Problems in

Electrical Circuit Theory 2

This book has been written as a sequel to *Problems in Electrical Circuit Theory, Book 1* to supplement courses in electrical theory and electronics in the latter part of degree and diploma courses in engineering and pure and applied science, by providing much-needed practice material. Each chapter is divided into three parts: theory summary, fully worked examples, and exercise problems (answers provided at the back of the book). Although the main emphasis is on the practice afforded by the examples, all necessary aspects of theory are covered. Modern notation and generalized methods are used throughout.

Dr Meadows, an experienced author, was formerly a scientific officer concerned with research for the Post Office. He is now Head of the Department of Electronic and Communication Engineering at the Polytechnic of North London, and was for some time UNESCO Professor at the University of Brasilia.

Problems in Electrical Circuit Theory 2

R. G. MEADOWS

B.Sc., M.Sc., Ph.D., C.Eng., M.I.E.E., M.Inst.P., A.R.C.S.

CASSELL · LONDON

Cassell & Co. Ltd—an imprint of
CASSELL & COLLIER MACMILLAN PUBLISHERS LTD
35 Red Lion Square, London WC1R 4SG
Sydney, Auckland
Toronto, Johannesburg

First published 1974

I.S.B.N. 0 304 29155 2

Printed in Great Britain by
Butler & Tanner Ltd, Frome and London

Preface

This book is the second of a two-book series designed to provide students who are following courses in engineering and science, with summarized notes and practice in solving problems in electrical circuit theory. To this end, each chapter is divided into three basic sections:
1. The first section gives a brief review of the relevant theory and summarizes the important results.
2. The second section contains fully worked examples.
3. The third section consists of exercise problems, to which answers are given at the back of the book.

The contents of Book 2 deal with electrical theory topics normally met by students in the latter half of University, Diploma, National Certificate or equivalent courses of study. Book 1 is intended for the first years of these courses.

R.G.M.

Contents

1 Laplace Transform Methods in Circuit Analysis

1.1. Theory Summary

1. INTRODUCTION

The equations governing the behaviour of a linear electrical circuit or system are generally in the form of linear differential or integro-differential equations specified in the time domain. The basic purpose of employing transform methods is to convert these equations into algebraic ones, solve them using algebraic methods—a procedure normally much simpler than the corresponding solution of the original differential equations—and then to convert these 'complex frequency domain' solutions back to the time domain, by means of an inverse transformation. Fig. 1.1 shows, in block diagram form, the basic steps in the transform solution process.

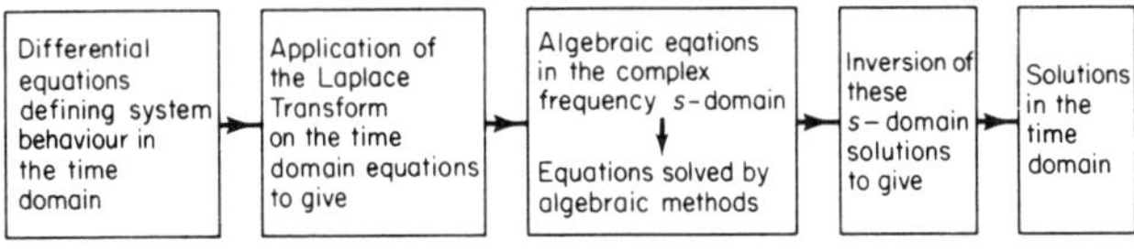

Fig. 1.1

2. THE LAPLACE TRANSFORM AND ITS PROPERTIES

(a) *The definition of the Laplace transform and its inverse*
The Laplace transform of a function of time $f(t)$ is defined as

$$\mathscr{L}\left[f(t)\right] = F(s) = \int_{0^-}^{\infty} f(t)\,\mathrm{e}^{-st}\,\mathrm{d}t$$

where $s = \sigma + j\omega$ is known in circuit theory as the complex frequency variable. σ is often known as the neper frequency and ω as the radian frequency.

Note that the lower limit of integration is taken at $t = 0^-$. In transient analysis this time represents the instant immediately prior to a switching operation, which is taken to occur at $t = 0$.

The condition for a function to possess a Laplace transform is

$$\int_{0-}^{\infty} |f(t)|\, e^{-\sigma t}\, dt < \infty \text{ for } \sigma \text{ real and positive}$$

The inverse Laplace transform is defined as

$$\mathcal{L}^{-1}[F(s)] \equiv f(t) = \frac{-j}{2\pi} \int_{\sigma'-j\omega}^{\sigma'+j\infty} F(s)\, e^{st}\, ds$$

where σ' is real and positive and greater than σ.

(b) *Linearity*
The Laplace transform of a sum of time functions is equal to the sum of the transforms of the individual functions, i.e.

$$\mathcal{L}\left[\sum_{r=1}^{n} f_r(t)\right] = \sum_{r=1}^{n} \mathcal{L}\left[f_r(t)\right] = \sum_{r=1}^{n} F_r(s)$$

(c) *Differentiation*

$$\mathcal{L}\left[\frac{df}{dt}\right] = sF(s) - f(0^-)$$

where $f(0^-)$ is the value of $f(t)$ at $t = 0^-$.

For example, the *i–v* relation for an iductor is $v = L\, di/dt$ and the transformed equation is

$$V(s) = LsI(s) - Li(0^-)$$

(d) *Integration*

$$\mathcal{L}\left[\int_{0-}^{t} f(t)\, dt\right] = \frac{F(s)}{s}$$

For example, the *i–v* relation for a capacitor is

$$v = \frac{1}{C}\int_{0-}^{t} i\, dt + v(0^-)$$

and the transformed equation is

$$V(s) = \frac{I(s)}{Cs} + \frac{v(0^-)}{s}$$

(e) *Translation and shifting property*

$$F(s-a) = \mathcal{L}\left[e^{at}f(t)\right], \; a \text{ is a complex constant.}$$

$$\mathcal{L}\left[f(t-T)\right] = e^{-Ts}F(s), \text{ where } f(t-T) = 0 \text{ for } t < T$$

(f) *Initial and final values*

$$\lim_{t \to 0+} [f(t)] = \lim_{s \to \infty} [sF(s)]$$

$$\lim_{t \to \infty} [f(t)] = \lim_{s \to 0} [sF(s)]$$

provided in the first case $f(t)$ is continuous or only contains a step discontinuity at $t = 0$; and in the second case the poles of $F(s)$ lie in the left half of the complex-frequency plane.

3. DETERMINATION OF LAPLACE TRANSFORMS

In order to obtain the Laplace transform or its corresponding inverse, recourse is usually made to a dictionary of known results. A table of some important results and the transform of some functions frequently encountered in circuit theory is given below.

(a) *Table of some standard results*

$f(t)$	$F(s)$
1. $f(t)$	$F(s) = \displaystyle\int_{0-}^{\infty} f(t)\,e^{-st}dt$
2. $\dfrac{df}{dt}$	$sF(s) - f(0^-)$
3. $\dfrac{d^2f}{dt^2}$	$s^2F(s) - sf(0^-) - \dfrac{df(0^-)}{dt}$
4. $\dfrac{d^nf}{dt^n}$	$s^nF(s) - s^{n-1}f(0^-) - s^{n-2}$ $\dfrac{df(0^-)}{dt} \cdots \dfrac{df^{n-1}(0^-)}{dt}$
5. $\displaystyle\int_{0-}^{t} f(t)\,dt$	$\dfrac{1}{s}F(s)$
6. $\displaystyle\int_{0-}^{t}\int_{0-}^{t} f(t_1)\,dt_1\,dt_2$	$\dfrac{1}{s^2}F(s)$
7. $-tf(t)$	$\dfrac{d}{ds}F(s)$

8. $(-t)^n f(t)$ $\dfrac{d^n}{ds^n} F(s)$

9. $e^{-\alpha t} f(t)$ $F(s+\alpha)$

10. $f(t-T)\,u(t-T)$ $e^{-Ts}\,F(s)$

11. Unit impulse $\delta(t)$ 1

12. Unit step function $u(t)$ $\dfrac{1}{s}$

13. Unit ramp function t $\dfrac{1}{s^2}$

14. Unit rectangular pulse (duration T) $\dfrac{1-e^{-Ts}}{s}$

15. $e^{-\alpha t}$ $\dfrac{1}{(s+\alpha)}$

16. $1-e^{-\alpha t}$ $\dfrac{\alpha}{s(s+\alpha)}$

17. $e^{-\alpha t}-e^{-\beta t}$ $\dfrac{\beta-\alpha}{(s+\alpha)(s+\beta)}$

18. $te^{-\alpha t}$ $\dfrac{1}{(s+\alpha)^2}$

19. $\dfrac{t^{n-1}}{(n-1)!}e^{-\alpha t}$ $\dfrac{1}{(s+\alpha)^n}$

20. $\sin \omega t$ $\dfrac{\omega}{s^2+\omega^2}$

21. $\cos \omega t$ $\dfrac{s}{s^2+\omega^2}$

22. $e^{-\alpha t}\sin \omega t$ $\dfrac{\omega}{(s+\alpha)^2+\omega^2}$

23. $e^{-\alpha t}\cos \omega t$ $\dfrac{s+\alpha}{(s+\alpha)^2+\omega^2}$

24. $t\sin \omega t$ $\dfrac{2\omega s}{(s^2+\omega^2)^2}$

25. $e^{-\alpha t}\left[\cos \omega t - \dfrac{\alpha}{\omega}\sin \omega t\right]$ $\qquad$ $\dfrac{s}{(s+\alpha)^2+\omega^2}$

26. $e^{-\alpha t}+\dfrac{\alpha}{\omega}\sin \omega t - \cos \omega t$ $\qquad$ $\dfrac{\alpha^2+\omega^2}{(s+\alpha)(s^2+\omega^2)}$

27. $\sinh \beta t$ $\qquad$ $\dfrac{\beta}{s^2-\beta^2}$

28. $\cosh \beta t$ $\qquad$ $\dfrac{s}{s^2-\beta^2}$

29. $e^{-\alpha t}\sinh \beta t$ $\qquad$ $\dfrac{\beta}{(s+\alpha)^2-\beta^2}$

30. $e^{-\alpha t}\cosh \beta t$ $\qquad$ $\dfrac{s+\alpha}{(s+\alpha)^2-\beta^2}$

(b) *Partial fraction expansions*

In circuit problems we are usually concerned with finding the inverse transform of functions which have the general form:

$$F(s) = \frac{b_m s^m + b_{m-1} s^{m-1} + \ldots b_1 s + b_0}{a_n s^n + a_{n-1} s^{n-1} + \ldots a_1 s + a_0}$$

where the use of the table of standard results is not immediately applicable. In these cases $F(s)$ may normally be expanded into a series of partial fractions and then recourse made to the standard results to find the inverse transform of the individual partial fraction terms. Some useful examples are reviewed below.

$$F(s) = \frac{as+b}{s^2+cs+d} = \frac{as+b}{(s-s_1)(s-s_2)} = \frac{K_1}{(s-s_1)} + \frac{K_2}{(s-s_2)}$$

$$F(s) = \frac{as^2+bs+c}{(s-s_1)(s^2+ds+e)} = \frac{K_1}{(s-s_1)} + \frac{K_2 s+K_3}{s^2+ds+e}$$

$$F(s) = \frac{as^2+bs+c}{(s-s_1)(s-s_2)^2} = \frac{K_1}{(s-s_1)} + \frac{K_2}{(s-s_2)} + \frac{K_3}{(s-s_2)^2}$$

When the degree of the numerator is equal to or greater than the degree of the denominator first divide out,

e.g.
$$\frac{4s^3+11s^2+8s-2}{s^3+2s^2+s} = 4 + \frac{3s^2+4s-2}{s(s+1)^2}$$

$$= 4 + \frac{K_1}{s} + \frac{K_2}{(s+1)} + \frac{K_3}{(s+1)^2}$$

The coefficients $K_1, K_2, K_3 \ldots$ etc. in the partial fraction expansions may be determined as follows.

(i) By using the identity formed by multiplying both sides by the denominator. For example,

$$\frac{3s^2+15s+68}{s^3+6s^2+34s} = \frac{3s^2+15s+68}{s(s^2+6s+34)} = \frac{K_1}{s} + \frac{K_2 s + K_3}{s^2+6s+34}$$

and on multiplying both sides by s^3+6s^2+34s and equating the numerators so obtained, we have

$$3s^2+15s+68 = K_1(s^2+6s+34)+s(K_2 s + K_3)$$

On substituting $s=0$, we have: $\quad 68 = 34K_1$, so $K_1 = 2$
On equating coefficients of s^2: $\quad 3 = K_1 + K_2$, so $K_2 = 1$
and equating coefficients of s: $\quad 15 = 6K_1 + K_3$, so $K_3 = 3$

Hence $F(s) = \dfrac{3s^2+15s+68}{s^3+6s^2+34s} = \dfrac{2}{s} + \dfrac{s+3}{s^2+6s+34}$

$$= \frac{2}{s} + \frac{s+3}{(s+3)^2+25}$$

and finally on referring to the transform table

$$f(t) = 2u(t) + e^{-3t} \cos \sqrt{25t}\, u(t)$$

(ii) If $F(s) = \dfrac{K_1}{(s-s_1)} + \dfrac{K_2}{(s-s_2)} + \cdots + \dfrac{K_i}{(s-s_i)} + \cdots$

then the partial fraction coefficients are given by

$$K_1 = [(s-s_1)F(s)]_{s=s_1}, \quad K_2 = [(s-s_2)F(s)]_{s=s_2}, \cdots$$
$$K_i = [(s-s_i)F(s)]_{s=s_i} \cdots$$

For example, $\dfrac{5s+6}{s^2+7s+12} = \dfrac{K_1}{(s+3)} + \dfrac{K_2}{(s+4)}$

where $K_1 = \left[(s+3)\dfrac{5s+6}{(s+3)(s+4)} \right]_{s=-3} = \left[\dfrac{5s+6}{s+4} \right]_{s=-3} = -9$

$K_2 = \left[(s+4)\dfrac{5s+6}{(s+3)(s+4)} \right]_{s=-4} = \left[\dfrac{5s+6}{s+3} \right]_{s=-4} = 14$

4. TRANSFORMED CIRCUITS

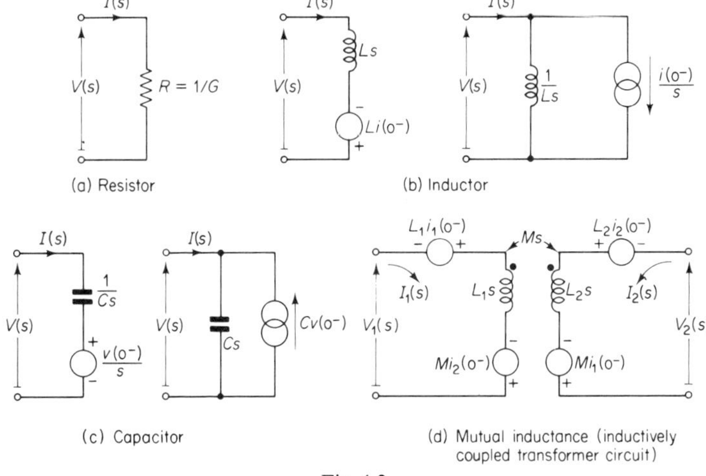

Fig. 1.2

Fig. 1.2. shows the complex frequency domain equivalent circuits for the passive circuit elements of R, L, C and M. These circuits are derived by taking the Laplace transform of the time domain equations:

Time domain equations	*s-domain equations*

$$v(t) = Ri(t) \qquad\qquad V(s) = RI(s)$$

$$i(t) = Gv(t) \qquad\qquad I(s) = GV(s)$$

$$v(t) = L\frac{di(t)}{dt} \qquad\qquad V(s) = Ls\,I(s) - Li(0^-)$$

$$i(t) = \frac{1}{L}\int_{0-}^{t} v(t)\,dt + i(\sigma) \qquad I(s) = \frac{1}{Ls}V(s) + \frac{i(0^-)}{s}$$

$$v(t) = \frac{1}{C}\int_{0-}^{t} i(t)\,dt + v(\sigma) \qquad V(s) = \frac{1}{Cs}I(s) + \frac{v(0^-)}{s}$$

$$i(t) = C\frac{dv(t)}{dt} \qquad\qquad I(s) = CsV(s) - Cv(0^-)$$

$$v_1 = L_1\frac{di_1}{dt} + M\frac{di_2}{dt} \qquad V_1 = L_1 sI_1 - L_1 i_1(0^-) + MsI_2 - Mi_2(0^-)$$

$$v_2 = M\frac{di_1}{dt} + L_2\frac{di_2}{dt} \qquad V_2 = MsI_1 - Mi_1(0^-) + L_2 sI_2 - L_2 i_2(0^-)$$

Note that the s-domain circuits contain the initial condition information represented in the form of voltage or current sources.

The process of solving a circuit problem may now be made in the following way.

(1) Replace all circuit elements by their transformed equivalent circuits. The initial conditions $v(0^-)$ and $i(0^-)$ are usually given in a problem or may be deduced from the data supplied. Replace also all applied sources by their transforms.

(2) Analyse the complete transformed circuit using Kirchhoff's laws to obtain the desired unknowns, $I(s)$ and/or $V(s)$.

(3) Use the transform table, partial fraction expansions and any other convenient means to transform the s-domain solutions back into the time domain.

5. THE TRANSFER OR SYSTEM FUNCTION

(a) *General*

If we consider linear systems which contain no stored energy at $t = 0^-$, i.e. all $v(0^-)$, $i(0^-)$ are zero, then the R, L and C relations in the s-domain reduce to:

$$V = RI \qquad V = LsI \qquad V = \frac{1}{Cs} I$$

$$I = GV \qquad I = \frac{1}{Ls} V \qquad I = CsV$$

In this case the system can be described by a transfer or system function $H(s)$ relating the response $R(s)$ and the excitation (driving function) $E(s)$ by the simple equation:

$$R(s) = H(s)E(s)$$

There are several different forms that the transfer function may take. For example, if $E(s) = V(s)$ is the voltage applied across a pair of terminals of a circuit and $R(s) = I(s)$ is the input current at these terminals, then

$$H(s) = \frac{I(s)}{V(s)} = Y(s)$$

represents a driving-point admittance.

If $\quad E(s) = I_1(s)$ is the current applied at one pair of terminals
and $\quad R(s) = V_2(s)$ is the voltage across another part of the circuit,

$$H(s) = \frac{V_2(s)}{I_1(s)} = Z_{21}(s)$$

represents a transfer impedance. If $V_2(s)$ corresponds to the same pair of terminals, then $H(s)$ is the driving-point impedance at these terminals.

If the excitation and response are both voltages, $H(s)$ is a voltage-ratio transfer function. Likewise if both are currents, $H(s)$ is a current-ratio transfer function.

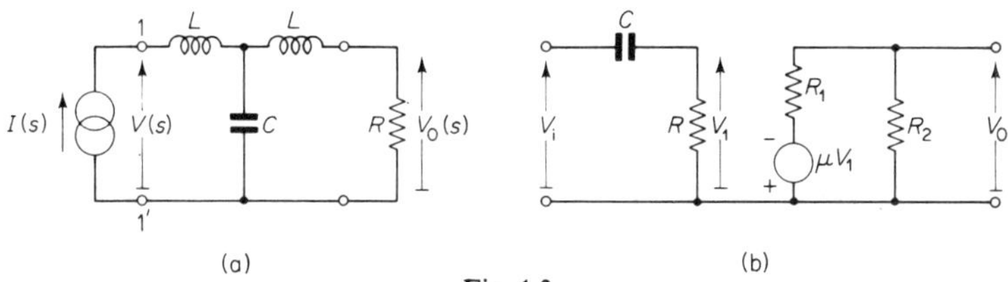

(a)   (b)

Fig. 1.3

In the circuit of fig. 1.3(a), for example

$$H(s) = Z(s) = \frac{V(s)}{I(s)} = Ls + \frac{1/(Cs)[R+Ls]}{R+Ls+1/(Cs)}$$

is the driving-point impedance at terminals $11'$, whereas

$$H(s) = \frac{V_0(s)}{I(s)} = \frac{1/(Cs) \times R}{R+Ls+1/(Cs)}$$

is a transfer impedance. In the circuit of fig. 1.3(b), the voltage-ratio transfer function,

$$H(s) = \frac{V_0}{V_i} = -\frac{\mu R_2 R}{(R_1+R_2)(R+1/(Cs))}$$

(b) *Poles and zeros of a transfer function: their significance in determining the natural response of a circuit*

The transfer function contains sufficient information to determine not only the response of a circuit (assuming zero initial conditions) but also the natural frequencies and the general solution of the force-free response as well.

In general a transfer function will consist of a ratio of two polynomials in s, i.e.

$$H(s) = \frac{R(s)}{E(s)} = \frac{b_m s^m + b_{m-1} s^{m-1} + \ldots + b_1 s + b_0}{a_n s^n + a_{n-1} s^{n-1} + \ldots + a_1 s + a_0}$$

and on factorizing, we have

$$H(s) = \left(\frac{b_m}{a_n}\right) \frac{(s-z_1)(s-z_2) \ldots (s-z_m)}{(s-p_1)(s-p_2) \ldots (s-p_n)}$$

where $z_1, z_2 \ldots z_m$ are the values of s for which $H(s) = 0$

9

and are therefore known as the zeros
of the transfer function

and $\quad p_1, p_2 \ldots p_n$ are the values of s for which $H(s) = \infty$;
these values are known as the poles
of the transfer function

In the case of force-free or natural response of a circuit $E(s) = R(s)/H(s) = 0$ and so for a non-trivial response, i.e. $R(s) \neq 0$, $H(s)$ should be infinite. Thus the natural frequencies of a circuit correspond to the poles of its transfer function. To be more specific, in the case of 1-port (two terminal) networks, each pole of the input admittance (or zero of the input impedance) defines a natural frequency of the circuit when the input terminals are short-circuited. Whilst each pole of the input impedance (or zero of the input admittance) defines a natural frequency of the circuit when the terminals are open-circuited.

The general solution of the force-free response has the form

$$r(t) = A_1 \, e^{p_1 t} + A_2 \, e^{p_2 t} + \ldots + A_n \, e^{p_n t}$$

provided the poles of $H(s)$ are all different. In the case of complex poles, they occur in conjugate pairs and give rise to damped oscillatory terms in the response, e.g. if $p_r = -\alpha_r + j\omega_r$, $p_r = -\alpha_r - j\omega_r$, then $A_r e^{p_r t} + A_{r+1} e^{p_r t}$ may be written as $B_r e^{-\alpha_r t} \sin(\omega_r t + \psi_r)$. If, however, some poles are equal, e.g. $p_1 = p_2 = \ldots p_r$, then

$$r(t) = (A_1 + A_2 t + \ldots + A_r t^{r-1}) e^{p_1 t} + \ldots + A_n \, e^{p_n t}$$

6. SOME ASPECTS OF POLE-ZERO ANALYSIS

(a) *Implication as to the nature of a function from plotting its poles on the s-plane*

The poles and zeros of a function $F(s)$ may be plotted on the s-plane, shown in fig. 1.4(a). A pole is denoted by a small cross and a zero by a small circle. The positions of the poles and zeros in this plane provide valuable information as to the time domain character of the function, as illustrated by the examples shown in fig. 1.4(b) to (g). In particular if any poles of $F(s)$ occur in the right half ($\sigma > 0$) of this plane they will give rise to an exponentially increasing f(t). Thus if a transfer function has any poles in the right hand s-plane, the system it defines is regarded as unstable. Likewise if the function has a pair of double poles on the $j\omega$ axis (e.g. see fig. 1.4(g)), the system is unstable.

10

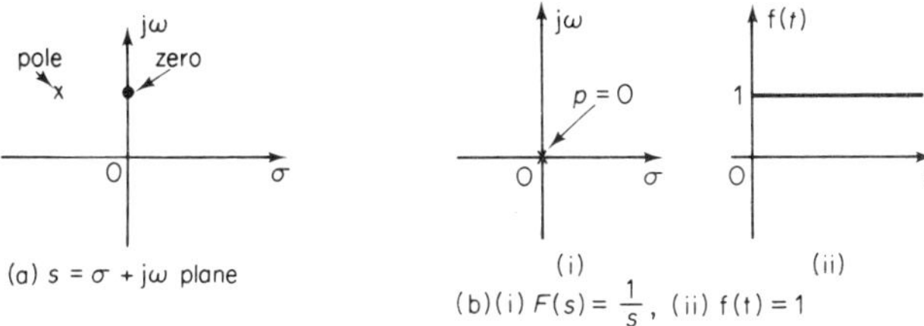

(a) $s = \sigma + j\omega$ plane

(b)(i) $F(s) = \dfrac{1}{s}$, (ii) $f(t) = 1$

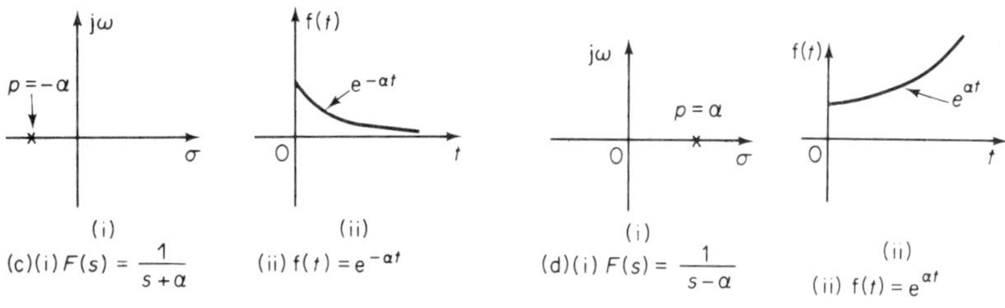

(c)(i) $F(s) = \dfrac{1}{s+\alpha}$ (ii) $f(t) = e^{-\alpha t}$

(d)(i) $F(s) = \dfrac{1}{s-\alpha}$ (ii) $f(t) = e^{\alpha t}$

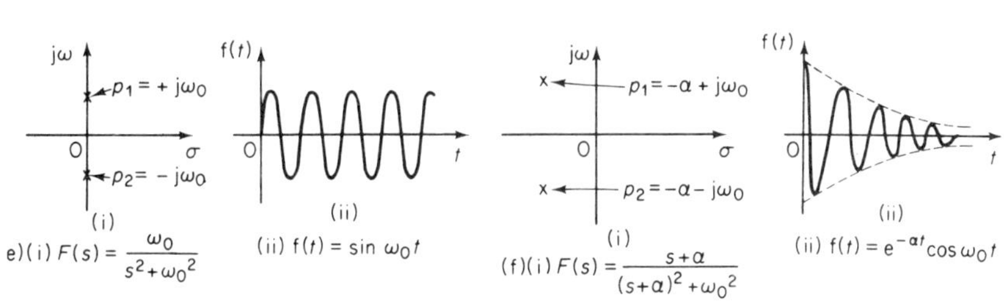

e)(i) $F(s) = \dfrac{\omega_0}{s^2+\omega_0^2}$ (ii) $f(t) = \sin \omega_0 t$

(f)(i) $F(s) = \dfrac{s+\alpha}{(s+\alpha)^2 + \omega_0^2}$ (ii) $f(t) = e^{-\alpha t}\cos \omega_0 t$

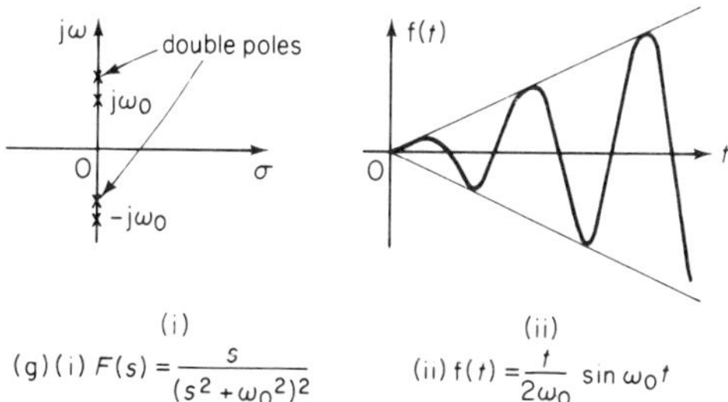

(g)(i) $F(s) = \dfrac{s}{(s^2 + \omega_0^2)^2}$ (ii) $f(t) = \dfrac{t}{2\omega_0}\sin \omega_0 t$

Fig. 1.4

(b) *A.C. (sinusoidal) steady state response*
Pole-zero analysis may also be used to study the a.c. sinusoidal steady
state response of a system by setting $s = j\omega$ in the transfer function:

$$H(s)_{s=j\omega} = \frac{b_m}{a_n} \frac{(j\omega - z_1)(j\omega - z_2) \ldots (j\omega - z_m)}{(j\omega - p_1)(j\omega - p_2) \ldots (j\omega - p_n)}$$

$$= |H(j\omega)| e^{j\phi(\omega)} \qquad \ldots (1)$$

where the magnitude response,

$$|H(j\omega)| = \frac{b_m}{a_n} \frac{|(j\omega - z_1)| \, |(j\omega - z_2)| \ldots |(j\omega - z_m)|}{|(j\omega - p_1)| \, |(j\omega - p_2)| \ldots |(j\omega - p_n)|}$$

and the phase response,

$$\phi(\omega) = [(\theta_{z_1} + \theta_{z_2} + \ldots + \theta_{z_m}) - (\theta_{p_1} + \theta_{p_2} + \ldots + \theta_{p_n})]$$

$\theta_{z_1}, \theta_{z_2} \ldots$ are the arguments of $(j\omega - z_1)$, $(j\omega - z_2) \ldots$ etc and
$\theta_{p_1}, \theta_{p_2} \ldots$ are the arguments of $(j\omega - p_1)$, $(j\omega - p_2) \ldots$ etc.
   In fact each of the terms $(j\omega - z_1)$, $(j\omega - z_2) \ldots (j\omega - p_1)$, $(j\omega - p_2) \ldots$
in $H(j\omega)$ correspond to the vector from the particular zero or pole to the
point $j\omega$ on the imaginary axis of the s-plane, as shown in fig. 1.5, and
thus for any particular value of ω:

$$|H(j\omega)| = \frac{\text{Product of vector magnitudes from zeros to the point } j\omega}{\text{Product of vector magnitudes from poles to the point } j\omega}$$

$$\phi(\omega) = \begin{array}{l} \text{Sum of angles of vectors from zeros to point } j\omega \\ - \text{sum of angles of vectors from poles to point } j\omega \end{array}$$

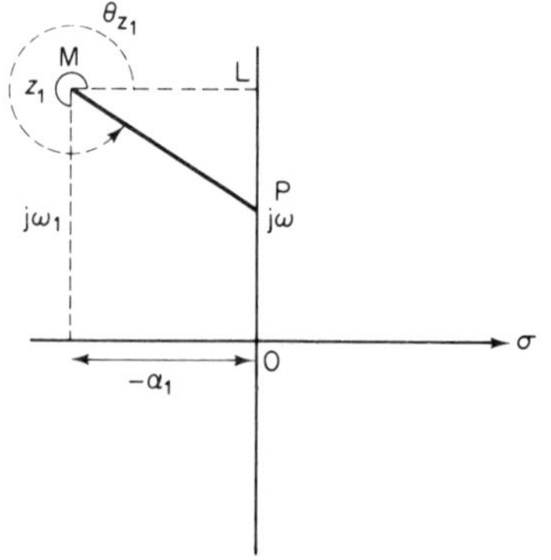

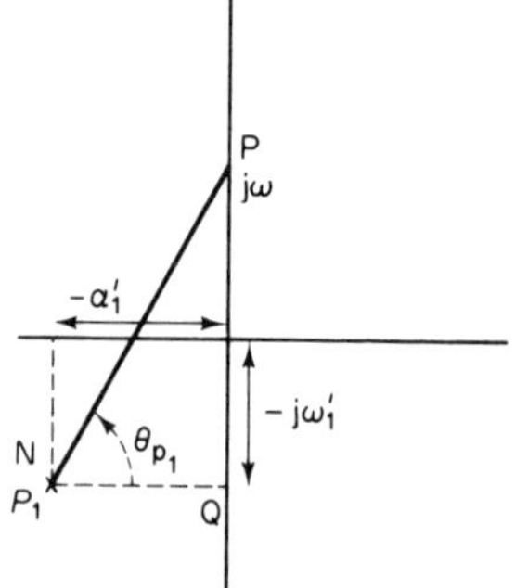

(a) $z_1 = -\alpha_1 + j\omega_1$, **MP** $= (j\omega - z_1)$, length
MP $= |(j\omega - z_1)|$, angle LMP $= \theta_{z_1}$

(b) $p_1 = -\alpha_1' - j\omega_1'$, **NP,** $= (j\omega - p_1)$, length
NP $= |(j\omega - p_1)|$, angle QNP $= \theta_{p_1}$

Fig. 1.5

One useful way of presenting the steady state response information is to plot graphs of $20\log_{10}|H(j\omega)|$ db and $\phi(\omega)$ versus frequency, the latter on a logarithmic scale. These graphs are known as *Bode plots* and are extensively used in network analysis and design. Examples of Bode plots for one and two-pole transfer functions are shown in fig. 1.6.

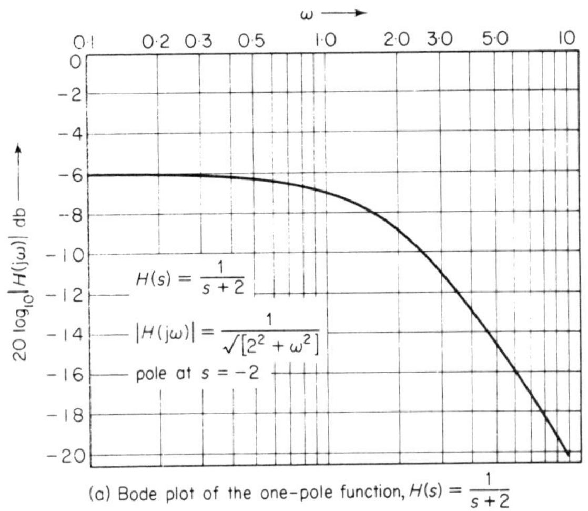

(a) Bode plot of the one-pole function, $H(s) = \dfrac{1}{s+2}$

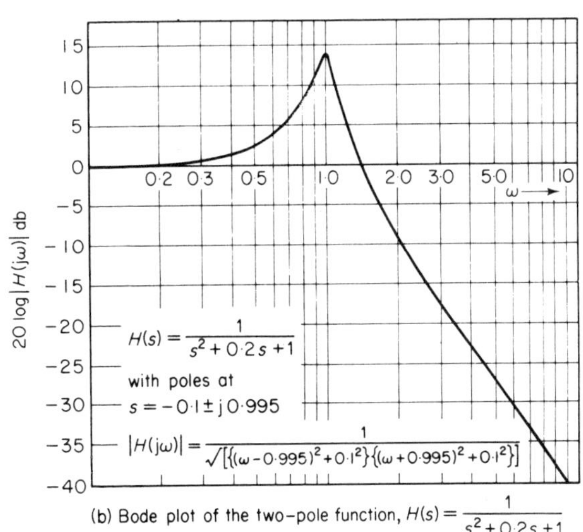

(b) Bode plot of the two-pole function, $H(s) = \dfrac{1}{s^2 + 0\cdot2s + 1}$

Fig. 1.6

1.2. Worked Problems

1. In the circuit of fig. 1.7 the switch S is suddenly changed from position 1 to 2 at $t = 0$. Just prior to switching the current i and the potential difference v have the values $i(0^-) = 4$ A, $v(0^-) = 5$ V. Draw the transformed circuit, determine the values of $I(s)$ and $V(s)$, and the variation of i and v for $t \geqslant 0$ for the cases (a) $R = 4\,\Omega$, $L = 1$ H, $C = \frac{1}{3}$F; (b) $R = 4\,\Omega$, $L = 2$ H, $C = \frac{1}{2}$F.

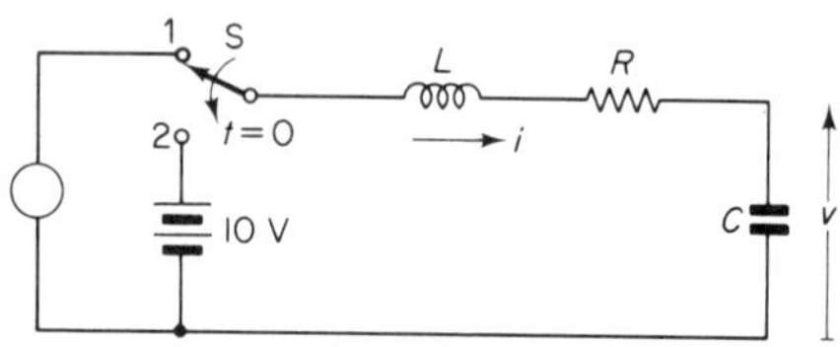

Fig. 1.7

Solution

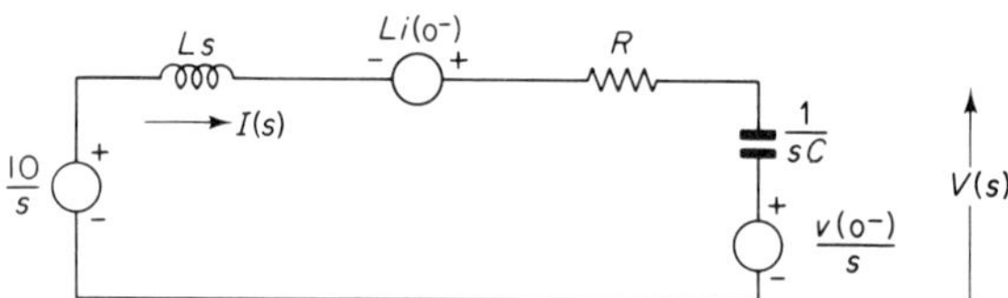

Fig. 1.8. The transformed circuit for $t \geqslant 0$

The transformed circuit is drawn in fig. 1.8. On applying Kirchhoff's voltage law to this circuit we obtain:

$$(Ls + R + 1/Cs)I(s) = \frac{10}{s} + Li(0^-) - \frac{v(0^-)}{s}$$

Hence
$$I(s) = \frac{10 + Li(0^-)s - v(0^-)}{Ls^2 + Rs + 1/C}$$

and
$$V(s) = \frac{1}{Cs}I(s) + \frac{v(0^-)}{s}$$

(a) On substituting $i(0^-) = 4$, $v(0^-) = 5$, $R = 4$, $L = 1$, $C = \frac{1}{3}$ we have

$$I(s) = \frac{5 + 4s}{s^2 + 4s + 3} = \frac{5 + 4s}{(s+1)(s+3)} = \frac{K_1}{(s+1)} + \frac{K_2}{(s+3)}$$

14

where the partial fraction coefficients K_1 and K_2 may be evaluated from

$$K_1 = [(s+1)F(s)]_{s=-1} , \quad K_2 = [(s+3)F(s)]_{s=-3} , \quad F(s) = \frac{5+4s}{(s+1)(s+3)}$$

Thus $\quad K_1 = \dfrac{1}{2}, \quad K_2 = \dfrac{7}{2}$

and $\quad I(s) = \dfrac{1}{2} \dfrac{1}{(s+1)} + \dfrac{7}{2} \dfrac{1}{(s+3)}$

On taking the inverse transform of $I(s)$, we obtain

$$i = \tfrac{1}{2}(e^{-t} + 7e^{-3t})u(t) \quad \text{A}$$

Likewise,

$$V(s) = \frac{3}{s} \frac{5+4s}{(s+1)(s+3)} + \frac{5}{s}$$

$$= 3\left[\frac{5}{3}\left(\frac{1}{s}\right) - \frac{1}{2}\left(\frac{1}{s+1}\right) - \frac{7}{6}\left(\frac{1}{s+3}\right)\right] + \frac{5}{s}$$

$$= \frac{10}{s} - \frac{3}{2}\left(\frac{1}{s+1}\right) - \frac{7}{2}\left(\frac{1}{s+3}\right)$$

and $\quad v = [10 - \tfrac{1}{2}(3e^{-t} + 7e^{-3t})]u(t) \quad \text{V}.$

(b) When $L = 2$, $C = \tfrac{1}{2}$

$$I(s) = \frac{5+8s}{2s^2 + 4s + 2} = \frac{5+8s}{2(s+1)^2} = \frac{1}{2}\left[\frac{K_1}{(s+1)} + \frac{K_2}{(s+1)^2}\right]$$

and evaluating the partial fraction coefficients this time by equating numerators, i.e.

$$5 + 8s \equiv K_1(s+1) + K_2$$

we find $K_2 = -3$, $K_1 = 8$ and therefore

$$I(s) = \frac{1}{2}\left[\frac{8}{(s+1)} - \frac{3}{(s+1)^2}\right]$$

On taking the inverse transform, we obtain

$$i = (4e^{-t} - \tfrac{3}{2}te^{-t})u(t) \quad \text{A}$$

Finally,

$$V(s) = \frac{5+8s}{s(s+1)^2} + \frac{5}{s}$$

and splitting the first term into partial fractions, we have

$$\frac{5+8s}{s(s+1)^2} = \frac{K_1}{s} + \frac{K_2}{(s+1)} + \frac{K_3}{(s+1)^2}$$

where the coefficients may be evaluated from the identity obtained by equating numerators of both sides, i.e.

$$5+8s \equiv K_1(s+1)^2 + K_2 s(s+1) + K_3 s$$

and when $s = 0$, we have $5 = K_1$; when $s = -1$, $-3 = -K_3$; when $s = 1$, $5+8 = 4K_1 + 2K_2 + K_3$, hence $K_2 = -5$.

Thus $\quad V(s) = \dfrac{5}{s} - \dfrac{5}{(s+1)} + \dfrac{3}{(s+1)^2} + \dfrac{5}{s}$

and $\qquad v = (10 - 5e^{-t} + 3t\,e^{-t})u(t)$ V

2. The circuit of fig. 1.9 contains no initial stored energy. Determine the mesh current i_1 if switch S is closed at $t = 0$.

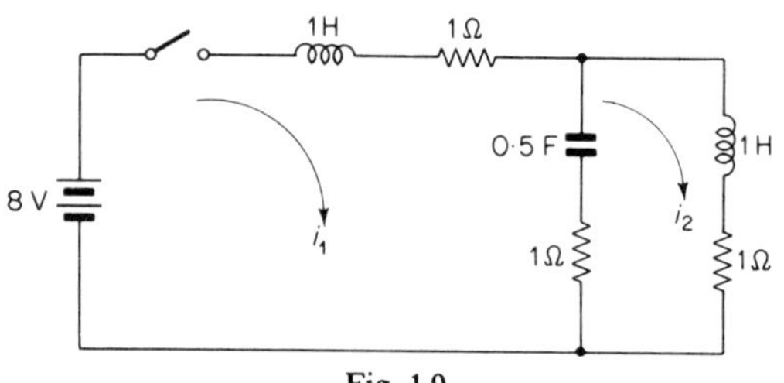

Fig. 1.9

Solution

The *s*-domain mesh equations for the circuit of fig. 1.9 are

$$(s+2+2/s)I_1(s) - (1+2/s)I_2(s) = \frac{8}{s}$$

$$-(1+2/s)I_1(s) + (s+2+2/s)I_2(s) = 0$$

and on solving for $I_1(s)$ we obtain,

$$I_1(s) = \frac{(8/s)(s+2+2/s)}{(s+2+2/s)^2 - (1+2/s)^2}$$

$$= \frac{8(s^2+2s+2)}{s(s+1)(s^2+3s+4)}$$

$$= 8\left[\frac{K_1}{s} + \frac{K_2}{s+1} + \frac{K_3 s + K_4}{s^2+3s+4}\right]$$

The partial fraction coefficient may be evaluated from the identity obtained by equating the numerators, i.e.

$$s^2 + 2s + 2 = K_1(s+1)(s^2+3s+4) + K_2 s(s^2+3s+4)$$
$$+ (K_3 s + K_4)s(s+1)$$

so when $s = 0$: $2 = K_1 \times 4$, $K_1 = \tfrac{1}{2}$

when $s = -1$: $1 = K_2 \times -2$, $K_2 = -\tfrac{1}{2}$

equating coefficients of s^3: $\qquad 0 = K_1 + K_2 + K_3$, so $K_3 = 0$

when $s = 1$: $\qquad 5 = K_1 \times 16 + K_2 \times 8 + K_4 \times 2$, so $K_4 = \tfrac{1}{2}$

Thus $I_1(s) = \dfrac{4}{s} - \dfrac{4}{s+1} + \dfrac{4}{s^2+3s+4} = \dfrac{4}{s} - \dfrac{4}{s+1} + \dfrac{4}{(s+\frac{3}{2})^2 + \frac{7}{4}}$

and $\quad i_1 = [4 - 4e^{-t} + \{4/\sqrt{(7/4)}\}e^{-3/2t}\sin\sqrt{(7/4)}t]u(t)$

$\qquad = [4 - 4e^{-t} + 3 \cdot 023 e^{-1 \cdot 5t}\sin(1 \cdot 323t)]u(t)$ A

3. Determine the current i in the series R–C circuit of fig. 1.10 when the following voltage sources are applied: (a) a voltage impulse $v = \delta(t)$, (b) a step voltage $v = u(t)$, (c) a ramp voltage $v = tu(t)$. The capacitor C is initially uncharged. Find also the current response of the circuit when the voltage waveform shown in fig. 1.11 is applied.

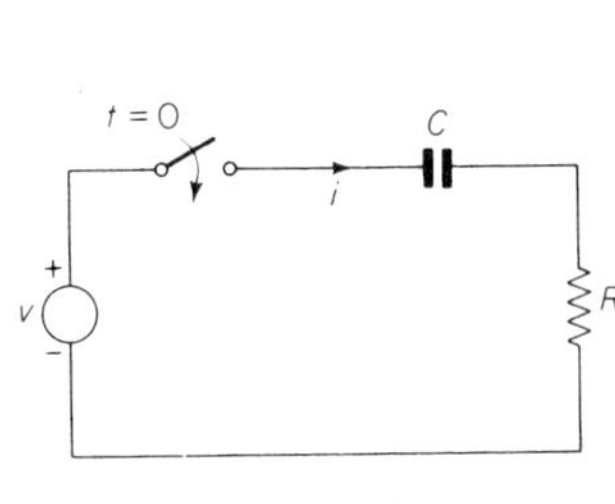

Fig. 1.10

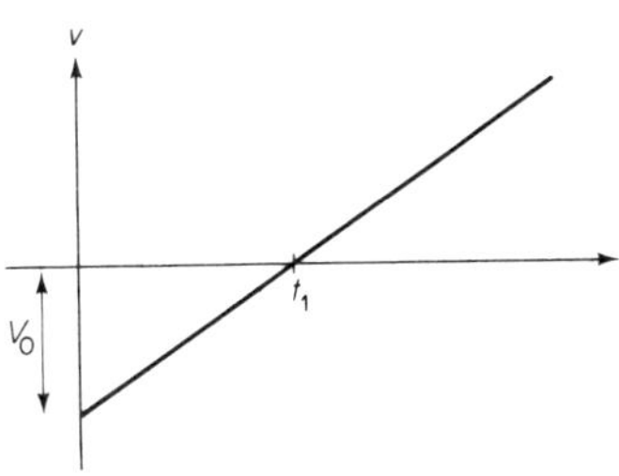

Fig. 1.11

Solution

The s-domain current $I(s)$ in an R–C circuit for an applied source $V(s)$ is

$$I(s) = \frac{V(s)}{R + \dfrac{1}{Cs}} = \frac{1}{R}\left(\frac{s}{s + \dfrac{1}{RC}}\right)V(s)$$

(a) For the voltage impulse $v = \delta(t)$, $V(s) = 1$ and therefore

$$I(s) = \frac{1}{R}\left(\frac{s}{s+\dfrac{1}{RC}}\right) = \frac{1}{R}\left[1 - \frac{1}{RC}\left(\frac{1}{s+\dfrac{1}{RC}}\right)\right]$$

$$i = \frac{1}{R}\left[\delta(t) - \frac{1}{RC}e^{-t/RC}u(t)\right] \text{amperes}$$

(b) For the step voltage $v = u(t)$, $V(s) = 1/s$ and so

$$I(s) = \frac{1}{R}\frac{1}{s+\dfrac{1}{RC}}$$

$$i = \frac{1}{R}e^{-t/RC}u(t)\,\text{amperes}$$

(c) For the ramp voltage $v = tu(t)$, $V(s) = 1/s^2$ and therefore

$$I(s) = \frac{1}{R}\left[\frac{1}{s\left(s+\dfrac{1}{RC}\right)}\right]$$

$$= \frac{1}{R}\left[\frac{RC}{s} - \frac{RC}{s+\dfrac{1}{RC}}\right]$$

$$i(t) = C[u(t) - e^{-t/RC}u(t)] = C[1 - e^{-t/RC}]u(t)\,\text{amperes}$$

The voltage waveform of fig. 1.11 has the equation,

$$v = -V_0 u(t) + \frac{V_0}{t_1}t\,u(t)$$

Hence the current i due to this applied voltage may be determined by adding $-V_0$ times the unit step response and V_0/t_1 times the ramp response. Thus

$$i = -V_0\left(\frac{1}{R}e^{-t/RC}\right)u(t) + CV_0/t_1[1 - e^{-t/RC}]u(t)$$

4. The circuit of fig. 1.12 has reached its steady state with the switch S in position 1. S is then placed in position 2 at $t = 0$. Find the initial values, $i_L(0^-)$ and $v_C(0^-)$, and draw the transformed circuit for $t \geq 0$. Evaluate v_C for the case $G = C = L = 1$, $V = 10$, $V_1 = 5$, $\omega = 2$.

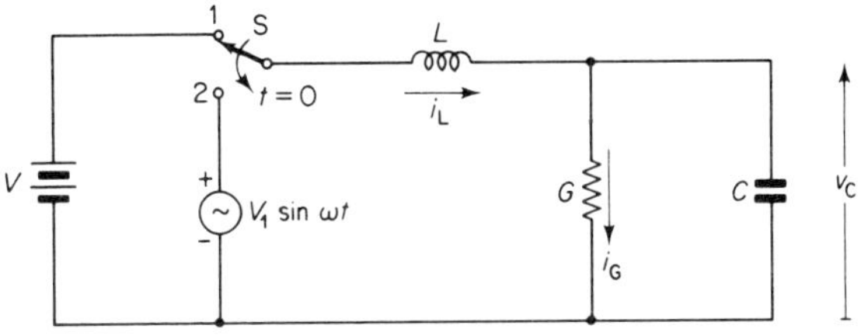

Fig. 1.12

Solution

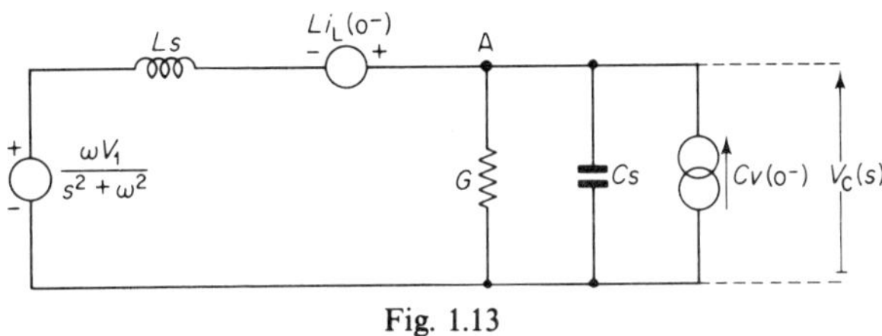

Fig. 1.13

In the steady state with an applied d.c. voltage L acts as an effective short-circuit and C as an open-circuit. Thus

$$v_C(0^-) = V \quad \text{and} \quad i_L(0^-) = i_G(0^-) = GV$$

The transformed circuit for $t \geqslant 0$ is drawn in fig. 1.13. On applying Kirchhoff's current law at node A, we obtain

$$\frac{1}{Ls}\left[V_C(s) - Li_L(0^-) - \frac{\omega V_1}{s^2 + \omega^2} \right] + (G + Cs)V_C(s) = Cv(0^-)$$

i.e.
$$V_C(s) = \frac{Cv(0^-) + \dfrac{i(0^-)}{s} + \dfrac{\omega V_1}{Ls(s^2 + \omega^2)}}{\left(\dfrac{1}{Ls} + G + Cs \right)}$$

$$= \frac{Ls(s^2 + \omega^2)Cv(0^-) + L(s^2 + \omega^2)i_L(0^-) + \omega V_1}{(s^2 + \omega^2)(1 + LGs + LCs^2)}$$

which, on substituting $G = C = L = 1$, $V_1 = 5$, $\omega = 2$, $v_C(0^-) = V = 10$, $i_L(0^-) = GV = 10$, becomes

$$V_C(s) = \frac{s(s^2 + 4)10 + (s^2 + 4)10 + 10}{(s^2 + 4)(s^2 + s + 1)}$$

19

$$= 10\left[\frac{s^3+s^2+4s+5}{(s^2+4)(s^2+s+1)}\right]=10\left[\frac{K_1s+K_2}{(s^2+4)}+\frac{K_3s+K_4}{(s^2+s+1)}\right]$$

On evaluating the partial fraction coefficients, we find $K_1=-\tfrac{1}{13}$, $K_2=-\tfrac{3}{13}$, $K_3=\tfrac{14}{13}$, $K_4=\tfrac{17}{13}$ so

$$V_C(s)=\frac{10}{13}\left[\frac{-s}{s^2+4}-\frac{3}{2}\frac{2}{s^2+4}+14\frac{s+\tfrac12}{(s+\tfrac12)^2+\tfrac34}+\frac{10}{\sqrt{(\tfrac34)}}\frac{\sqrt{(\tfrac34)}}{(s+\tfrac12)^2+\tfrac34}\right]$$

$$v_C(t)=-\frac{10}{13}\cos 2t-\frac{15}{13}\sin 2t+\frac{140}{13}e^{-1/2t}\sin\sqrt{(\tfrac34)}t$$

$$+\frac{100}{13\sqrt{(\tfrac34)}}e^{-1/2t}\sin\sqrt{(\tfrac34)}t$$

5. Find the poles and zeros of the transfer function V/I_G for the circuit shown in fig. 1.14. Hence deduce
   (a) if $I_G = I\sin\omega t$, the values of ω for which $v = 0$
   (b) the nature of the force-free response when switch S is opened.

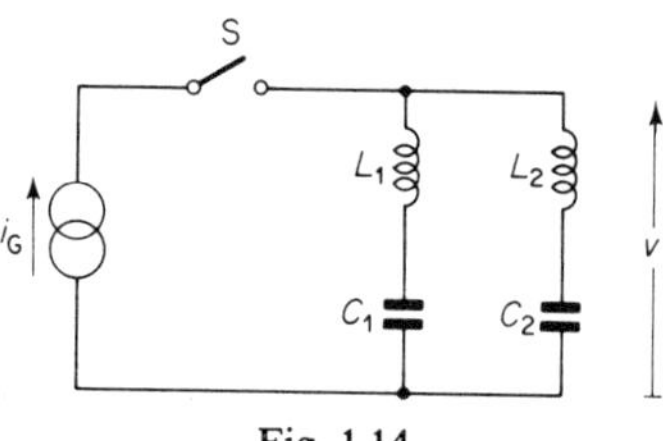

Fig. 1.14

Solution

The transfer function V/I_G is the impedance of the L–C network,

i.e. $$\frac{V(s)}{I_G(s)}=Z(s)=Y(s)^{-1}$$

where $$Y(s)=\frac{1}{L_1 s+1/(C_1 s)}+\frac{1}{L_2 s+1/(C_2 s)}$$

$$=\frac{C_1 s(1+L_2 C_2 s^2)+C_2 s(1+L_1 C_1 s^2)}{(1+L_1 C_1 s^2)(1+L_2 C_2 s^2)}$$

$$=\frac{L_1+L_2}{L_1 L_2}\frac{s[s^2+(C_1+C_2)/\{C_1 C_2(L_1+L_2)\}]}{[s^2+1/(L_1 C_1)][s^2+1/(L_2 C_2)]}$$

20

so
$$Z(s) = \frac{L_1 L_2}{L_1 + L_2} \frac{[s^2 + 1/(L_1 C_1)][s^2 + 1/(L_2 C_2)]}{s[s^2 + (C_1 + C_2)/\{C_1 C_2(L_1 + L_2)\}]}$$

and hence the zeros of $Z(s)$ are $\pm j1/\sqrt{(L_1 C_1)}, \pm j1/\sqrt{(L_2 C_2)}$ and the poles of $Z(s)$ are $0, \pm j\sqrt{[(C_1 + C_2)/\{C_1 C_2(L_1 + L_2)\}]}$
(a) For the sinusoidal source, $V(s) = 0$ when $Z(j\omega) = 0$, i.e. at the zeros $j\omega_1 = j1/\sqrt{(L_1 C_1)}$ and $j\omega_2 = j1/\sqrt{(L_2 C_2)}$, so the values of ω for which $V(j\omega) = v = 0$ are

$$\omega_1 = 1/\sqrt{(L_1 C_1)} \text{ rad/s}, \quad \omega_2 = 1/\sqrt{(L_2 C_2)} \text{ rad/s}.$$

(b) The qualitative nature of the force-free response of the L–C network may be found directly from the poles of $Z(s)$, since if $V(s) = Z(s) I_G(s)$ is to be non-zero when $I_G(s) = 0$, $Z(s)$ must tend to infinity. The pole at $s = 0$ leads to a possible d.c. term, the poles $\pm j1/\sqrt{[(L_1 + L_2)C']}$, where $C' = (1/C_1 + 1/C_2)^{-1}$ leads to an oscillatory component at the radian frequency $\omega = 1/\sqrt{\{(L_1 + L_2)C'\}}$. Depending on the initial state of the circuit at switching, there may be also an impulse change at this instant.

6. Determine the transfer function $V_2(s)/V_G(s)$ in the circuit of fig. 1.15. If $L_1 = 2{\cdot}0236$ H, $C_2 = 0{\cdot}9941$ F, $L_3 = 2{\cdot}0236$ H show that the transfer function has a pole at $s = -0{\cdot}494$ and find its other poles. Plot also its a.c. steady state magnitude response on a Bode diagram for $\omega = 0{\cdot}1$ to $\omega = 1{\cdot}4$ rad/s.

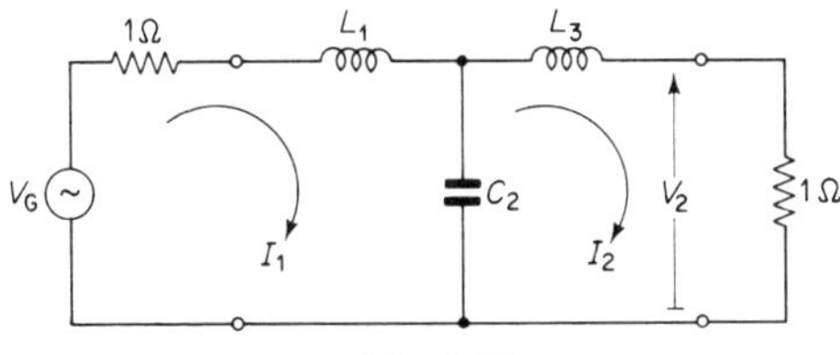

Fig. 1.15

Solution

The transfer function $H(s) = V_2/V_G$ may be found by writing down the mesh equations of the circuit:

$$[1 + L_1 s + 1/(C_2 s)]I_1 \qquad - [1/(C_2 s)]I_2 = V_G$$

$$- [1/(C_2 s)]I_1 + [1/(C_2 s) + L_3 s + 1]I_2 = 0$$

and solving for I_2 (note $V_2 = I_2 \times 1$),

$$I_2 = V_2 = \frac{V_G}{C_2 s[\{1 + L_1 s + 1/(C_2 s)\}\{1/(C_2 s) + L_3 s + 1\} - \{1/(C_2 s)\}^2]}$$

hence $H(s) =$

$$V_2/V_G = \frac{1}{L_1 L_3 C_2 s^3 + (L_1 + L_3)C_2 s^2 + (L_1 + L_3 + C_2)s + 2}$$

$$= \frac{0 \cdot 246}{s^3 + 0 \cdot 988\, s^2 + 1 \cdot 238\, s + 0 \cdot 491} \qquad \cdots (1)$$

on substituting $L_1 = L_3 = 2 \cdot 0236$, $C = 0 \cdot 9941$

The poles of $H(s)$ are the roots of $s^3 + 0 \cdot 988\, s^2 + 1 \cdot 238\, s + 0 \cdot 491 = 0$. On substituting $s = -0 \cdot 494$ in the latter we find the equation is satisfied so $-0 \cdot 494$ is a pole of $H(s)$. Also on dividing the denominator of $H(s)$ by the factor $(s + 0 \cdot 494)$ and solving the resulting quadratic we may find the two other poles:

$$-0 \cdot 249 + j0 \cdot 972 \quad \text{and} \quad -0 \cdot 249 - j0 \cdot 972$$

Thus $H(s)$ may also be written in the form:

$$H(s) = \frac{0 \cdot 246}{(s + 0 \cdot 494)(s + 0 \cdot 249 - j0 \cdot 972)(s + 0 \cdot 249 + j0 \cdot 972)} \qquad \cdots (2)$$

The a.c. steady state magnitude of $H(s)$ is

$$|H(j\omega)| = \frac{0 \cdot 246}{|(0 \cdot 491 - 0 \cdot 988\, \omega^2) + j(1 \cdot 238\, \omega - \omega^3)|} \quad \text{using (1)}$$

$$= \frac{0 \cdot 246}{|(0 \cdot 494 + j\omega)\{0 \cdot 249 + j(\omega - 0 \cdot 972)\}\{0 \cdot 249 + j(\omega + 0 \cdot 972)\}|}$$

$$\text{using (2)}$$

and hence

$$20 \log_{10} |H(j\omega)| = 20 \log_{10} 0 \cdot 246 - 10 \log_{10} \big[(0 \cdot 491 - 0 \cdot 988\, \omega^2)^2$$
$$+ (1 \cdot 238\, \omega - \omega^3)^2\big]\, \text{dB}$$

$$\text{or} \qquad = 20 \log_{10} 0 \cdot 246 - 10 \log_{10} (0 \cdot 494^2 + \omega^2)$$

$$- 10 \log_{10} \big[0 \cdot 249^2 + (\omega - 0 \cdot 972)^2\big]$$

$$- 10 \log_{10} \big[0 \cdot 249^2 + (\omega + 0 \cdot 972)^2\big]\, \text{dB}$$

The Bode plot, i.e. $20 \log_{10} |H(j\omega)|$ versus ω, is drawn in fig. 1.16.

1.3. Exercise Problems

1. The switch S in the circuit of fig. 1.17 is closed at $t = 0$. If C is initially uncharged, determine the subsequent variation of capacitor voltage v_C.

2. Determine the output voltage v_2 across the resistor R in the C–R network of fig. 1.18 for an applied voltage pulse, $v_1 = K$ for $0 \leqslant t \leqslant a$, assuming C is initially uncharged.

3. Determine the current i in a series R–L network when the following

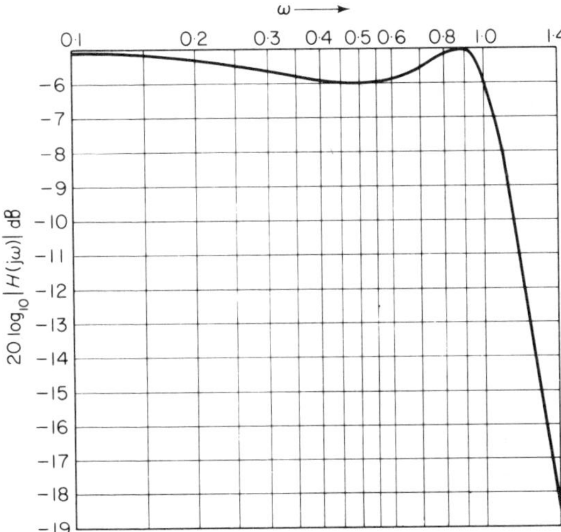

Fig. 1.16. Bode plot of magnitude response

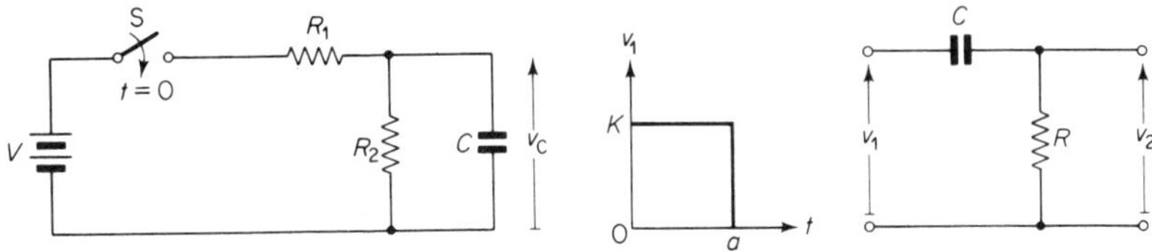

Fig. 1.17. Circuit for problem 1   Fig. 1.18. Network for problem 2

voltage sources are impressed across its terminals:
(a) $v = \delta(t)$, a voltage impulse; (b) $v = u(t)$, a step voltage;
(c) $v = tu(t)$, a ramp voltage. Assume $i(0^-) = 0$ in each case

4. Draw the transformed circuit for the 3 mesh circuit of fig. 1.19 for $t \geqslant 0$ and write down its s-domain equations in terms of $I_1(s)$, $I_2(s)$, and $I_3(s)$.

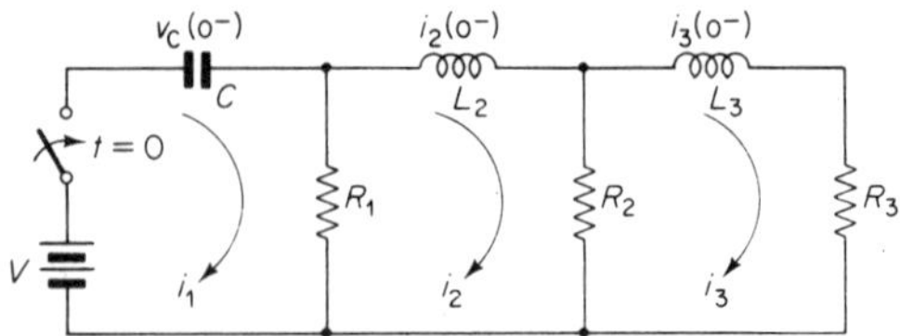

Fig. 1.19. Circuit for problem 4

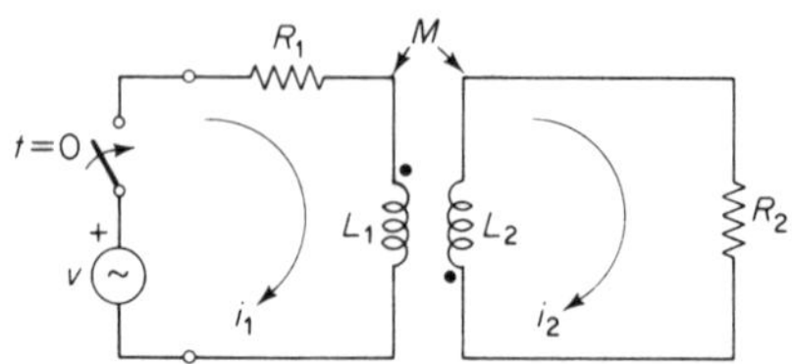

Fig. 1.20. Circuit for problem 5

5. A voltage generator $v = V \cos \omega t\, u(t)$ is applied across the input of the transformer circuit of fig. 1.20. Write down the s-domain mesh equations of the circuit and determine the output current $I_2(s)$. Assume $i_1(0^-) = i_2(0^-) = 0$.

6. Draw the transformed circuit for the circuit of fig. 1.21 and write down the s-domain nodal equations in terms of the node-pair voltages $V_1(s)$ and $V_2(s)$.

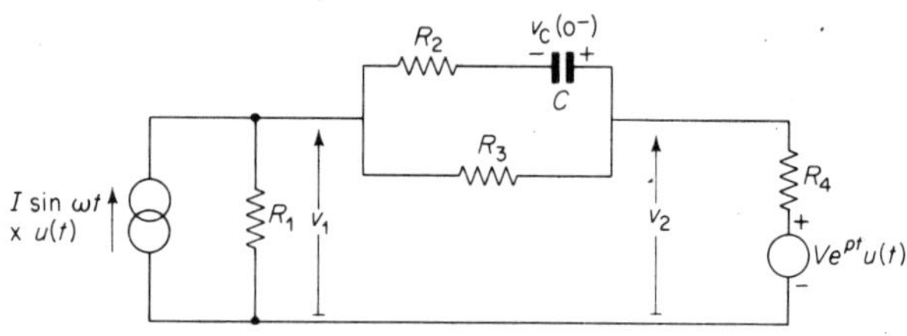

Fig. 1.21. Circuit for problem 6

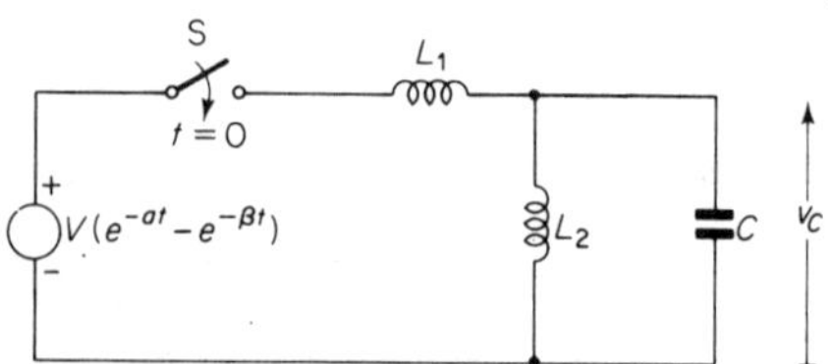

Fig. 1.22. Circuit for problem 7

7. The switch S in the circuit of fig. 1.22 is closed at $t = 0$ applying the voltage $v = V(e^{-\alpha t} - e^{-\beta t})$ across the L_1–L_2–C network, which contains no initial stored energy. Determine the subsequent variation in the capacitor voltage v_C.

8. In the series L–R–C circuit of fig. 1.23 switch S is changed from position 1 to 2 at $t = 0$. If the current in L and the potential difference across C just prior to switching are $i(0^-) = 2\,\text{A}$, $v_C(0^-) = 20\,\text{V}$, determine the subsequent variation in v_C.

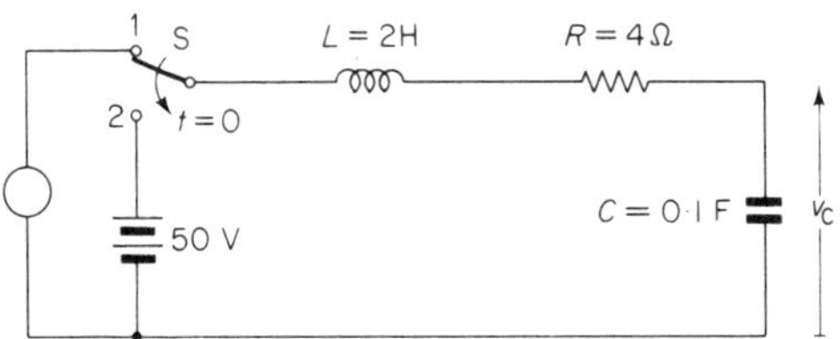

Fig. 1.23. Circuit for problem 8

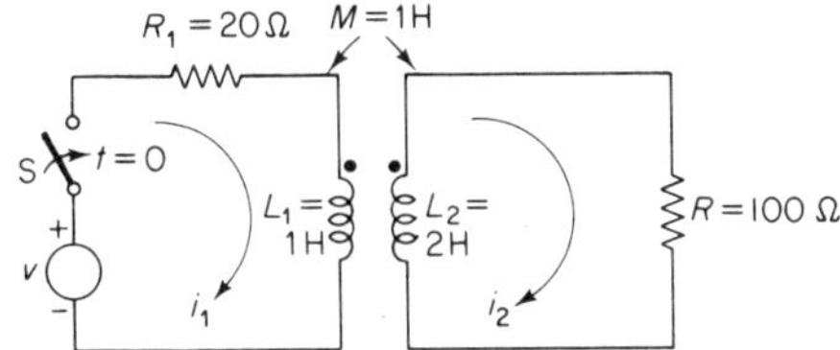

Fig. 1.24. Circuit for problem 9

9. In the circuit of fig. 1.24 switch S is closed at $t = 0$ applying the voltage $v = 40\,e^{-5t}$. Assuming $i_1(0^-) = i_2(0^-) = 0$, determine the subsequent variation of i_1 and i_2.

10. The circuit of fig. 1.25 has reached its steady state prior to the closing of switch S at $t = 0$. Draw the transformed circuit for $t \geqslant 0$ and find the solution for i_1 and i_2.

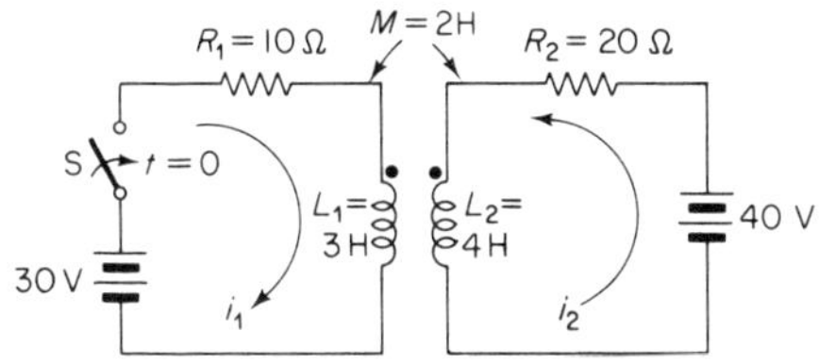

Fig. 1.25. Circuit for problem 10

11. The circuit of fig. 1.26 is in its steady state with switch S closed. If S
is opened at $t=0$, find i_1 for $t \geqslant 0$.

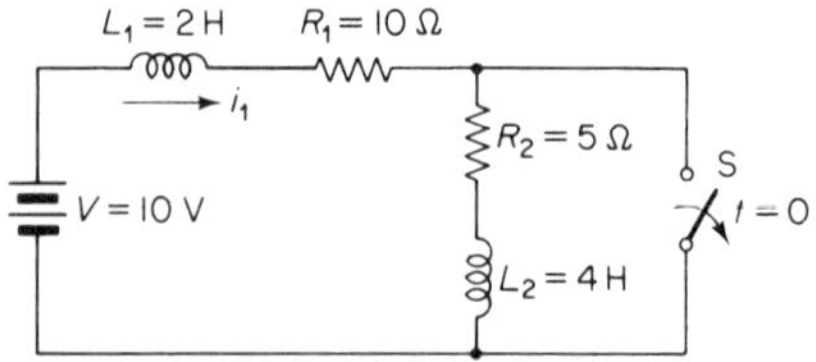

Fig. 1.26. Circuit for problem 11

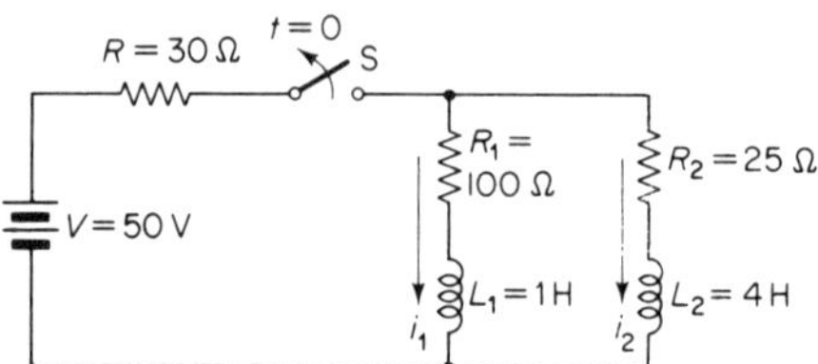

Fig. 1.27. Circuit for problem 12

12. The circuit of fig. 1.27 is in its steady state with switch S closed. If S is
opened at $t = 0$, determine i_1 and i_2 for $t \geqslant 0$.
13. A trapezoidal pulse, shown in fig. 1.28, is applied across the terminals of a series C–R network. Determine the voltage across the
resistor and sketch its variation with time for the case $C = 1000\,\text{pF}$,
$R = 10\,\text{k}\Omega$, C being initially uncharged.

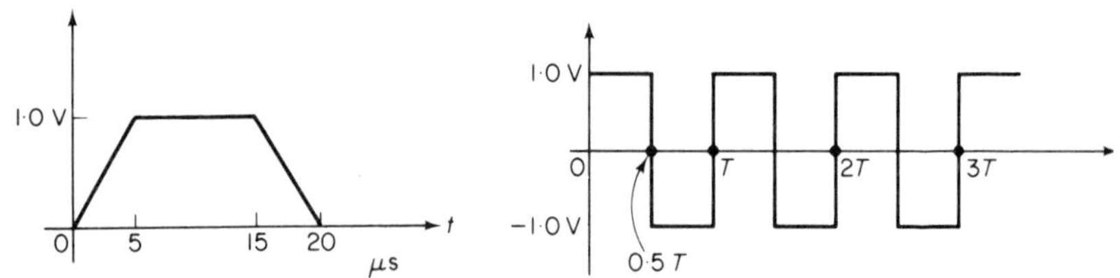

Fig. 1.28. Input for problem 13   Fig. 1.29. Input waveform for problem 14

14. The square wave, shown in fig. 1.29, is applied across the terminals
of a series C–R network. Determine the voltage across the resistor
and sketch its variation with time for the case when $CR = 0.1\,T$, C
being initially uncharged.

26

15. At $t = 0$ a 100 V d.c. supply is connected across terminals **AB** of the transformer circuit of fig. 1.30. Derive expressions for the currents i_1 and i_2 (assuming $i_1(0^-) = i_2(0^-) = 0$) and find the time at which i_2 reaches its maximum value.

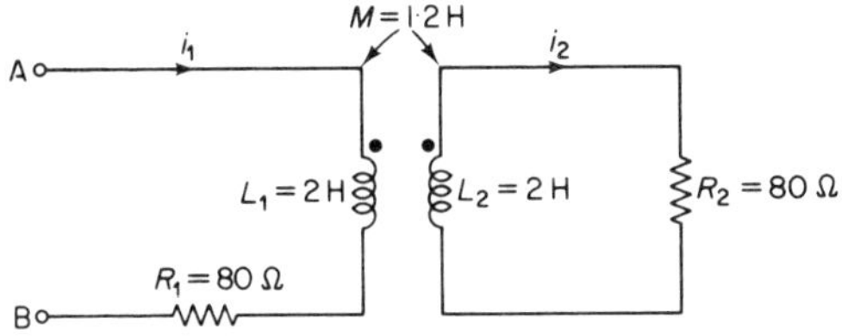

Fig. 1.30. Circuit for problem 15

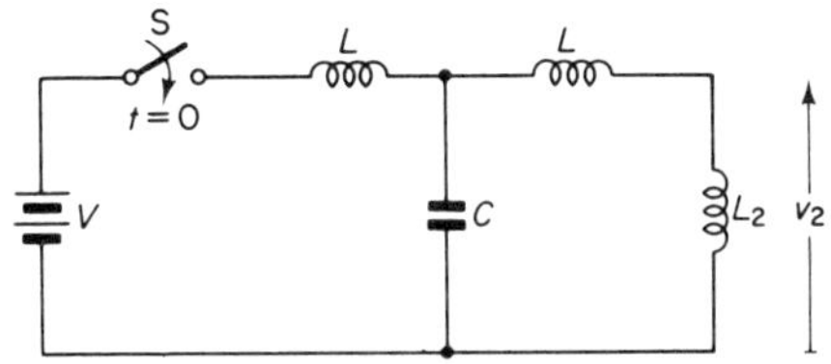

Fig. 1.31. Circuit for problem 16

16. If the switch S in the circuit of fig. 1.31 is closed at $t = 0$ and the circuit contains no initial stored energy, determine the output voltage v_2 for $t \geqslant 0$.
17. Determine the generalized impedance, $Z(s) = V(s)/I(s)$ of the parallel C–G–L network of fig. 1.32(a). Assuming the network contains no initial stored energy, determine the transform response $V(s)$ of the voltage $v(t)$ for the following current sources applied across terminals **AB**:

(a) $i = \cos \omega t u(t)$, (b) the pulse, $i = I$ for $0 \leqslant t \leqslant T$ (see fig. 1.32(b))
(c) the sawtooth, $i = It$ for $0 \leqslant t \leqslant T$, $i = 0$ for $t > T$ (see fig. 1.32(c)).

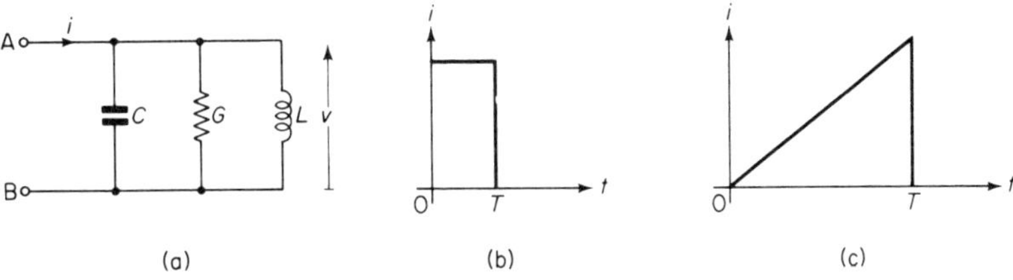

Fig. 1.32. For problem 17

18. Determine the transfer admittance $Y_{21}(s) = I_2(s)/V_1(s)$ for the circuit of fig. 1.33. Find also the current $I_2(s)$ if the switch S is switched from position 1 to 2 at $t = 0$, the circuit having reached its steady state in position 1 prior to switching.

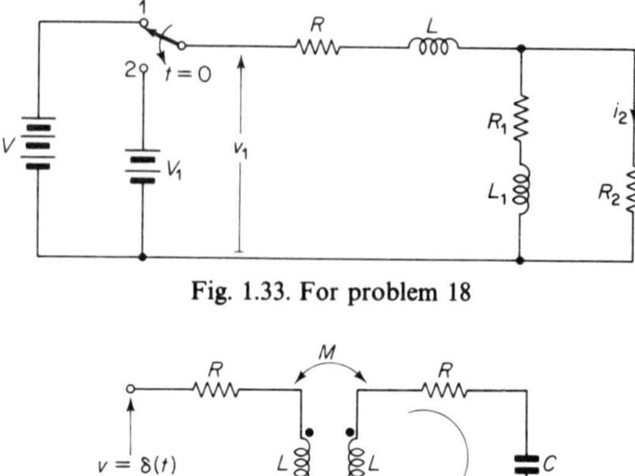

Fig. 1.33. For problem 18

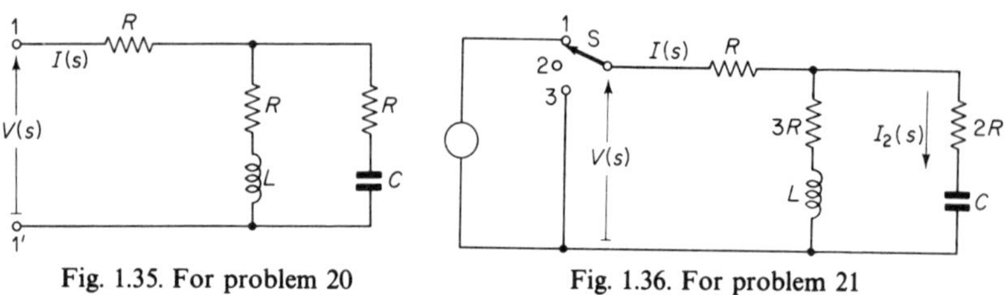

Fig. 1.34. For problem 19

19. A voltage impulse $v = \delta(t)$ is applied across the input terminals of the transformer network of fig. 1.34. Assuming that the network contains no initial stored energy, determine the transform secondary current $I_2(s)$. Also, given that $M = 1/\sqrt{2}$, $R = L = 1$, $C = \frac{1}{2}$, determine the time domain solution $i_2(t)$.

20. Determine the input admittance $Y(s) = I(s)/V(s)$ of the network of fig. 1.35. Write down the poles of $Y(s)$ and hence determine the condition such that the natural response of the network should be oscillatory when terminals 11' are short-circuited. If this condition is satisfied evaluate the radian frequency and the exponential damping factor of the natural response. If $R = L = C = 1$ is the natural response oscillatory?

Fig. 1.35. For problem 20

Fig. 1.36. For problem 21

21. Determine the input impedance $Z(s) = V(s)/I(s)$ of the circuit of fig. 1.36, and write down its poles and zeros. Hence find the general form of the solution for the response (a) $i_2(t)$ when the circuit is open-circuited (switch S in position 2), (b) $i(t)$ when the circuit is short-circuited (S in position 3).

22. Determine the voltage transfer function $V_0(s)/V_i(s)$ for the circuit of fig. 1.37. Hence or otherwise determine the output voltage v_0 for a d.c. input $v_i = u(t)$, assuming zero initial conditions.

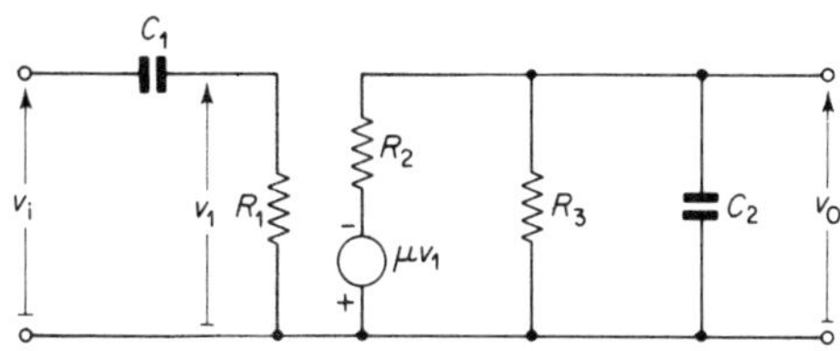

Fig. 1.37. For problem 22

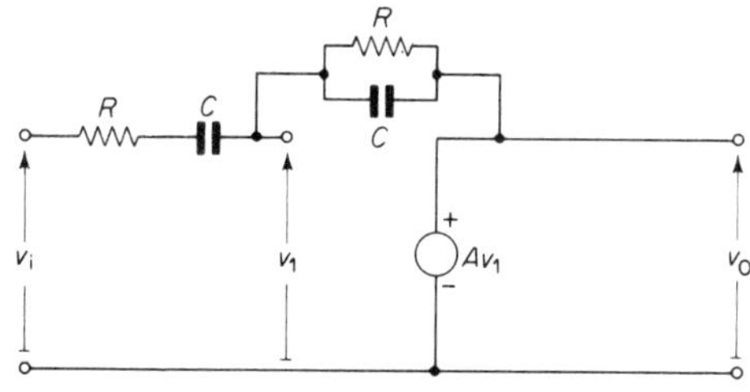

Fig. 1.38. For problem 23

23. Determine the voltage transfer function $V_0(s)/V_i(s)$ for the circuit of fig. 1.38. Assuming that $A \gg 1$ and zero initial conditions determine the impulse response, i.e. the output $v_0(t)$ for a voltage input $v_i(t) = \delta(t)$.

24. Determine the transfer function $H(s) = Y(s)/X(s)$ of the feedback system shown in fig. 1.39, and find the output response $y(t)$ for a step function input $x(t) = u(t)$.

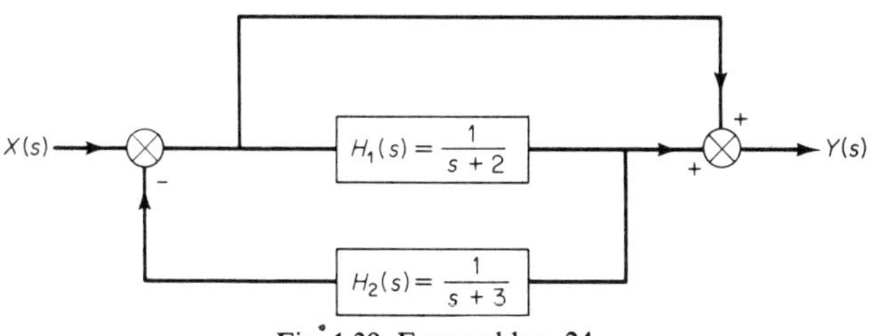

Fig. 1.39. For problem 24

25. Determine the transfer function $Y(s)/X(s)$ of the system of fig. 1.40.

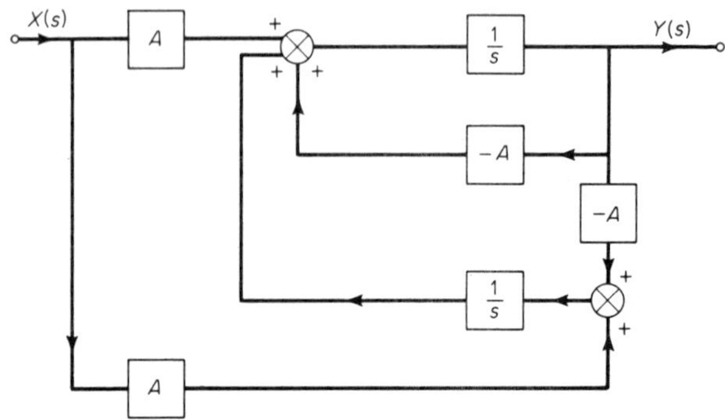

Fig. 1.40. For problem 25

26. Determine the output function $Y(s)$ and $y(t)$ for the inputs $x_1(t) = u(t)$ and $x_2(t) = \delta(t)$ in the system of fig. 1.41.

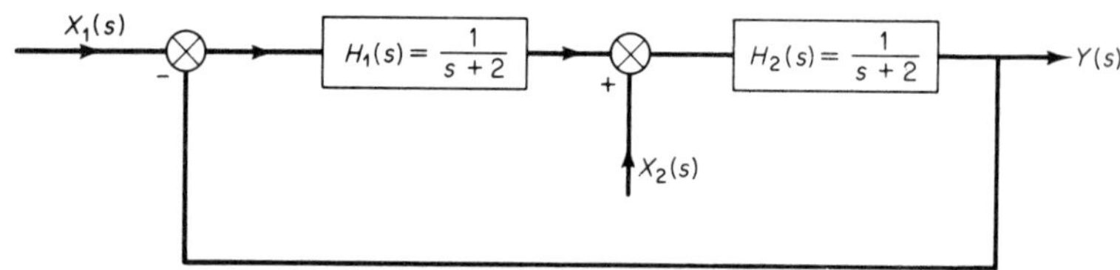

Fig. 1.41. For problem 26

27. Find the poles and zeros of (a) the impedance $Z(s) = V(s)/I(s)$ of the network of fig. 1.42(a), (b) the admittance $Y(s) = I(s)/V(s)$ of the network of fig. 1.42(b).

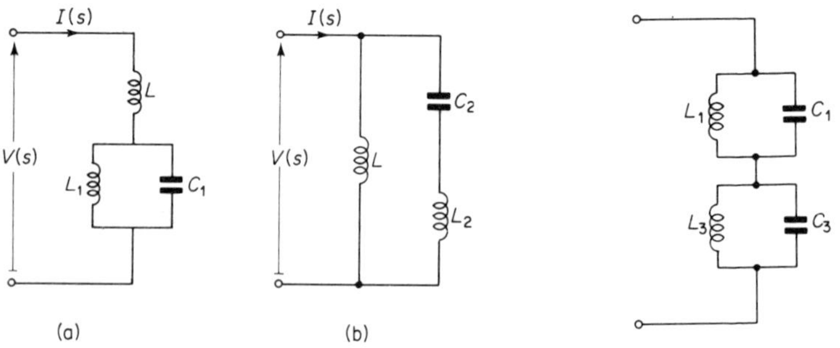

(a)

(b)

Fig. 1.42. For problem 27

Fig. 1.43. For problem 28

30

28. Find the poles and zeros of the impedance $Z(s)$ of the 4-element L–C network of fig. 1.43. Assuming $L_1 C_1 > L_3 C_3$, plot the poles and zeros in the s-plane. What deduction may be made concerning the relative positions of the poles and zeros on this diagram?

29. Determine the transfer function $H(s) = V_0(s)/V_G(s)$ for the tuned circuit of fig. 1.44 and find its poles assuming $R < 2\sqrt{(L/C)}$. Adopting the notation $\alpha = R/(2L)$, $\beta = \sqrt{[(1/LC) - (R^2/4L)]}$ show that $H(s)$ may be written in the form

$$H(s) = \frac{\alpha^2 + \beta^2}{(s + \alpha + j\beta)(s + \alpha - j\beta)}$$

and hence find the frequency at which $|H(j\omega)|$ is a maximum.

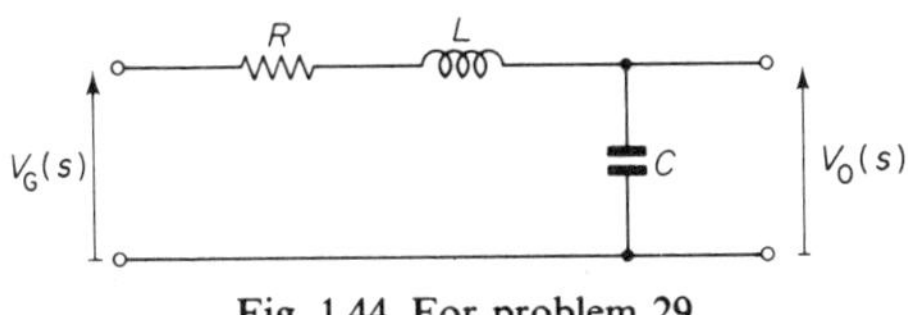

Fig. 1.44. For problem 29

30. Plot the magnitude and phase Bode diagrams for the transfer function

$$H(s) = \frac{1}{s^2 + 0.1\,s + 1} \quad \text{for } 0.1 \leqslant \omega \leqslant 2.0.$$

What is the maximum amplitude of $|H(j\omega)|$ and determine also the low and high frequency asymptotes of $|H(j\omega)|$.

2 Fourier Methods in Waveform and Circuit Analysis

2.1. Theory Summary

Section 1: Fourier Series

1. INTRODUCTION

The a.c. steady state theory developed for the analysis of circuits driven by constant amplitude sinusoidal sources may be applied to solve linear network problems for the more general case of non-sinusoidal periodic waveforms.

By means of Fourier analysis most of the periodic waveforms encountered in practice may be resolved into a convergent series composed of purely sinusoidal terms. This process provides immediately valuable information concerning the frequency component structure of the waveform. Also if the waveform is applied to a linear network each sinusoid of the series can be considered as acting separately and phasor-complex methods can be applied to determine the steady state response for each frequency component in turn. Finally on applying the principle of superposition the complete steady state solution is obtained by summing the individual solutions. For example, suppose that the driving function in the circuit of fig. 2.1 may be expanded as the series

$$v(t) = a_0 + a_1 \cos \omega t + a_2 \cos 2\omega t + \ldots a_n \cos n\omega t + \ldots$$

then the complex current response is given by

$$I = I_0 + \sum_{n=1}^{\infty} I_n$$

where $\quad I_0 = \dfrac{a_0}{Z(0)} \qquad$ is the d.c. current component

$\qquad I_n = \dfrac{a_n}{Z(jn\omega)} \qquad$ is the phasor current component due to the source $a_n \cos n\omega t$

$\qquad Z(jn\omega) = \qquad$ impedance of the network at the radian frequency $n\omega$

and the time domain solution:

$$i(t) = I_0 + \sum_{n=1}^{\infty} \text{Re}\left[I_n\, e^{jn\omega t}\right]$$

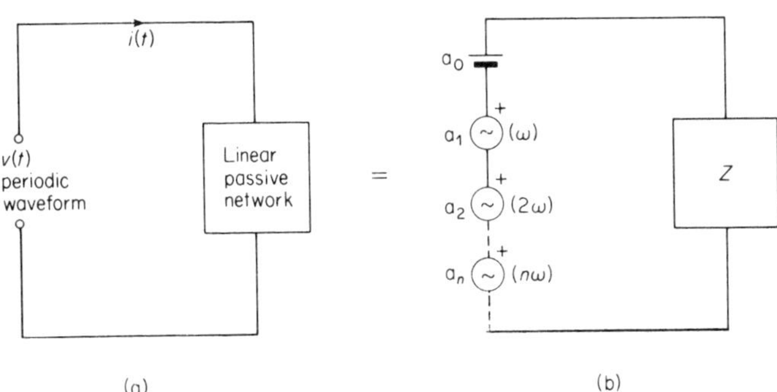

Fig. 2.1

2. REAL FORM OF THE FOURIER SERIES

A periodic waveform $f(t)$ of period T may be expanded as a uniformily convergent series consisting of the sum of a number of pure sinusoids of various amplitude but harmonically related frequencies $f, 2f, 3f \ldots nf \ldots$ in the general form of*

$$f(t) = a_0 + a_1 \cos \omega t + a_1 \cos 2\omega t$$

$$+ a_1 \cos 3\omega t + \ldots + a_n \cos n\omega t + \ldots$$

$$+ b_1 \sin \omega t + b_2 \sin 2\omega t$$

$$+ b_3 \sin 3\omega t + \ldots b_n \sin n\omega t + \ldots$$

$$= \sum_{n=0}^{\infty} (a_n \cos n\omega t + b_n \sin n\omega t)$$

where $\omega = 2\pi f = 2\pi/T$,

* Provided $f(t)$ satisfies the following conditions:
(i) $f(t)$ is truly periodic, i.e. $f(t) = f(t + nT)$, $n = \pm 1, \pm 2, \ldots$
(ii) $f(t)$ is a single-valued function.
(iii) In any period the magnitude of any discontinuity is finite and the number of discontinuities are finite.
(iv) $\displaystyle\int_{t}^{t+T} |f(t)|\, dt$ exists.

and the coefficients in the Fourier series (1) are:

$$a_0 = \frac{1}{T} \int_t^{t+T} f(t)\,dt$$

$$a_n = \frac{2}{T} \int_t^{t+T} f(t)\cos n\omega t\,dt$$

$$b_n = \frac{2}{T} \int_t^{t+T} f(t)\sin n\omega t\,dt \qquad \dots (2)$$

Note that the lower limit of integration is arbitrarily specified at any value of t. For convenience these limits are normally selected from $t = 0$ to T or $t = -\frac{1}{2}T$ to $+\frac{1}{2}T$.

By combining sinusoids of the same frequency (1) can be written in the form:

$$f(t) = \sum_{n=0}^{\infty} c_n \cos(n\omega t - \phi_n) \qquad \dots (3)$$

where $c_0 \cos \phi_0 = a_0$ ... the amplitude of the d.c. component

$$c_n = \sqrt{(a_n^2 + b_n^2)} \dots \text{the amplitude of the } n^{\text{th}} \text{ harmonic}$$

$$\phi_n = \tan^{-1}(b_n/a_n) \qquad \dots (4)$$

3. COMPLEX FORM OF THE FOURIER SERIES

The Fourier series of a function may also be expressed in complex form by substituting

$$\cos n\omega t = \tfrac{1}{2}(e^{jn\omega t} + e^{-jn\omega t}), \quad \sin n\omega t = \frac{1}{2j}(e^{jn\omega t} - e^{-jn\omega t})$$

in (1), i.e. $f(t) = \sum_{n=0}^{\infty} [\tfrac{1}{2}(a_n - jb_n)e^{jn\omega t} + \tfrac{1}{2}(a_n + jb_n)e^{-jn\omega t}]$

$$= \sum_{n=0}^{\infty} [d_n e^{jn\omega t} + d_{-n} e^{-jn\omega t}]$$

$$= \sum_{n=-\infty}^{\infty} d_n e^{jn\omega t} \qquad \dots (5)$$

where $d_n = \frac{1}{T} \int_t^{t+T} f(t)e^{-jn\omega t}\,dt \qquad \dots (6)$

Note also $d_n = \frac{1}{2}(a_n - jb_n)$, $d_{-n} = \frac{1}{2}(a_n + jb_n)$
so $|d_n| = |d_{-n}| = \frac{1}{2}\sqrt{(a_n^2 + b_n^2)} = \frac{1}{2}c_n$
and the amplitude of the nth harmonic, $c_n = 2|d_n|$.

4. CONSIDERATIONS OF WAVEFORM SYMMETRY IN SIMPLIFYING THE FOURIER ANALYSIS OF CERTAIN WAVEFORMS

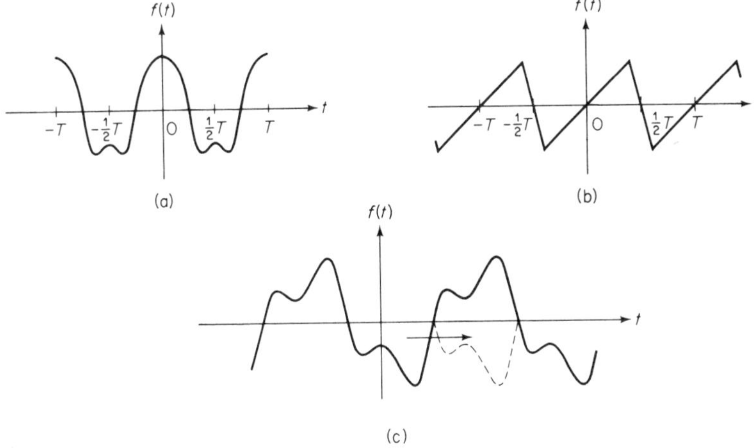

Fig. 2.2. Waveforms possessing symmetry: (a) an even function; (b) an odd function; (c) half-wave or mirror symmetry

(a) *Even functions, $f(t) = f(-t)$*

In this case $f(t) = \displaystyle\sum_{n=0}^{\infty} a_n \cos n\omega t$

$$a_n = \frac{4}{T} \int_0^{1/2T} f(t) \cos n\omega t \, dt \qquad \ldots (7)$$

(b) *Odd functions, $f(t) = -f(-t)$*

$$f(t) = \sum_{n=1}^{\infty} b_n \sin n\omega t, \quad b_n = \frac{4}{T} \int_0^{1/2T} f(t) \sin n\omega t \, dt \qquad \ldots (8)$$

(c) *Half-wave or mirror symmetry, $f(t) = -f(t + \frac{1}{2}T)$*
If a function possesses half-wave symmetry, then no even harmonic terms appear in its Fourier series. The Fourier coefficients for n odd are:

$$a_n = \frac{4}{T} \int_0^{1/2T} f(t) \cos n\omega t \, dt, \quad b_n = \frac{4}{T} \int_0^{1/2T} f(t) \sin n\omega t \, dt,$$

35

$$d_n = \frac{2}{T} \int_0^{1/2T} f(t)\,e^{-jn\omega t}\,dt$$

5. PARSEVAL'S THEOREM AND THE R.M.S. VALUE OF A PERIODIC WAVEFORM

Consider $f(t)$ as either a periodic voltage or current applied to a resistance of one ohm, then the instantaneous power dissipated in the resistor, $p = [f(t)]^2$

and the mean power,
$$\begin{aligned}
P &= \frac{1}{T} \int_0^T [f(t)]^2\,dt \\[6pt]
&= \frac{1}{T} \int_0^T f(t) \left[\sum_{-\infty}^{\infty} d_n\,e^{+jn\omega t} \right] dt \\[6pt]
&= \sum_{-\infty}^{\infty} \left[d_n \times \frac{1}{T} \int_0^T f(t)\,e^{-j(-n)\omega t}\,dt \right] \\[6pt]
&= \sum_{-\infty}^{\infty} d_n\,d_{-n} = \sum_{-\infty}^{\infty} |d_n|^2 \\[6pt]
&= c_0^2 + \sum_{n=1}^{\infty} \tfrac{1}{2}c_n^2 \qquad\qquad \dots (9)
\end{aligned}$$

This relation relating mean power to the Fourier series component amplitudes is known as Parseval's theorem. By definition the above expression is the mean square value of $f(t)$, thus the root-mean-square (r.m.s.) value of $f(t)$ is

$$f_{\text{rms}} = \sqrt{\left[\sum_{-\infty}^{\infty} |d_n|^2 \right]} = \sqrt{\left[c_0^2 + \sum_{n=1}^{\infty} \tfrac{1}{2}c_n^2 \right]} = \sqrt{\left[a_0^2 + \sum_{n=1}^{\infty} (a_n^2 + b_n^2) \right]}$$
$$\dots (10)$$

6. POWER, VOLT-AMPERES AND POWER FACTOR OF A CIRCUIT DRIVEN BY A PERIODIC WAVEFORM

If the Fourier series for the terminal voltage and current of a linear network are

$$v(t) = V_0 + \sum_{n=1}^{\infty} V_n \cos(n\omega t - \phi_{vn})$$

$$i(t) = I_0 + \sum_{n=1}^{\infty} I_n \cos{(n\omega t - \phi_{I_n})}$$

then the mean power supplied and dissipated in the circuit is

$$P = V_0 I_0 + \tfrac{1}{2} V_1 I_1 \cos{\phi_1} + \tfrac{1}{2} V_2 I_2 \cos{\phi_2} + \ldots + \tfrac{1}{2} V_n I_n \cos{\phi_n} + \ldots$$

where $\phi_1, \phi_2 \ldots \phi_n = (\phi_{V_n} - \phi_{I_n}) \ldots$ are the phase angle differences between the individual harmonic components of voltage and current; volt-amperes = (r.m.s. voltage amplitude) × (r.m.s. current amplitude)

$$= \sqrt{\left[\left(V_0^2 + \tfrac{1}{2} \sum_{n=1}^{\infty} V_n^2 \right) \left(I_0^2 + \tfrac{1}{2} \sum_{n=1}^{\infty} I_n^2 \right) \right]}$$

and the power factor $= \dfrac{\text{Mean Power}}{\text{Volt-amperes}}$

7. LINE SPECTRA REPRESENTATION OF PERIODIC SIGNALS AND WAVEFORMS

This is essentially a graphical means of representing the frequency components contained in the Fourier series of a periodic signal. Normally a plot of component amplitude versus frequency and phase versus frequency are drawn. For example the line spectra of the square wave of fig. 2.3(a) are shown for the real Fourier series in fig. 2.3(b) and for the complex Fourier series in fig. 2.3(c).

Section 2: Fourier Transforms

8. DEVELOPMENT OF THE FOURIER TRANSFORM FROM THE COMPLEX FOURIER SERIES

(a) *Introductory comment*

By the use of Fourier series expansions periodic waveforms may be resolved into a series of harmonic components. A.C. steady state analysis may be then applied to analyse the performance of systems driven by such functions. Such a procedure is a particular example of transforming from the time to the frequency domain to effect a solution.

In many problems we are concerned with signals which are not periodic, but continually change with time or have pulse type characteristics. To extend Fourier methods to these types of signals we may consider the 'period' T of the waveform as being very much longer than all the variations contained in the signal. The signal can then be 'Fourier analysed' in the standard way. The resulting frequency spectra would

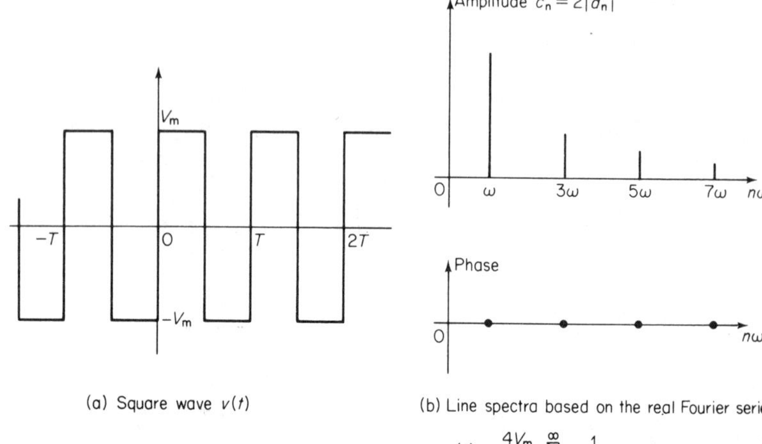

(a) Square wave $v(t)$

(b) Line spectra based on the real Fourier series

$$v(t) = \frac{4V_m}{\pi} \sum_{r=0}^{\infty} \frac{1}{2r+1} \sin(2r+1)\,\omega t$$

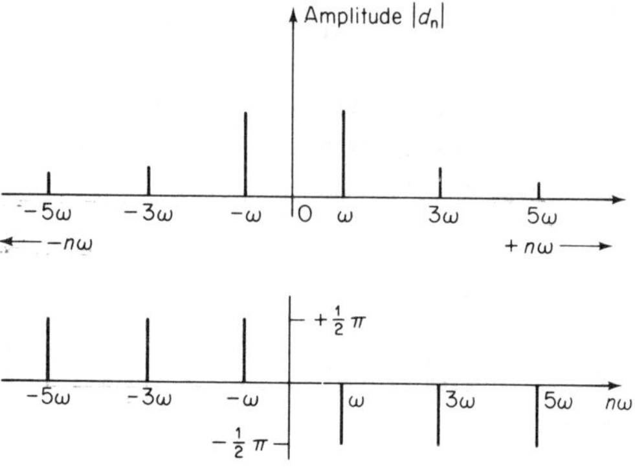

(c) Line spectra based on the complex Fourier series

$$v(t) = \sum_{-\infty}^{\infty} \frac{2V_m}{j\pi(2r+1)}\, e^{j(2r+1)\omega t}$$

Fig. 2.3

then consist of a vast number of closely spaced lines, spaced at intervals of $1/T$ hertz. In the limit as T approaches infinity the line spectrum would merge into a continuous spectrum. This limiting process is used below to develop the Fourier transform and the Fourier integral representation of a non-periodic signal.

38

(b) *Mathematical development: the Fourier transform and integral*
Using the complex Fourier results (5) and (6) we have for a periodic
function $f(t)$:

$$f(t) = \sum_{n=-\infty}^{\infty} d_n \, e^{jn\omega_0 t}$$

where

$$d_n = \frac{1}{T} \int_{-1/2T}^{1/2T} f(t) e^{-jn\omega_0 t} \, dt$$

and so

$$f(t) = \sum_{-\infty}^{\infty} \left[\frac{1}{T} \int_{-1/2T}^{1/2T} f(t) e^{-jn\omega_0 t} \right] e^{jn\omega_0 t} \qquad \dots (11)$$

Assume next that the period $T = 2\pi/\omega_0$ is considered as very large
compared to any variation in $f(t)$ (i.e. $f(t)$ tends to a non-periodic
function), and that ω_0 is replaced by $\delta\omega$, $n\omega_0$ by ω, and T by $2\pi/\delta\omega$, then
(11) becomes:

$$f(t) = \sum_{-\infty}^{\infty} \left[\frac{1}{2\pi} \int_{-\pi/\delta\omega}^{+\pi/\delta\omega} f(t) e^{-j\omega t} \, dt \right] e^{j\omega t} \delta\omega$$

Further on taking the limit as $\delta\omega \to 0$ we have

$$f(t) = \frac{1}{2\pi} \int_{-\infty}^{\infty} \left[\int_{-\infty}^{\infty} f(t) e^{-j\omega t} \, dt \right] e^{j\omega t} \, d\omega$$

$$= \frac{1}{2\pi} \int_{-\infty}^{\infty} F(\omega) e^{j\omega t} \, d\omega \qquad \dots (12)$$

where
$$F(\omega) = \int_{-\infty}^{\infty} f(t) e^{-j\omega t} \, dt \qquad \dots (13)$$

is the **FOURIER TRANSFORM** (or Fourier integral) of $f(t)$ and
relation (12) is the Fourier integral (inverse Fourier transform of $F(\omega)$)
representation of $f(t)$.

9. RAYLEIGH'S ENERGY THEOREM: THE PHYSICAL SIGNIFICANCE OF THE FOURIER TRANSFORM

Physical significance can be given to the Fourier transform of electrical
signals by interpreting $F(\omega)$ as a spectrum density function quantifying
the continuous spectrum characteristic of a non-periodic function.

Consider a non-periodic voltage or current $f(t)$ applied to a one ohm
resistor, then the total energy dissipated in the resistor,

$$W = \int_{-\infty}^{\infty} [f(t)]^2 \, dt$$

$$= \int_{-\infty}^{\infty} f(t) \left[\frac{1}{2\pi} \int_{-\infty}^{\infty} F(\omega) \, e^{j\omega t} \, d\omega \right] dt$$

$$= \frac{1}{2\pi} \int_{-\infty}^{\infty} F(\omega) \left[\int_{-\infty}^{\infty} f(t) \, e^{-j(-\omega)t} \, dt \right] d\omega$$

$$= \frac{1}{2\pi} \int_{-\infty}^{\infty} F(\omega) F(-\omega) \, d\omega = \frac{1}{2\pi} \int_{-\infty}^{\infty} |F(\omega)|^2 \, d\omega$$

$$= \frac{2}{2\pi} \int_{0}^{\infty} |F(\omega)|^2 \, d\omega = 2 \int_{0}^{\infty} |F(\omega)|^2 \, df \qquad \ldots (14)$$

where $f = \omega/2\pi$. Expression (14) is a statement of Rayleigh's energy theorem and compares directly with Parseval's power theorem for the Fourier series of a periodic function. $F(\omega)$ describes the distribution of signal voltage or current with frequency, whilst $|F(\omega)|^2$ describes the distribution of energy with frequency of a non-periodic signal. $|F(\omega)|^2$ is often referred to as the energy spectral density function. In particular the fraction of the total energy carried by the signal in the frequency range f_1 to f_2 is

$$W_{12}/W = \left[2 \int_{f_1}^{f_2} |F(\omega)|^2 \, df \right] / W \qquad \ldots (15)$$

10. PROPERTIES AND TABLES OF FOURIER TRANSFORMS

Listed below, in the tables of fig. 2.4 and 2.5 are some of the more common theorems and transform pairs. Also, tabulated in fig. 2.6, are some numerical values of the sinc and Si functions.

	$f(t)$	$F(\omega)$
Linearity	$a_1 f_1(t) + a_2 f_2(t)$	$a_1 F_1(\omega) + a_2 F_2(\omega)$
Time shift	$f(t - t_0)$	$F(\omega)\, e^{-j\omega t_0}$
Scaling	$f(at)$	$\frac{1}{\lvert a \rvert} F(\frac{\omega}{a})$
Duality	$f(t)$ $F(t)$	$F(\omega)$ $2\pi f(-\omega)$
Frequency translation	$f(t)\, e^{j\omega_c t}$	$F(\omega - \omega_c)$
Differentiation	$\frac{d}{dt} f(t)$	$j\omega F(\omega)$
	$\frac{d^n}{dt^n} f(\omega)$	$(j\omega)^n F(\omega)$
Integration	$\int_{-\infty}^{t} f(t)\, dt$	$\frac{1}{j\omega} F(\omega)$
Multiplication	$f_1(t) f_2(t)$	$\frac{1}{2\pi} \int_{-\infty}^{\infty} F_1(\omega - x) F_2(x)\, dx$
(Convolution theorem)	$\int_{-\infty}^{\infty} f(t - x) f_2(x)\, dx$	$F_1(\omega) F_2(\omega)$

Fig. 2.4

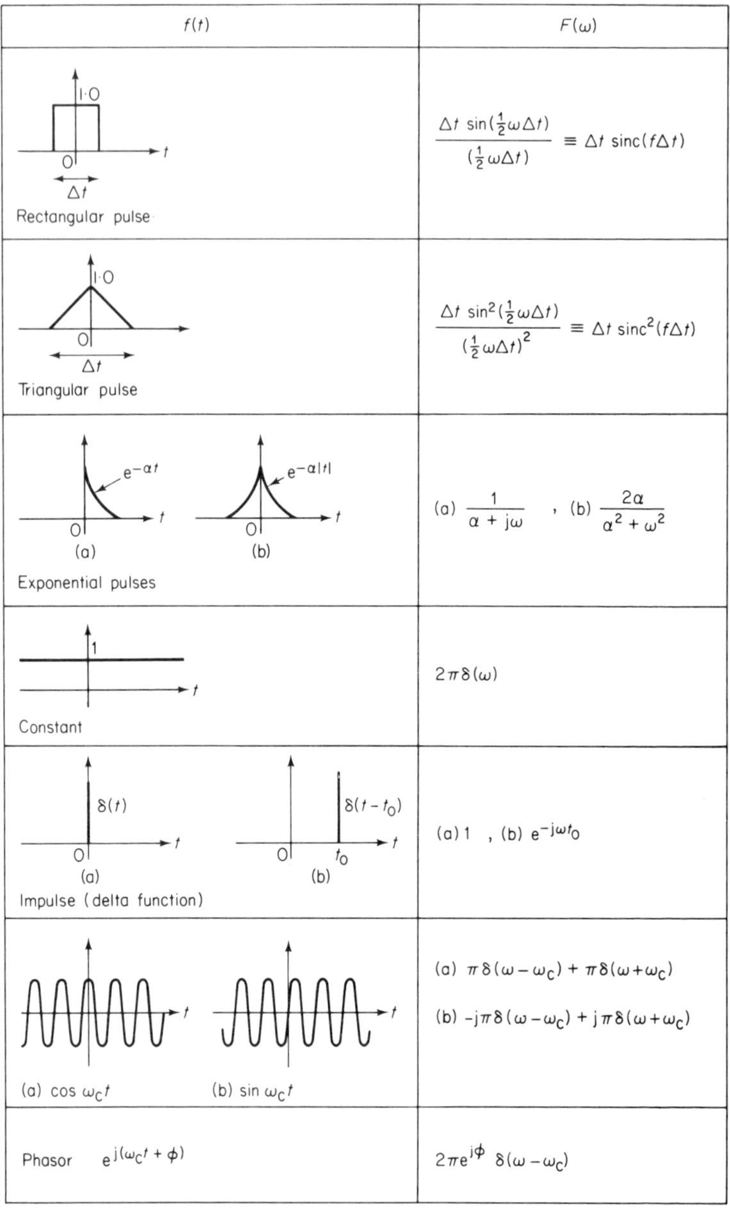

Fig. 2.5. Table of some transform pairs

x	$\mathrm{sinc}\ x$ $= \dfrac{\sin \pi x}{\pi x}$	$\mathrm{sinc}^2\ x$ $= \left(\dfrac{\sin \pi x}{\pi x}\right)^2$	u	$\dfrac{\sin u}{u}$	$\mathrm{Si}\ (x) =$ $\displaystyle\int_0^x \dfrac{\sin u}{u}\,du$
0·0	1	1	0·0	1	0
0·2	0·935	0·875	0·2	0·993	0·200
0·4	0·757	0·573	0·4	0·974	0·397
0·6	0·505	0·255	0·6	0·941	0·588
0·8	0·234	0·055	0·8	0·897	0·772
1·0	0	0	1·0	0·842	0·946
1·2	−0·156	0·024	1·2	0·777	1·108
1·4	−0·216	0·047	1·4	0·704	1·256
1·6	−0·189	0·036	$\pi/2$	0·637	1·370
1·8	−0·104	0·011	1·6	0·625	1·389
2·0	0	0	1·8	0·541	1·506
2·2	0·085	0·007	2·0	0·455	1·605
2·4	0·126	0·016	2·2	0·368	1·688
2·6	0·116	0·014	2·4	0·281	1·753
2·8	0·067	0·004	2·6	0·198	1·800
3·0	0	0	2·8	0·120	1·832
3·2	−0·058	0·003	3·0	0·047	1·849
3·4	−0·089	0·008	π	0	1·851
3·6	−0·084	0·007	2π	0	1·419
3·8	−0·049	0·002	3π	0	1·675
4·0	0	0	4π	0	1·492
∞	0	0	∞	0	$\pi/2 \approx 1\cdot571$

Fig. 2.6

11. DETERMINATION OF NETWORK RESPONSE USING FOURIER TRANSFORMS

Using the general philosophy of transform methods the response of a system or network may be found in the following manner, providing of course the Fourier transform pairs exist and can be evaluated.

(i) The Fourier transform $E(\omega)$ of the input signal $e(t)$ is first found.

(ii) The complex transfer function of the network, $H(j\omega)$, is evaluated. This is accomplished using standard a.c. theory (i.e. R, L, C are represented by the impedance elements R, $j\omega L$, $1/(j\omega C)$).

(iii) The frequency and time domain responses are then given, respectively, by:

$$R(\omega) = H(j\omega)E(\omega)$$

$$r(t) = \mathscr{F}^{-1}[H(j\omega)E(\omega)]$$

where $\mathscr{F}^{-1}$ denotes the inverse transform operation.

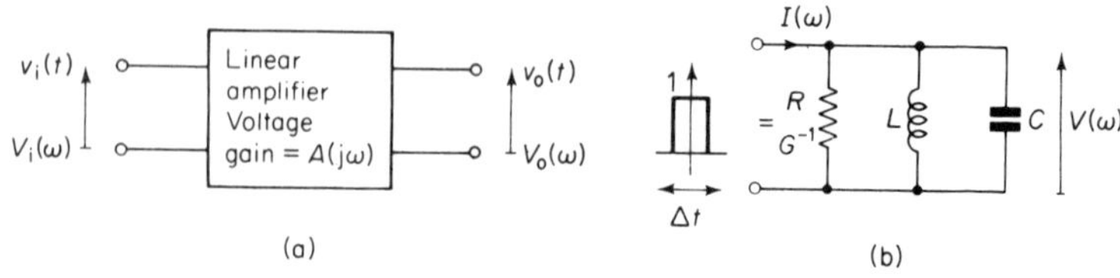

(a) (b)

Fig. 2.7

For example in the amplifier circuit of fig. 2.7(a),

$$V_0(\omega) = A(j\omega)V_i(\omega), \quad v_0(t) = \mathscr{F}^{-1}[A(j\omega)V_i(\omega)]$$

whilst for the R–L–C network of fig. 2.7(b):
the Fourier transform of the input current pulse is $I(\omega) = \Delta t \sin (\tfrac{1}{2}\omega\Delta t)/(\tfrac{1}{2}\omega\Delta t)$, the a.c. transfer function (in this case the network impedance) is

$$H(j\omega) = \frac{V(\omega)}{I(\omega)} = \frac{1}{G + j\omega C + 1/(j\omega L)},$$

and the transform output voltage is then given by

$$V(\omega) = H(j\omega)I(\omega) = \frac{\Delta t \sin (\tfrac{1}{2}\omega\Delta t)/(\tfrac{1}{2}\omega\Delta t)}{G + j\omega C + 1/(j\omega L)}$$

2.2. Worked Problems

1. Determine the Fourier series expansion for the half wave rectifier waveform shown in fig. 2.8(a).

   Calculate (a) the current supplied to, and (b) the peak voltage amplitudes developed across terminals AB at 50 Hz and 100 Hz, in the network of fig. 2.8(b) when it is fed by a half wave rectifier voltage of peak amplitude $V_m = 200$ volts at 50 Hz.

Solution
The half wave rectifier voltage of fig. 2.8(a) is defined by:

$$f(t) = V_m \sin \omega t \text{ from } t = 0 \text{ to } \tfrac{1}{2}T,$$

44

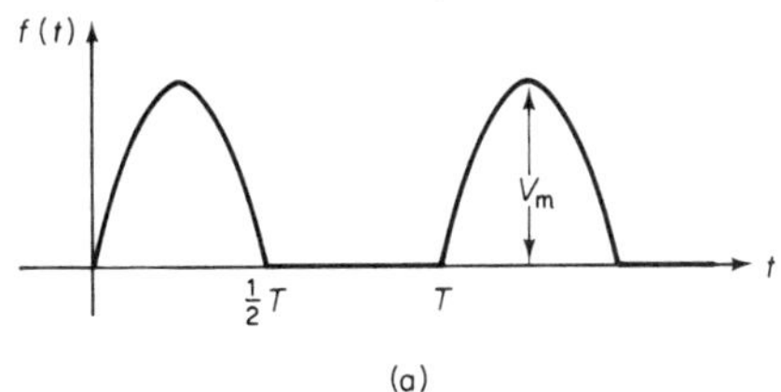

(a)

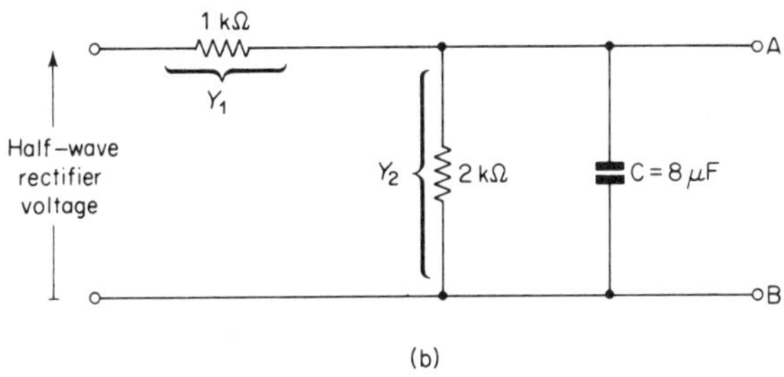

(b)

Fig. 2.8

and $f(t) = 0$ from $t = \frac{1}{2}T$ to T.

Thus $a_0 = \dfrac{1}{T} \displaystyle\int_0^T f(t)\,\mathrm{d}t$

$$= \frac{1}{T}\left[\int_0^{1/2T} V_{\mathrm{m}} \sin \omega t\,\mathrm{d}t + \int_{1/2T}^T 0\,\mathrm{d}t\right] = \frac{V_{\mathrm{m}}}{\pi}$$

$$a_{\mathrm{n}} = \frac{2}{T}\int_0^T f(t)\cos n\omega t\,\mathrm{d}t = \frac{2}{T}\int_0^{1/2T} V_{\mathrm{m}}\sin \omega t \cos n\omega t\,\mathrm{d}t$$

$$= \frac{2V_{\mathrm{m}}}{T}\left[\int_0^{1/2T}\{\sin(n+1)\omega t - \sin(n-1)\omega t\}\,\mathrm{d}t\right]$$

$$= \frac{2V_{\mathrm{m}}}{T}\left[\frac{-\cos(n+1)\omega t}{(n+1)\omega} + \frac{\cos(n-1)\omega t}{(n-1)\omega}\right]_0^{1/2T}$$

$$= 0 \text{ if } n \text{ is odd}$$

$$= \frac{2V_{\mathrm{m}}}{T\omega}\left(\frac{1}{n+1} - \frac{1}{n-1}\right) = \frac{-2V_{\mathrm{m}}}{\pi(n^2-1)} \text{ if } n \text{ is even.}$$

$$b_{\mathrm{n}} = \frac{2}{T}\int_0^T f(t)\sin n\omega t\,\mathrm{d}t = \frac{2}{T}\int_0^{1/2T} V_{\mathrm{m}}\sin \omega t \sin n\omega t\,\mathrm{d}t$$

45

$$= \begin{cases} \frac{1}{2}V_m, & n=1 \\ 0, & n=2,3,4 \ldots \end{cases}$$

and $f(t) = \sum_{n=0}^{\infty} (a_n \cos n\omega t + b_n \sin n\omega t)$

$$= V_m \left[\frac{1}{\pi} + \frac{1}{2} \sin \omega t - \frac{2}{\pi} \sum_{n=2,4,6 \ldots}^{\infty} \frac{\cos n\omega t}{n^2 - 1} \right]$$

$$= V_m \left(\frac{1}{\pi} + \frac{1}{2} \sin \omega t - \frac{2}{3\pi} \cos 2\omega t \right.$$

$$\left. - \frac{2}{15\pi} \cos 4\omega t - \frac{2}{35\pi} \cos 6\omega t \ldots \right) \qquad \ldots (1)$$

(a) The current supplied to the network of fig. 2.8(b) may be found by computing the total admittance at each of the frequencies 0, 50, 100, 200 Hz etc. present in (1) and calculating therefrom the amplitude and phase of the individual currents by considering each voltage term in (1) as acting separately.

At d.c. C acts as an open circuit, hence the d.c. component is:

$$i_{dc} = \frac{\frac{1}{\pi}V_m}{(1+2)10^3} = \frac{200}{3 \cdot 10^3 \pi} = 0 \cdot 0212 \text{ A.}$$

The admittance of the circuit at any harmonic frequency $n\omega$ is

$$Y(n\omega) = \frac{Y_1 Y_2}{Y_1 + Y_2}$$

where $Y_1 = 10^{-3}$, $Y_2 = (0 \cdot 5 \times 10^{-3} + jn\omega C)$, $C = 8 \times 10^{-6}$, $\omega = 100\pi$.

Hence $Y(n\omega) = \dfrac{10^{-3}(0 \cdot 5 + jn\omega C \times 10^3)}{(1 \cdot 5 + jn\omega C \times 10^3)}$,

and on substituting $\omega C = 2 \cdot 513 \times 10^{-3}$

$$Y(n\omega) = \frac{10^{-3}[(0 \cdot 75 + n^2 6 \cdot 315) + jn2 \cdot 513]}{2 \cdot 25 + n^2 6 \cdot 315}$$

from which $|Y(n\omega)| = \dfrac{10^{-3}[(0 \cdot 75 + n^2 6 \cdot 315)^2 + n^2 6 \cdot 315]^{1/2}}{2 \cdot 25 + n^2 6 \cdot 315}$

and the argument of $Y(n\omega)$, $\phi = \tan^{-1} \left[\dfrac{2 \cdot 513n}{0 \cdot 75 + 6 \cdot 315n^2} \right]$

At $f = 50\,\text{Hz}$, $|Y| = 0.875 \times 10^{-3}$, $\phi = 19° 35'$ so the current component at this frequency is

$$i_1 = |Y|\tfrac{1}{2}V_m \sin(\omega t + \phi) = 0.0875 \sin(\omega t + 19° 35')$$

At $f = 100\,\text{Hz}$, $|Y| = 0.963 \times 10^{-3}$, $\phi = 10° 56'$ so the current is

$$i_2 = |Y|\left(-\frac{2}{3\pi}V_m\right)\cos(2\omega t + \phi) = -0.0409 \cos(2\omega t + 10° 56')$$

At $f = 200\,\text{Hz}$, $|Y| = 0.991 \times 10^{-3}$, $\phi = 5° 38'$ so the $200\,\text{Hz}$ current is

$$i_4 = |Y|\left(-\frac{2}{15\pi}V_m\right)\cos(4\omega t + \phi) = -0.00841 \cos(4\omega t + 5° 38')$$

The resultant current i is the sum of all the individual components, i.e.

$$i = i_{dc} + i_1 + i_2 + i_4 + \dots$$
$$= 0.0212 + 0.0875 \sin(\omega t + 19° 35') - 0.0409 \cos(2\omega t + 10° 56')$$
$$- 0.00841 \cos(4\omega t + 5° 38') - \dots$$

(b) The peak voltage amplitude developed across AB at $n\omega$ is

$$|V_2(n\omega)| = (\text{amplitude of current at } n\omega) \times |Z_2(n\omega)|$$

where $|Z_2(n\omega)| = \dfrac{1}{|Y_2(n\omega)|} = \dfrac{1}{|0.5 \times 10^{-3} + jn\omega C|}$

At $f = 50\,\text{Hz}$, $|Z_2| = 390.3\,\Omega$; and at $f = 100\,\text{Hz}$, $|Z_2| = 198.0\,\Omega$
Therefore the respective peak amplitudes are

$$|V_2(50\,\text{Hz})| = 390.3 \times 0.0875 = 34.14\,\text{V}$$

$$|V_2(100\,\text{Hz})| = 198.0 \times 0.0409 = 8.10\,\text{V}$$

2. Determine the complex Fourier series for the square waveform shown in fig. 2.9(a). Express the result also in real form. Using the last result write down the Fourier series for the waveform of fig. 2.9(b).

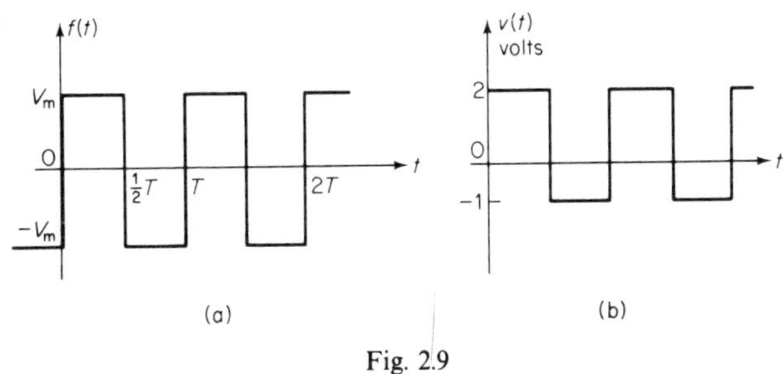

Fig. 2.9

Solution

The square waveform of fig. 2.9(a) is defined by

$$f(t) = +V_m \text{ from } t = 0 \text{ to } t = \tfrac{1}{2}T$$

$$= -V_m \text{ from } t = \tfrac{1}{2}T \text{ to } t = T$$

Thus $d_n = \dfrac{1}{T} \displaystyle\int_0^T f(t) e^{-jn\omega t}\, dt$

$$= \frac{1}{T}\left[\int_0^{1/2T} V_m\, e^{-jn\omega t}\, dt + \int_{1/2T}^T -V_m\, e^{-jn\omega t}\, dt \right]$$

$$= \frac{V_m}{T}\left[\left\{\frac{e^{-jn\omega t}}{-jn\omega}\right\}_0^{1/2T} - \left\{\frac{e^{-jn\omega t}}{-jn\omega}\right\}_{1/2T}^T \right]$$

$$= \frac{-V_m}{jn\omega T}\left(e^{-jn\pi} - 1 - e^{-jn2\pi} + e^{-jn\pi}\right) = \frac{V_m}{jn\pi}\left(1 - e^{-jn\pi}\right)$$

So $d_n = 0$ if n is even,

$$d_n = \frac{2V_m}{jn\pi} \text{ if } n \text{ is odd;}$$

i.e. $d_n = \dfrac{2V_m}{j(2r+1)\pi}$, where $(2r+1)$ replaces n and r can

take all integral values from $-\infty$ to $+\infty$. Thus:

$$f(t) = \sum_{n=-\infty}^{+\infty} d_n\, e^{jn\omega t} = \frac{2V_m}{j\pi} \sum_{r=-\infty}^{+\infty} \frac{1}{(2r+1)} e^{j(2r+1)\omega t}$$

Expressing the Fourier series in the real form of

$$f(t) = \sum_{n=0}^{\infty} (a_n \cos n\omega t + b_n \sin n\omega t)$$

we have

$$a_n = d_n + d_{-n} = \frac{2V_m}{jn\pi} + \frac{2V_m}{-jn\pi} = 0$$

and

$$b_n = j(d_n - d_{-n}) = \frac{4V_m}{n\pi} = \frac{4V_m}{(2r+1)\pi}$$

so

$$f(t) = \frac{4V_m}{\pi} \sum_{r=0}^{\infty} \frac{1}{2r+1} \sin(2r+1)\omega t \qquad \dots (1)$$

Fig. 2.9(b) also depicts a square wave but with a d.c. vertical shift. The peak to peak amplitude of this wave is $2V_m = 2 - (-1) = 3$ V and its d.c. vertical shift $= 2 - V_m = 2 - 1\cdot5 = 0\cdot5$ V.

48

Thus the new waveform may be represented as a combination of a symmetrical square wave of amplitude $V_m = 1\cdot5\,V$ plus a d.c. voltage of $0\cdot5\,V$. Its Fourier series may then be found by adding $0\cdot5$ to expansion (1) and setting $V_m = 1\cdot5$, thus

$$v(t) = 0\cdot5 + \frac{6}{\pi} \sum_{r=0}^{\infty} \frac{1}{2r+1} \sin(2r+1)\omega t \text{ volts}$$

3. Determine the Fourier series for each of the waveforms shown in fig. 2.10.

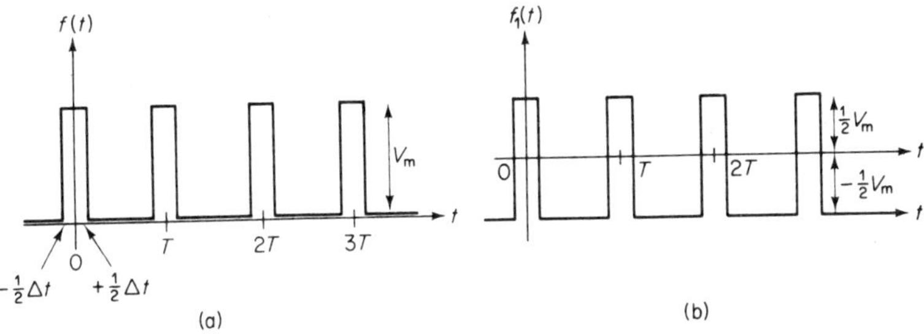

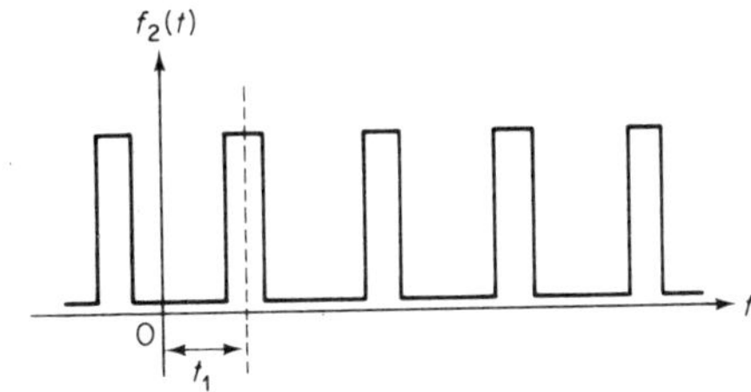

Fig. 2.10

Solution
The waveform of fig. 2.10(a) is an even function and therefore contains no sine terms. Over the interval $t = 0$ to $\tfrac{1}{2}T$ the wave is defined by $f(t) = V_m\ 0 \leqslant t \leqslant \tfrac{1}{2}\Delta t,\ f(t) = 0\ \tfrac{1}{2}\Delta t < t < \tfrac{1}{2}T$. Thus

$$f(t) = \sum_{n=0}^{\infty} a_n \cos n\omega t$$

where

$$a_0 = \frac{1}{T} \int_{-1/2T}^{1/2T} f(t)\,dt = \frac{2}{T} \int_{0}^{1/2T} f(t)\,dt$$

$$= \frac{2}{T} \int_0^{1/2\Delta t} V_m \, dt = \frac{\Delta t}{T} V_m$$

and
$$a_n = \frac{2}{T} \int_{-1/2T}^{1/2T} f(t) \cos n\omega t \, dt$$

$$= \frac{4}{T} \int_0^{1/2T} f(t) \cos n\omega t \, dt$$

$$= \frac{4}{T} \int_0^{1/2\Delta t} V_m \cos n\omega t \, dt = \frac{4}{T} \left[V_m \frac{\sin n\omega t}{n\omega} \right]_0^{1/2\Delta t}$$

$$= \frac{4V_m}{T} \frac{\sin\left(\frac{1}{2}n\omega\Delta t\right)}{n\omega} = \frac{2\Delta t V_m}{T} \frac{\sin\left(\frac{1}{2}n\omega\Delta t\right)}{\left(\frac{1}{2}n\omega\Delta t\right)}$$

Hence
$$f(t) = \frac{\Delta t}{T} V_m + \frac{2\Delta t}{T} V_m \sum_{n=1}^{\infty} \frac{\sin\left(\frac{1}{2}n\omega\Delta t\right)}{\left(\frac{1}{2}n\omega\Delta t\right)} \cos n\omega t \qquad \ldots (1)$$

The waveform of fig. 2.10(b) is identical to that of (a) except for a change in the d.c. level of $-\frac{1}{2}V_m$. The series (1) therefore applies provided the first term is replaced by the new d.c. value of

$$a_0 = -\tfrac{1}{2}V_m + \frac{\Delta t}{T} V_m = -\tfrac{1}{2}V_m\left(1 - \frac{2\Delta t}{T}\right)$$

In fig. 2.10(c) the origin has been shifted to the left by t_1 relative to (a). Thus if the t variable in (1) is replaced by $(t - t_1)$ the Fourier series for the waveform of (c) is

$$f_2(t) = \frac{\Delta t}{T} V_m + \frac{2\Delta t}{T} V_m \sum_{n=1}^{\infty} \frac{\sin\left(\frac{1}{2}n\omega\Delta t\right)}{\left(\frac{1}{2}n\omega\Delta t\right)} \cos\{n\omega(t - t_1)\}$$

4. Determine the frequency spectrum of the r.f. pulse waveform shown in fig. 2.11, and plot its line spectrum amplitude v. frequency graph. The wave train of fig. 2.11 is passed through an ideal bandpass filter which has cut-off frequencies of 0·955 MHz and 1·045 MHz. Assuming matched conditions find the percentage of power in the output wave train relative to that in the input. If the output is then detected, sketch the resulting waveform as a function of time.

Solution
The r.f. pulse wave train is defined in one repetition period as
$$v(t) = \begin{cases} 5\sin\omega_c t & \text{for } -10\,\mu s \leqslant t \leqslant 10\,\mu s \\ 0 & \text{for rest of period } T \end{cases}$$

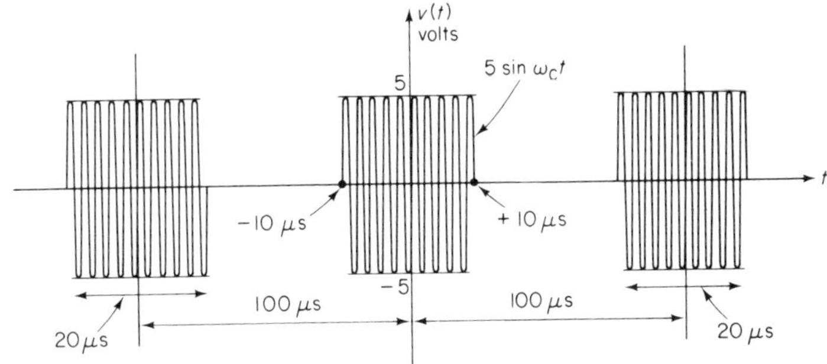

Fig. 2.11. R.F. pulse waveform. R.F. carrier frequency $f_c = \omega_c/2\pi = 1$ MHz, pulse duration 20 μs, pulse repetition frequency 10 kHz

where $T = 100\,\mu$s, the pulse repetition frequency $f_0 = \omega_0/(2\pi) = 1/T = 10$ kHz. Since $v(t)$ is an odd function its Fourier series contains only sine terms. Thus

$$v(t) = \sum_{n=1}^{\infty} b_n \sin n\omega_0 t$$

where $b_n = \dfrac{4}{T} \int_0^{1/2T} v(t) \sin n\omega_0 t = \dfrac{4}{T} \int_0^{1/2\Delta t} 5 \sin \omega_c t \sin n\omega_0 t \, dt$

$$= \frac{4}{T} \int_0^{1/2\Delta t} 5 \times \tfrac{1}{2}[\cos (\omega_c - n\omega_0)t - \cos (\omega_c + n\omega_0)t] \, dt$$

$$= \frac{10}{T} \left[\frac{\sin (\omega_c - n\omega_0)t}{(\omega_c - n\omega_0)} - \frac{\sin (\omega_c + n\omega_0)t}{(\omega_c + n\omega_0)} \right]_0^{1/2\Delta t}$$

$$= \frac{5\Delta t}{T} \left[\frac{\sin [(\omega_c - n\omega_0)\tfrac{1}{2}\Delta t]}{[(\omega_c - n\omega_0)\tfrac{1}{2}\Delta t]} - \frac{\sin [(\omega_c + n\omega_0)\tfrac{1}{2}\Delta t]}{[(\omega_c + n\omega_0)\tfrac{1}{2}\Delta t]} \right]$$

$$= \frac{5\Delta t}{T} \left[\frac{\sin [(f_c - nf_0)\Delta t\pi]}{[(f_c - nf_0)\Delta t\pi]} - \frac{\sin [(f_c + nf_0)\Delta t\pi]}{[(f_c + nf_0)\Delta t\pi]} \right]$$

$$= \frac{5\Delta t}{T} \left[\operatorname{sinc} [\pi(f_c - nf_0)\Delta t] - \frac{5\Delta t}{T} \operatorname{sinc} [\pi(f_c + nf_0)\Delta t] \right.$$

where the sinc function is defined as $\operatorname{sinc} x = \sin (\pi x)/(\pi x)$. A graph of the latter is shown in fig. 2.12(a). Note that when $x = 0$ $\operatorname{sinc} x = 1$ and is at a maximum; and for large values of x, $\operatorname{sinc} x$ approaches zero.

$|b_n|$ is the voltage amplitude of the nf_0 harmonic frequency component, and on considering the properties of the sinc function we see that

$$b_n \approx \frac{5\Delta t}{T} [\operatorname{sinc} [\pi(f_c - nf_0)\Delta t]]$$

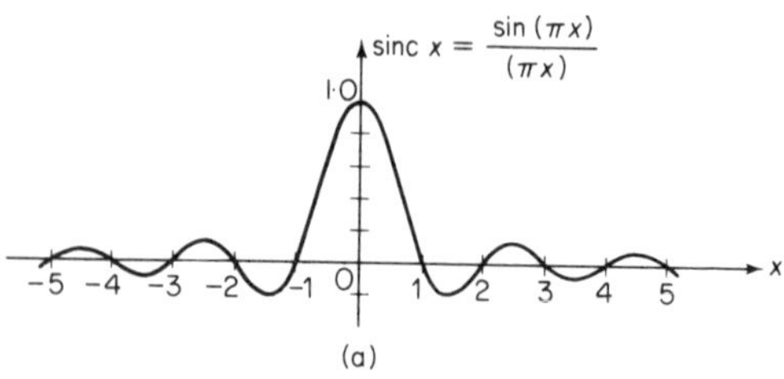

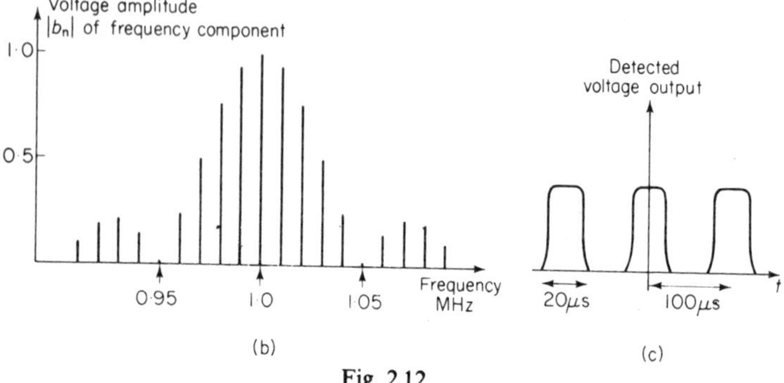

Fig. 2.12

with a maximum amplitude occurring when $f_c - nf_0 = 0$, i.e. at the carrier frequency of 1 MHz when $n = 100$. The sinc $[\pi(f_c + nf_0)\Delta t]$ term is omitted since for all positive values of n it is negligible in value. The amplitudes of the most significant frequency components are listed in the following table and the corresponding frequency spectrum of $|b_n|$ versus frequency is drawn in fig. 2.12(b).

n	Frequency $f = nf_0$ kHz	Voltage Amplitude $|b_n|$	Relative Power $|b_n|^2$
100	1000	1	1
99, 101	1010, 990	0·9355	0·875
98, 102	1020, 980	0·7569	0·573
97, 103	1030, 970	0·505	0·255
96, 104	1040, 960	0·234	0·055
95, 105	950, 1050	0	0

The band-pass filter will pass only those frequency components greater than 955 kHz and less than 1045 kHz. Thus the components transmitted through the filter are those at

960, 970, 980, 990, 1000, 1010, 1020, 1030, 1040 kHz

and the power associated with these components is proportional to (equal to if we consider a $1\,\Omega$ load),

$$\tfrac{1}{2}\sum_{n=96}^{104}|b_n|^2 = \tfrac{1}{2}[1+2(0{\cdot}875+0{\cdot}573+0{\cdot}255+0{\cdot}055)]$$
$$= 2{\cdot}258$$

The power contained in the input wave is proportional to

$$\frac{1}{T}\int_{-1/2T}^{1/2T} v^2(t)\,dt = \frac{1}{T}\int_{-1/2\Delta t}^{1/2\Delta t} 5^2\sin^2\omega_c t = \frac{1}{2}\times 5^2\frac{\Delta t}{T} = 2{\cdot}5$$

Hence the percentage of output power relative to the input is

$$\frac{2{\cdot}258}{2{\cdot}5}\times 100 = 90{\cdot}3\%$$

The detected voltage output is sketched in fig. 2.12(c). Since not all the frequency components in the input wave are passed by the filter the 'edges' of the output wave are not sharp but slightly rounded.

5. Determine the Fourier transform of the cosine-squared pulse,
$$f(t) = \cos^2(\pi t/T): -\tfrac{1}{2}T \leqslant t \leqslant \tfrac{1}{2}T$$
$$= 0 \qquad : |t| > \tfrac{1}{2}T$$
and sketch its magnitude and phase frequency spectra.

Solution
The Fourier transform of the cosine-squared pulse is

$$F(\omega) = \int_{-\infty}^{\infty} f(t)\,e^{-j\omega t}\,dt = \int_{-1/2T}^{1/2T}\cos^2(\pi t/T)\,e^{-j\omega t}\,dt$$

On substituting $\cos^2(\pi t/T) = [\tfrac{1}{2}(e^{j(\pi t/T)}+e^{-j(\pi t/T)})]^2$ we obtain

$$F(\omega) = \tfrac{1}{4}\int_{-1/2T}^{1/2T}(e^{j2(\pi t/T)}+e^{-j2(\pi t/T)}+2)\,e^{-j\omega t}\,dt$$

$$= \tfrac{1}{4}\left[\frac{e^{j(2\pi/T-\omega)t}}{j(2\pi/T-\omega)}-\frac{e^{-j(2\pi/T+\omega)t}}{j(2\pi/T+\omega)}-\frac{2}{j\omega}e^{-j\omega t}\right]_{-1/2T}^{1/2T}$$

$$= \tfrac{1}{2}\left[\frac{\sin\left[(2\pi/T-\omega)\tfrac{1}{2}T\right]}{(2\pi/T-\omega)}+\frac{\sin\left[(2\pi/T+\omega)\tfrac{1}{2}T\right]}{(2\pi/T+\omega)}+\frac{2}{\omega}\sin(\omega\tfrac{1}{2}T)\right]$$

$$= \frac{\sin(\omega T/2)}{2}\left[\frac{2}{\omega}+\frac{(2\pi/T+\omega)-(2\pi/T-\omega)}{4\pi^2/T^2-\omega^2}\right]$$

$$= \frac{4\pi^2\sin(\tfrac{1}{2}\omega T)}{\omega(4\pi^2/T^2-\omega^2 T^2)}$$

The magnitude and phase of $F(\omega)$ is plotted in fig. 2.13

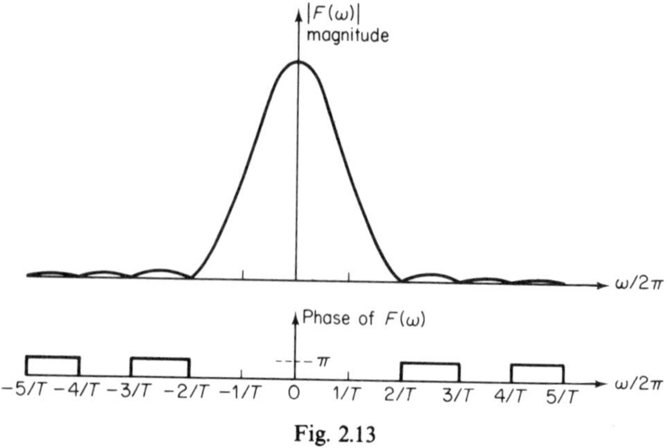

Fig. 2.13

6. Determine the Fourier transform of the exponentially decaying pulse:

$$v(t) = 0 \text{ for } t < 0, \ v(t) = e^{-\alpha t} \text{ for } t \geqslant 0.$$

This pulse is applied to the input of a C–R network as shown in fig. 2.14. Determine the Fourier transform and the time domain solutions, $V_0(\omega)$ and $v_0(t)$, of the output voltage.

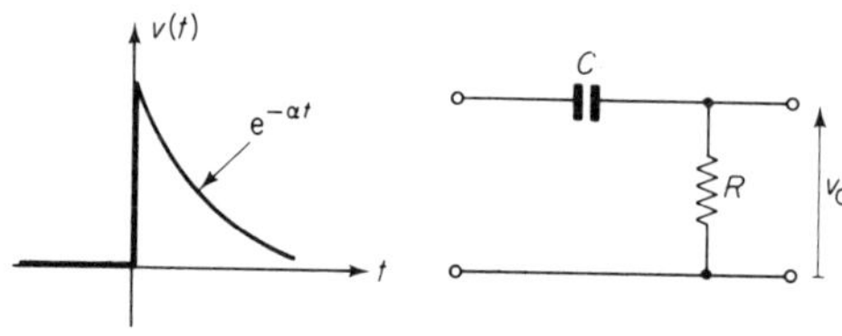

Fig. 2.14

Solution

The Fourier transform of the exponential pulse is

$$V(\omega) = \int_{-\infty}^{\infty} v(t) e^{-j\omega t} \, dt = \int_{0}^{\infty} e^{-\alpha t} e^{-j\omega t} \, dt$$

$$= \left[\frac{e^{-(\alpha + j\omega)t}}{-(\alpha + j\omega)} \right]_{0}^{\infty} = \frac{1}{\alpha + j\omega}$$

The voltage transfer function of the C–R network is

$$H(j\omega) = \frac{V_0(\omega)}{V(\omega)} = \frac{R}{R + 1/(j\omega C)} = \frac{j\omega CR}{1 + j\omega CR}$$

54

and the output voltage transform,

$$V_0(\omega) = H(j\omega)V(\omega) = \frac{j\omega CR}{1+j\omega CR}\frac{1}{\alpha+j\omega}$$

$$= \frac{j\omega CR}{CR[(1/(CR)+j\omega)(\alpha+j\omega)]}$$

$$= j\left[\frac{j}{(CR\alpha-1)(1/(CR)+j\omega)} - \frac{jCR\alpha}{(CR\alpha-1)(\alpha+j\omega)}\right] \qquad \dots (1)$$

on expanding $V_0(\omega)$ as a sum of two partial fractions.

On taking the inverse transform of (1) we have

$$v_0(t) = \frac{-1}{(CR\alpha-1)}e^{-t/(CR)} + \frac{CR\alpha}{(CR\alpha-1)}e^{-\alpha t} \text{ for } t \geqslant 0.$$

7. The rectangular pulse of fig. 2.15 is applied to an idealized low-pass filter whose transfer function is

$$H(j\omega) = e^{j\omega td} \qquad |\omega| \leqslant \omega_c$$

$$= 0 \qquad |\omega| > \omega_c$$

Determine the output voltage in the frequency and time domains, i.e. $V_0(\omega)$ and $v_0(t)$. Using the table of fig. 2.6 determine the maximum amplitude of the output waveform when $\omega_c = 2\pi/\Delta t$ and when $\omega_c = \pi/\Delta t$. Sketch these waveforms over the range $|t| \leqslant \frac{1}{2}\Delta t$

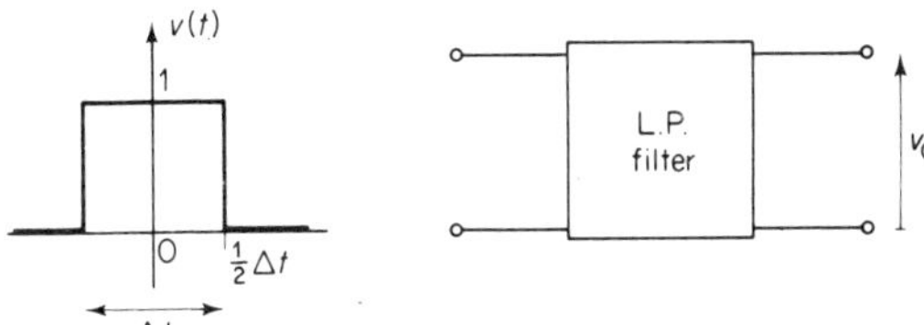

Fig. 2.15

Solution
The Fourier transform of the input pulse is

$$V(\omega) = \int_{-\infty}^{\infty} v(t)e^{-j\omega t}\,dt = \int_{-1/2\Delta t}^{1/2\Delta t} 1 \times e^{-j\omega t}\,dt$$

$$= \frac{\Delta t \sin\left(\frac{1}{2}\omega\Delta t\right)}{\left(\frac{1}{2}\omega\Delta t\right)}$$

The Fourier transform of the output signal is

$$V_0(\omega) = H(\mathrm{j}\omega)V(\omega)$$

$$= \frac{\Delta t \sin\left(\tfrac{1}{2}\omega\Delta t\right)}{\left(\tfrac{1}{2}\omega\Delta t\right)}\,\mathrm{e}^{-\mathrm{j}\omega t_\mathrm{d}}$$

for $|\omega| \leqslant \omega_\mathrm{c}$, $= 0$ for $|\omega| > \omega_\mathrm{c}$, whilst the inverse transform gives the time domain solution,

$$v_0(t) = \frac{1}{2\pi}\int_{-\infty}^{\infty} V_0(\omega)\,\mathrm{e}^{\mathrm{j}\omega t}\,\mathrm{d}\omega$$

$$= \frac{\Delta t}{2\pi}\int_{-\omega_\mathrm{c}}^{\omega_\mathrm{c}} \frac{\sin\left(\tfrac{1}{2}\omega\Delta t\right)}{\left(\tfrac{1}{2}\omega\Delta t\right)}\,\mathrm{e}^{\mathrm{j}\omega(t-t_\mathrm{d})}\,\mathrm{d}t$$

$$= \frac{\Delta t}{2\pi}\int_{-\omega_\mathrm{c}}^{\omega_\mathrm{c}} \frac{\sin\left(\tfrac{1}{2}\omega\Delta t\right)}{\left(\tfrac{1}{2}\omega\Delta t\right)}\,\mathrm{e}^{\mathrm{j}\omega t'}\,\mathrm{d}t' \text{ where } t' = t - t_\mathrm{d}$$

Now substituting $\mathrm{e}^{\mathrm{j}\omega t'} = \cos\omega t' + \mathrm{j}\sin\omega t'$,

$$v_0(t) = \frac{\Delta t}{2\pi}\left[\int_{-\omega_\mathrm{c}}^{\omega_\mathrm{c}} \frac{\sin\left(\tfrac{1}{2}\omega\Delta t\right)}{\left(\tfrac{1}{2}\omega\Delta t\right)}\cos\omega t'\,\mathrm{d}\omega + \mathrm{j}\int_{-\omega_\mathrm{c}}^{\omega_\mathrm{c}} \frac{\sin\left(\tfrac{1}{2}\omega\Delta t\right)}{\left(\tfrac{1}{2}\omega\Delta t\right)}\sin\omega t'\,\mathrm{d}\omega\right]$$

$$= \frac{2\Delta t}{2\pi}\int_{0}^{\omega_\mathrm{c}} \frac{\sin\left(\tfrac{1}{2}\omega\Delta t\right)\cos\omega t'}{\tfrac{1}{2}\omega\Delta t}\,\mathrm{d}\omega + 0$$

since the $\sin\left(\tfrac{1}{2}\omega\Delta t\right)/\left(\tfrac{1}{2}\omega\Delta t\right)$ is an even function.

Further using $\sin\left(\tfrac{1}{2}\omega\Delta t\right)\cos\omega t' = \tfrac{1}{2}\left[\sin\omega(t' + \tfrac{1}{2}\Delta t) - \sin\omega(t' - \tfrac{1}{2}\Delta t)\right]$ we obtain,

$$v_0(t) = \frac{1}{\pi}\left[\int_{0}^{\omega_\mathrm{c}} \left\{\frac{\sin\omega(t' + \tfrac{1}{2}\Delta t)}{\omega} - \frac{\sin\omega(t' - \tfrac{1}{2}\Delta t)}{\omega}\right\}\mathrm{d}\omega\right]$$

$$= \frac{1}{\pi}\left[\int_{0}^{\omega_\mathrm{c}(t' + 1/2\Delta t)} \frac{\sin x}{x}\,\mathrm{d}x - \int_{0}^{\omega_\mathrm{c}(t' - 1/2\Delta t)} \frac{\sin x}{x}\,\mathrm{d}x\right]$$

$$= \frac{1}{\pi}\left[\mathrm{Si}\{\omega_\mathrm{c}(t' + \tfrac{1}{2}\Delta t)\} - \mathrm{Si}\{\omega_\mathrm{c}(t' - \tfrac{1}{2}\Delta t)\}\right]$$

Using the table of Si function results, we find that the maximum amplitude occurs when $t' = 0$ and is given by

$$v_{0\max} = \frac{2}{\pi}\mathrm{Si}[\tfrac{1}{2}\omega_\mathrm{c}\Delta t]$$

$$= \frac{2}{\pi}\mathrm{Si}[\pi] = \frac{2}{\pi}\times 1\cdot851 = 1\cdot178 \text{ when } \omega_\mathrm{c} = 2\pi/\Delta t$$

$$= \frac{2}{\pi} \text{Si}\left[\frac{\pi}{2}\right] = \frac{2}{\pi} \times 1 \cdot 370 = 0 \cdot 872 \text{ when } \omega_c = \pi/\Delta t$$

The sketches of output voltage waveforms are shown in fig. 2.16

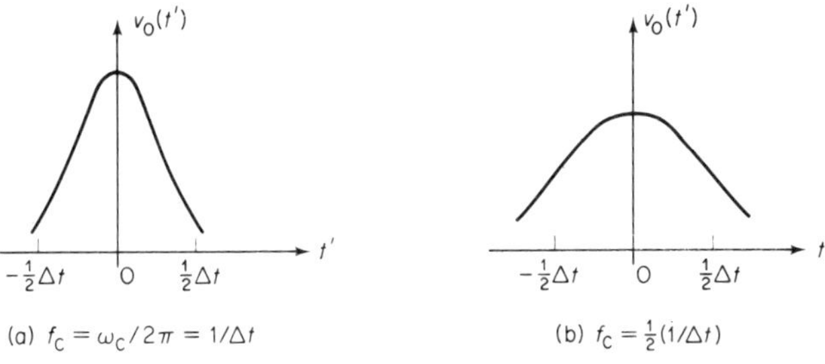

Fig. 2.16

2.3. Exercise Problems

1. Determine the Fourier series for the waveforms shown in fig. 2.17. Calculate also their root mean square (r.m.s.) values.

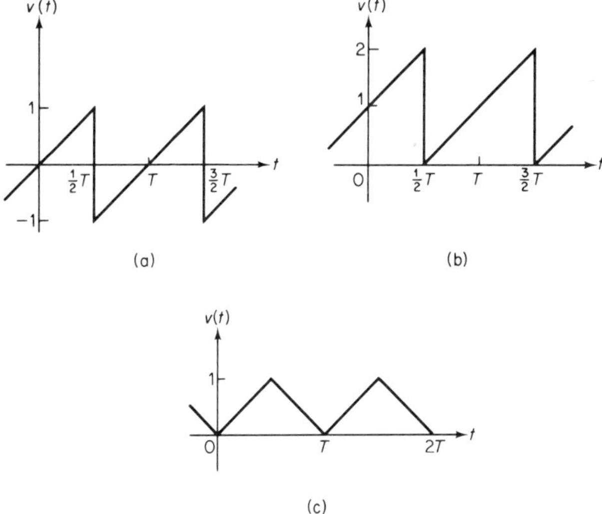

Fig. 2.17. For problem 1

2. Determine the Fourier series for the full wave rectified voltage waveform shown in fig. 2.18(a). This waveform is applied across the

input of the L–C smoothing network shown in fig. 2.18(b). Estimate the amplitude of the voltage output across R at the second and fourth harmonics, 2ω and 4ω, assuming $\omega L \gg 1/(\omega C)$ and $\omega C \gg 1/R$.

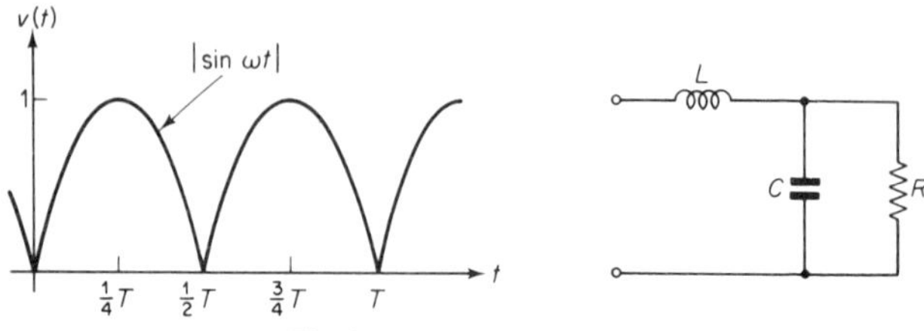

Fig. 2.18. For problem 2

3. Determine the Fourier series of the train of impulses shown in fig. 2.19 and sketch the frequency spectrum of the train.

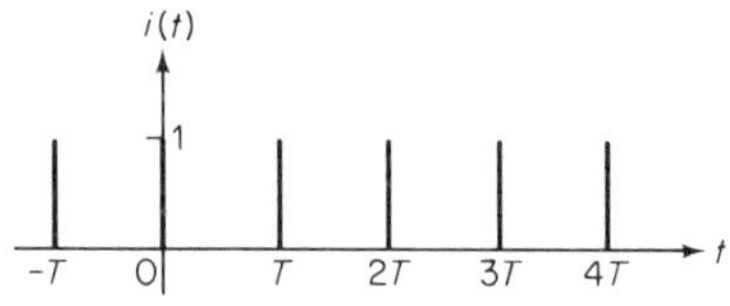

Fig. 2.19. For problem 3

4. Determine the Fourier series for the square wave voltage of fig. 2.20(a). This waveform is applied to the input of the R–C network of fig. 2.20(b). Determine the output voltage $v_0(t)$ and sketch its form when $RC \gg T$.

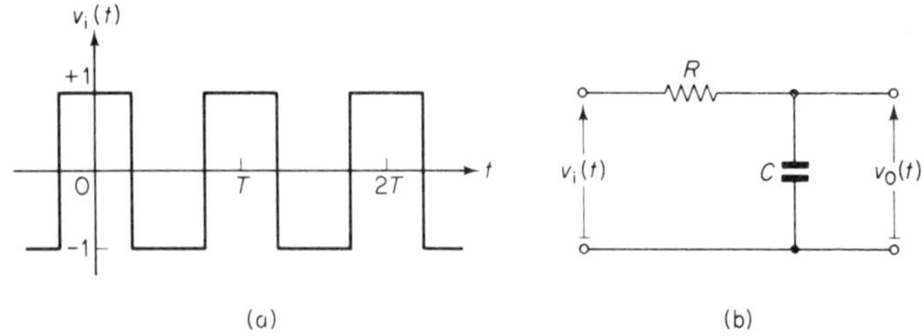

Fig. 2.20. For problem 4

5. Determine the Fourier series of the triangular current wave shown in fig. 2.21(a). If this wave is connected directly across a resistor of R ohms, determine the mean power dissipated in the resistor. If the wave is now passed via an ideal low pass filter, which transmits all

frequencies from 0 to $5/T$ hertz with zero loss, as shown in fig. 2.21(b), evaluate the mean power dissipated in R.

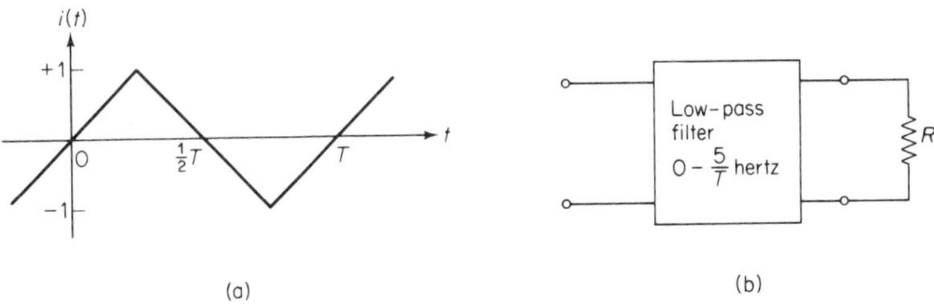

Fig. 2.21. For problem 5

6. Determine the amplitude frequency spectrum of the rectangular pulse train shown in fig. 2.22, and evaluate as a fraction of the total power the power contained in the frequency range 0 to 100 kHz.

   Calculate also the voltage amplitude of the outputs at 10 kHz and 40 kHz if the train is fed through a d.c. amplifier of voltage gain, $G_V = 100/[1+j10^{-2}f]$ where f = frequency in kHz.

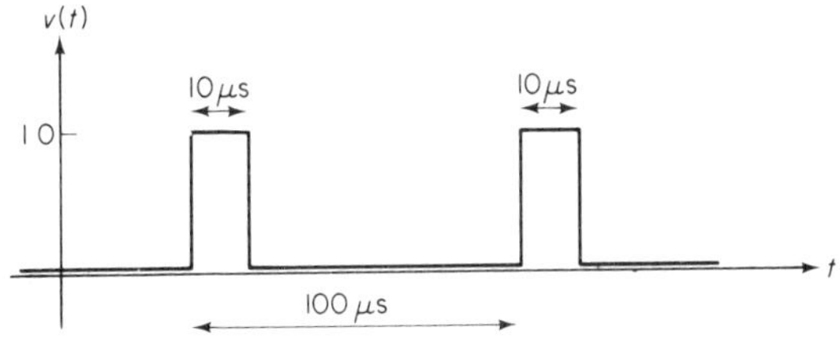

Fig. 2.22. For problem 6

7. If the terminal voltage and current of a circuit are given by

$$v = 63{\cdot}7 + \sqrt{2} \times 70{\cdot}7 \sin \omega t - \sqrt{2} \times 30{\cdot}0 \cos 2\omega t \text{ V}$$

$$i = 10{\cdot}0 + \sqrt{2} \times 5{\cdot}0 \sin (\omega t + 40°) - \sqrt{2} \times 3{\cdot}0 \cos (2\omega t - 27°) \text{ A}$$

   evaluate (a) the r.m.s. values of voltage and current, (b) the mean power absorbed by the circuit, (c) the power factor.

8. In fig. 2.23 the signal $v_i = V_m(\cos \omega t + \cos 20\omega t)$ volts is fed into the 2-port device which has the non-linear transfer characteristic of $v_0 = av_i + bv_i^2$. Determine the frequency components present in the output and the power dissipated by each in the output resistor of R ohms.

9. A double-sideband, amplitude modulated current wave, $i = A(1 + m \cos \omega_m t) \cos \omega_c t$ is applied to a parallel circuit tuned to ω_c. The impedance of this circuit is given by

$Z = R[1+j2Q(\omega-\omega_c)/\omega_c]$. Determine the amplitude of the frequency components in the current wave and those of the voltage output across the tuned circuit.

10. Fig. 2.24 shows a diagram of a single-sideband, amplitude modulation detector in which a signal $v_i = A\cos(\omega_c+\omega_m)t$ is added to a local carrier $v_1 = 2{\cdot}5\cos\omega_c t$ and then the combined signal is passed through an envelope detector. The latter gives an output voltage proportional to the amplitude of the combined signal, excluding any frequency variation at the carrier or like frequencies. Determine the detector output and the amplitude of second harmonic $(2\omega_m)$ content in the latter.

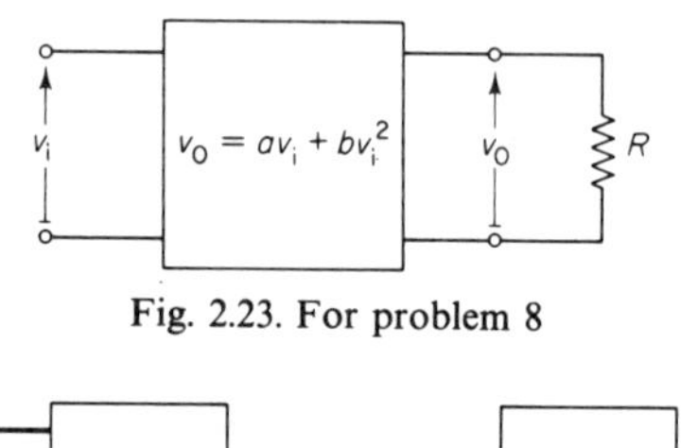

Fig. 2.23. For problem 8

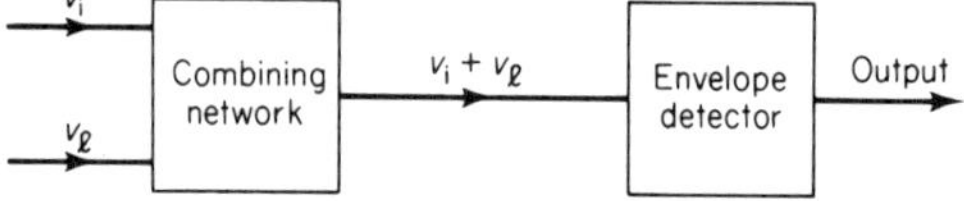

Fig. 2.24. For problem 10

11. Fig. 2.25 shows a pulse amplitude modulated waveform, with pulses of $5\,\mu s$ duration at a pulse repetition frequency of 20 kHz. The pulse heights vary according to $(1+0{\cdot}5\sin\omega_m t)$ volts, where $\omega_m = 2\pi \times 5 \times 10^3$ rad/s. Determine the voltage amplitudes of the frequency components of the waveform over the range 0 to 50 kHz. If the waveform is passed through a filter sketch the output waveform if the filter has (a) a low-pass characteristic with a 0 to 6 kHz pass-band, (b) a band-pass characteristic from 14 to 26 kHz.

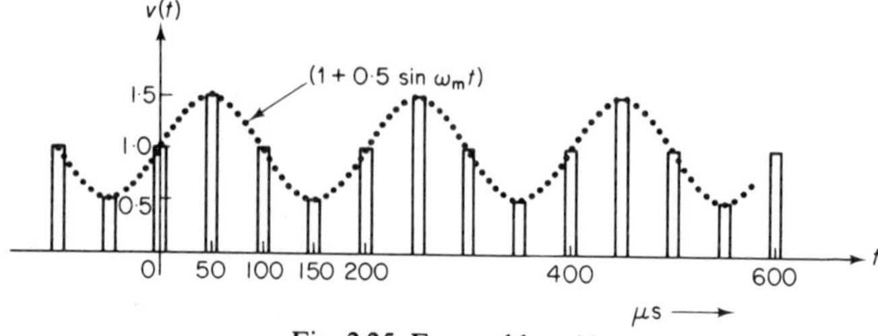

Fig. 2.25. For problem 11

60

12. Derive from first principles an expression for the r.m.s. value of a non-sinusoidal current or voltage.

An r.m.s. current of 7·65 A containing only a fundamental, third and fifth harmonic components flows in a coil which has an inductance of 0·01 H and a resistance of 10 Ω. The r.m.s. voltage across the coil is 89·26 V and the fundamental frequency is 50 Hz. Calculate the magnitude of the fundamental, third and fifth harmonic components of the current if the peak amplitude of the fundamental component of voltage is 100 V.

13. A voltage of

$$v = 200 \sin 2\pi f_0 t + 100 \sin 6\pi f_0 t + 50 \sin 2\pi f t,$$

where f_0 is the fundamental frequency and f is a harmonic frequency, is applied to a circuit consisting of a coil of resistance 50 Ω and inductance 2 mH in series with a variable capacitor C. Selective resonance is obtained when $C = 20\,\mu F$ at the fundamental frequency, when $C = 0·8\,\mu F$ at f Hz. Determine f and find expressions for the instantaneous value of the total current and power dissipated in the circuit when $C = 10\,\mu F$.

14. (a) What meaning is attached to the term 'power-factor' when applied to a circuit in which the voltages and currents contain harmonics? State briefly why the power-factor cannot usually be improved to unity by means of shunt capacitors in circuits of this type.

(b) When a voltage $v = \sqrt{2} \times 300 \sin 1000t$ is applied to a non-linear impedance a current

$$i = \sqrt{2} \times 15 \sin (1000t - 36·9°) + \sqrt{2} \times 9 \sin 3000t$$

flows. Calculate the overall power-factor.

(c) A capacitor is now connected in parallel and it takes a current of 9 A r.m.s. What will be the new value of overall power-factor under these conditions?

15. A low-frequency alternator, of negligible impedance, is connected to an ideal full-wave rectifier, as shown in Fig. 2.26. The rectified output voltage is given by:

$$v = 100\left(1 - \frac{2}{3}\cos 2\omega t - \frac{2}{15}\cos 4\omega t\right) \text{volts}$$

(neglecting higher harmonics) where $\omega = 50$ rad/s.

This output is fed to a smoothing circuit, comprising R_1 and C, which supplies an inductive load, represented by R_2 and L.

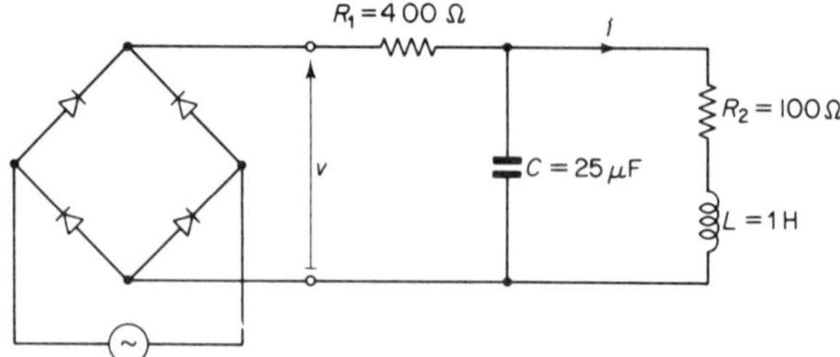

Fig. 2.26. For problem 15

Derive an expression for the instantaneous load current i. Hence, determine its r.m.s. value and the power dissipated in the load.

16. An air-cored coil has a resistance of $5\,\Omega$ and an inductance of $30\,\text{mH}$. The r.m.s. voltage across the coil is $100\,\text{V}$ and the r.m.s. current through it is $7\,\text{A}$.

    Calculate (a) the peak amplitudes of the fundamental and harmonic components of the current if it contains a $50\,\text{Hz}$ fundamental and a third harmonic, (b) the total power absorbed.

17. Determine the Fourier transform of the rectangular pulses shown in fig. 2.27(a) and (b) and sketch the magnitude and phase spectrums over the range $-3/\Delta t < f < 3/\Delta t$.

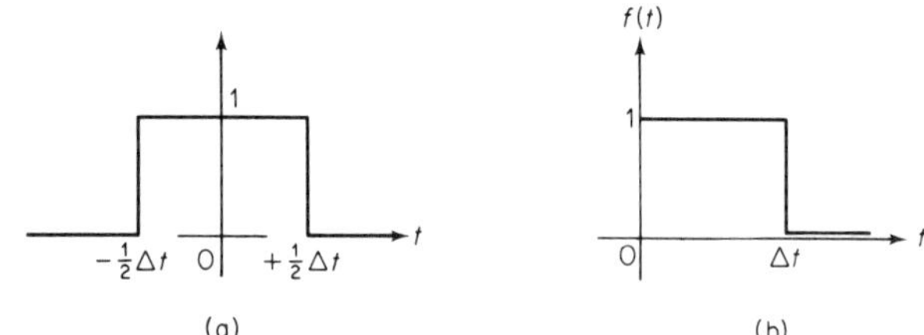

Fig. 2.27. For problem 17

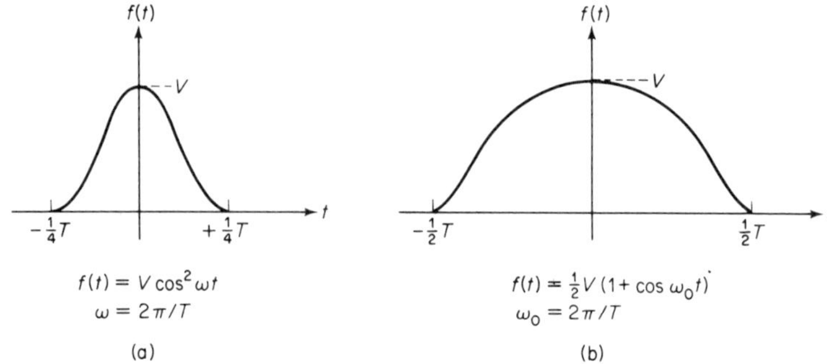

Fig. 2.28. For problem 18

18. Determine the Fourier transform of the non-periodic waveforms shown in fig. 2.28 and sketch the magnitude and phase diagrams.
19. Determine the Fourier transform of the single triangular and the double pulse waveforms shown in fig. 2.29. Sketch their magnitude versus frequency spectrums.

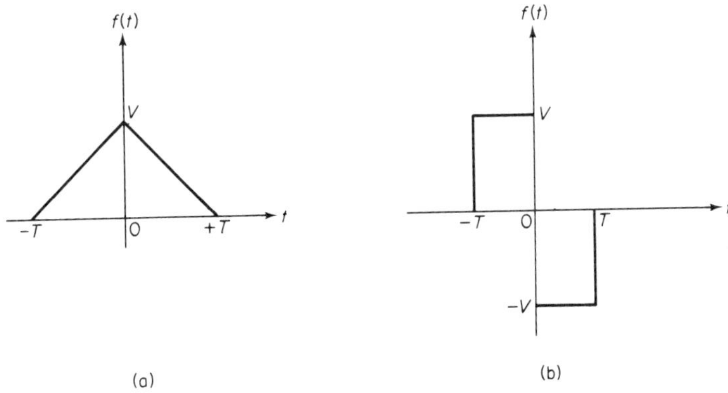

Fig. 2.29. For problem 19

20. Derive an expression for the fraction of the total energy of a 1 volt rectangular pulse of duration Δt seconds contained in the frequency range 0 to $1/\Delta t$ hertz.

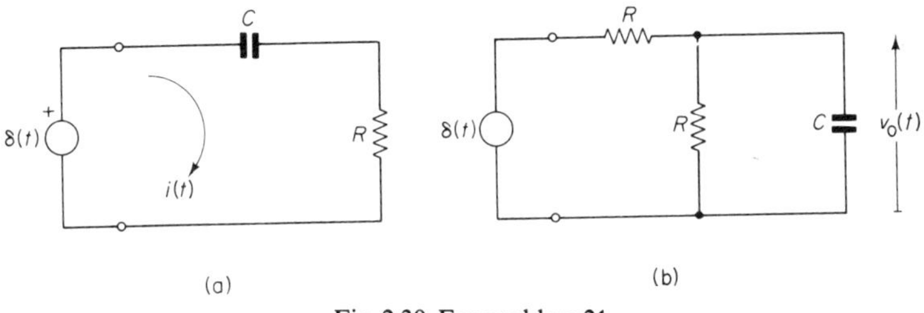

Fig. 2.30. For problem 21

21. Using Fourier methods determine the impulse response of the current $i(t)$ in the circuit of fig. 2.30(a) and the impulse response of the voltage $v_0(t)$ in the circuit of fig. 2.30(b).
22. A rectangular pulse of amplitude V volts and duration Δt seconds is applied to the input of an ideal low pass filter with a cut-off frequency $\omega_c = 1 \cdot 6/\Delta t$ rad/s. Using the Si function table of fig. 2.6 determine the maximum height of the output pulse and estimate the heights of this pulse at times $\pm 0 \cdot 3\Delta t$ seconds displaced from the maximum.

3 Transmission Line Theory

3.1. Theory Summary

1. INTRODUCTION

In lumped circuit theory we consider that the resistive, inductive and capacitative properties of an element can be regarded as concentrated into individual lumped units or mathematically equivalent models of R, L and C, although in the practical case these properties are distributed throughout the component and associated with the electric and magnetic fields within the component and the medium surrounding it. It is also assumed that the effect of an applied source is propagated instantaneously throughout the circuit, whereas the transfer of electromagnetic energy cannot exceed the velocity of light. Provided the physical size of the circuit components and the system dimensions are very much smaller than the wavelength λ associated with the transmitted electromagnetic waves the representation of components as lumped elements gives excellent agreement with the experimentally observed behaviour. However where the circuit dimensions become comparable with or much greater than λ, we must revise lumped circuit theory to take into account the true distributed nature of the components. This is particularly so in the transmission of electrical energy down 'long' lines in which there is a continuous variation of voltage and current from one point to the next. Distinction between 'long' and 'short' transmission lines is made as follows. Lines are termed long, if at the frequency of the transmitted signal, the line length is at least a considerable fraction of λ; whilst short if the line length is less than $0 \cdot 1\lambda$ or so. Most r.f. lines fall into the long line category, whilst power and telephone lines may fall into the short line range. However, the review of theory given below applies quite generally.

2. THE TRANSMISSION LINE EQUATIONS AND THEIR A.C. STEADY STATE SOLUTIONS

(a) *The distributed parameters of a line*

In the circuit theory approach to transmission line theory it is assumed that a uniform line can be characterized by four parameters which are

64

distributed uniformily along the line length. The distributed parameters, also known as the line primary constants, are:

R = the series a.c. resistance per unit length of line (including both conductors)

L = the series inductance per unit length of line (including both conductors)

C = the shunt capacitance between the line conductors per unit line length

G = the shunt conductance between the line conductors per unit line length

Some values for some common 2-wire lines are given in the table of fig. 3.1.

(b) *The transmission line equations*

The representation of a uniform line as a cascade connection of a finite number of small elements of length $\delta z \ll \lambda$ is shown in fig. 3.2. The transmission line equations may be derived from any of the elemental sections shown (by using Kirchhoff's laws and taking the limit as $\delta z \to 0$) and are given in general by

$$\frac{\partial v}{\partial z} = -\left(Ri + L\frac{\partial i}{\partial t} \right) \qquad \ldots (1)$$

$$\frac{\partial i}{\partial z} = -\left(Gv + C\frac{\partial v}{\partial t} \right) \qquad \ldots (2)$$

which in the a.c. steady state, i.e. assuming $v = V(z)e^{j\omega t}$, $i = I(z)e^{j\omega t}$ reduce to

$$\frac{dV}{dz} = -ZI \qquad \ldots (3)$$

$$\frac{dI}{dz} = -YV \qquad \ldots (4)$$

where v, V = voltage between line conductors at plane z

i, I = current flowing in line conductors through plane z

$Z = (R + j\omega L)$ = series impedance of line per unit length

$Y = (G + j\omega C)$ = shunt admittance of line per unit length

On eliminating I from (3) and (4), or V from (3) and (4) we obtain:

$$\frac{d^2 V}{dz^2} - ZYV = 0 \qquad \ldots (5)$$

$$\frac{d^2 I}{dz^2} - ZYI = 0 \qquad \ldots (6)$$

	Coaxial line	Parallel wire line	$p=\dfrac{s}{d}$, $q=\dfrac{s}{D}$	Parallel plate or stripline
Resistance $R\ \Omega\text{m}^{-1}$	$\dfrac{R_s}{2\pi}\left(\dfrac{1}{a}+\dfrac{1}{b}\right)$	$\dfrac{2R_s}{\pi d}\left[\dfrac{s/d}{\sqrt{\{(s/d)^2-1\}}}\right]$	$\dfrac{2R_{s_2}}{\pi d}\left[1+\dfrac{1+2p^2}{4p^4}(1-4q^2)\right]$ $+\dfrac{8R_{s_1}}{\pi D}q^2\left[1+q^2-\dfrac{1+4q^2}{8p^4}\right]$	$\dfrac{2R_s}{b}$
Inductance $L\ \text{Hm}^{-1}$	$\dfrac{\mu}{2\pi}\log_e\!\left(\dfrac{b}{a}\right)$	$\dfrac{\mu}{\pi}\cosh^{-1}\!\left(\dfrac{s}{d}\right)$		$\mu\dfrac{a}{b}$
Conductance $G\ \text{Sm}^{-1}$	$\dfrac{2\pi\sigma}{\log_e\!\left(\dfrac{b}{a}\right)}$	$\dfrac{\pi\sigma}{\cosh^{-1}\!\left(\dfrac{s}{d}\right)}$		$\sigma\dfrac{b}{a}$
Capacitance $C\ \text{Fm}^{-1}$	$\dfrac{2\pi\epsilon}{\log_e\!\left(\dfrac{b}{a}\right)}$	$\dfrac{\pi\epsilon}{\cosh^{-1}\!\left(\dfrac{s}{d}\right)}$		$\epsilon\dfrac{b}{a}$
Characteristic impedence, $Z_C\ \Omega$	$\dfrac{1}{2\pi}\sqrt{\dfrac{\mu'}{\epsilon'}}\log_e\!\left(\dfrac{b}{a}\right)$	$\dfrac{1}{\pi}\sqrt{\dfrac{\mu'}{\epsilon'}}\cosh^{-1}\!\left(\dfrac{s}{d}\right)$	$\dfrac{1}{\pi}\sqrt{\dfrac{\mu'}{\epsilon'}}\left[\log_e\!\left\{2p\left(\dfrac{1-q^2}{1+q^2}\right)\right\}\right.$ $\left.-\dfrac{1+4p^2}{16p^4}(1-4q^2)\right]$	$\sqrt{\dfrac{\mu'}{\epsilon'}}\dfrac{a}{b}$

Fig. 3.1. $R_s=(\omega\mu_0/2\sigma_m)^{1/2}$ ... skin effect surface resistivity, $\sigma_m=$ conductivity of metal line conductors, $\varepsilon=\varepsilon'-j\varepsilon''$ ... complex permittivity of dielectric, $\mu=\mu'-j\mu''$ ... complex permeability of dielectric between line conductors

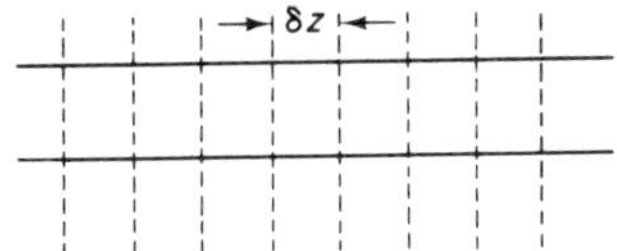 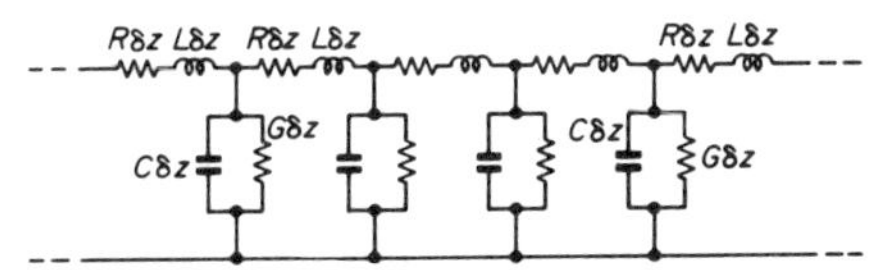

(a) Transmission line showing its division into elemental sections of length $\delta z \ll \lambda$

(b) An approximate lumped circuit representation of a line as an unbalanced ladder network. For representation as a balanced network $R\delta z$ and $L\delta z$ should be split into two and $\frac{1}{2}R\delta z$ and $\frac{1}{2}L\delta z$ located in both top and bottom rails

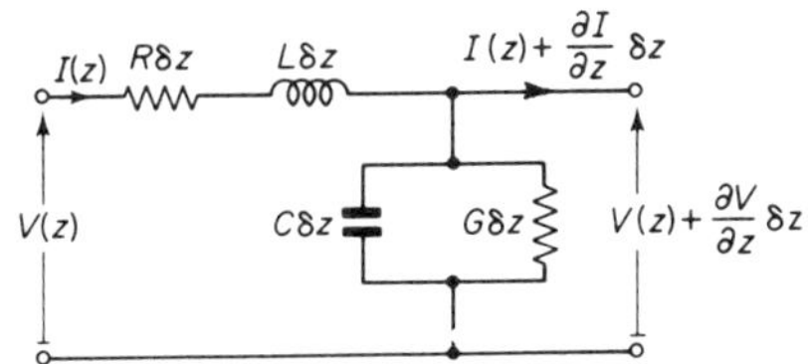 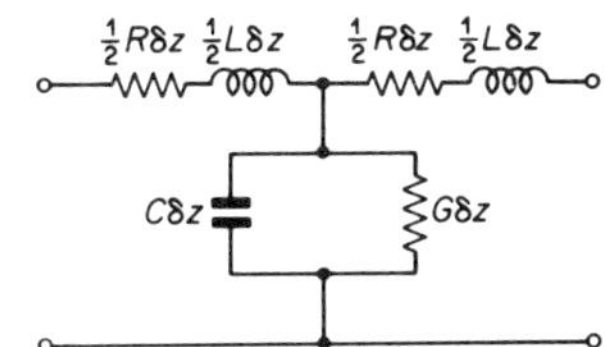

(c) Approximate equivalent circuit of an element of line, length $\delta z \ll \lambda$

(d) Approximate equivalent T circuit

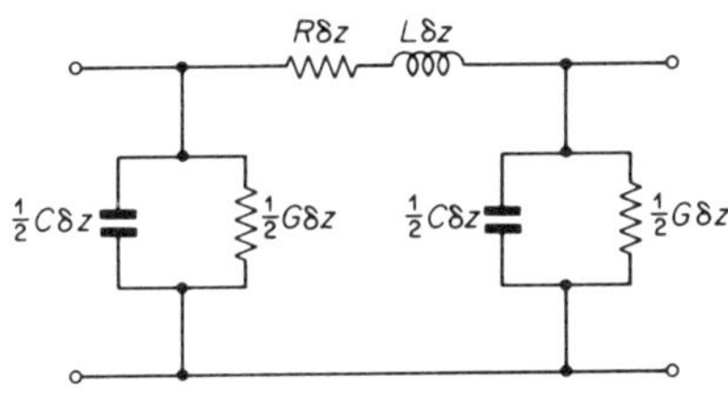

(e) Approximate equivalent π circuit

Fig. 3.2

(c) *The a.c. steady state solutions*

The general solutions of the steady state transmission line equations are

$$V = V^+ e^{-\gamma z} + V^- e^{+\gamma z} \qquad \cdots (7)$$

$$I = I^+ e^{-\gamma z} + I^- e^{-\gamma z} = \frac{V^+}{Z_C} e^{-\gamma z} - \frac{V^-}{Z_C} e^{+\gamma z} \qquad \cdots (8)$$

where $\gamma = \sqrt{(ZY)}$

is known as the propagation constant of the line,

$$Z_C = \sqrt{\left(\frac{Z}{Y}\right)} = \frac{V^+}{I^+} = \frac{V^-}{-I^-}$$

is known as the line characteristic impedance,

V^+, V^-, I^+, I^- are constants which may be evaluated when the conditions at the load and generator ends are given.

γ and Z_C may be expressed in terms of the line parameters as follows:

$$\gamma = \sqrt{[(R+j\omega L)(G+j\omega C)]} \equiv \alpha + j\beta$$

$$\alpha = \frac{1}{\sqrt{2}}\sqrt{[(RG-\omega^2 LC)+\sqrt{\{(R^2+\omega^2 L^2)(G^2+\omega^2 C^2)\}}]}$$

is known as the attenuation constant; units, nepers per unit length

$$\beta = \frac{1}{\sqrt{2}}\sqrt{[(\omega^2 LC-RG)+\sqrt{\{(R^2+\omega^2 L^2)(G^2+\omega^2 C^2)\}}]}$$

is known as the phase constant; units radians per unit length

$$Z_C = \sqrt{\left(\frac{R+j\omega L}{G+j\omega C}\right)}, \; |Z_C| = \left(\frac{R^2+\omega^2 L^2}{G^2+\omega^2 C^2}\right)^{1/4},$$

$$\angle Z_C = \frac{1}{2}\left[\tan^{-1}\frac{\omega L}{R}-\tan^{-1}\frac{\omega C}{G}\right]$$

For low-loss r.f. lines $\omega L \gg R, \omega C \gg G$ and the following approximations apply:

$$\alpha = \frac{1}{2}\left[R\sqrt{\frac{C}{L}}+G\sqrt{\frac{L}{C}}\right], \; \beta = \omega\sqrt{(LC)}, \; Z_C = \sqrt{\frac{L}{C}}$$

The full time varying solutions, corresponding to cosine or sinusoidal sources, can be found by taking the real or imaginary parts of $V(z)\,e^{j\omega t}$, $I(z)\,e^{j\omega t}$, e.g. for a cosine variation

$$v(t, z) = \mathrm{Re}(V(z)\,e^{j\omega t})$$

$$= \mathrm{Re}[V^+\,e^{-\alpha z}\,e^{j(\omega t - \beta z)} + V^-\,e^{\alpha z}\,e^{j(\omega t + \beta z)}]$$

$$= |V^+|\,e^{-\alpha z}\cos(\omega t - \beta z + \phi^+)$$

$$+ |V^-|\,e^{\alpha z}\cos(\omega t + \beta z + \phi^-) \qquad \ldots (9)$$

assuming $V^+ = |V^+|\,e^{j\phi^+}$ and $V^- = |V^-|\,e^{j\phi^-}$

The V^+ and V^- terms in (7) and (9) may be interpreted as the incident and reflected wave components making up the total voltage across the line. Likewise with the I^+ and I^- components in (8). The incident wave component in (9) propagates down the line in the positive z direction at a velocity v_p, known as the phase velocity:

$$v_p = \frac{\omega}{\beta} = \frac{2\pi f}{2\pi/\lambda} = f\lambda \qquad \ldots (10)$$

with an amplitude decaying with distance according to $e^{-\alpha z}$. The reflected wave propagates in the negative z axis direction at the same velocity and with the same exponential decay in amplitude.

In the case of the loss free line and the 'distortionless' line (condition $RC = LG$) $v_p = 1/\sqrt{(LC)}$ and is constant for all frequencies. However, in general, v_p depends on frequency and signals at different frequencies will travel at different velocities on the line. Thus if a group of waves whose frequencies lie between ω and $\omega + d\omega$ is considered, then the resultant amplitude envelope of the group, which carries the energy contained in the signals, travels down the line at a velocity v_g different from v_p. v_g is known as the group velocity:

$$v_g(\omega) = \frac{d\omega}{d\beta} \qquad \ldots (11)$$

(d) Hyperbolic forms of the steady state solutions

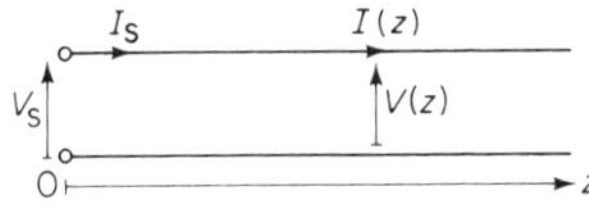

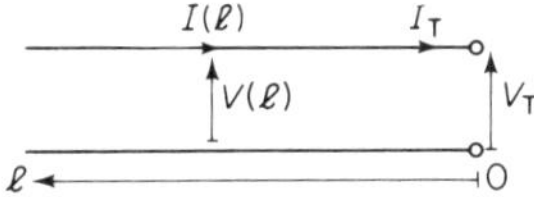

Fig. 3.3

If we are given the sending end values of voltage and current, i.e. V_S and I_S at $z = 0$, then on substituting into (7) and (8) we have

$$V_S = V^+ + V^-, \quad I_S = (V^+ - V^-)/Z_C$$

On solving these equations and substituting back into (7) and (8) for V^+ and V^-, the solutions may be expressed in the hyperbolic forms of

$$V(z) = V_S \cosh \gamma z - Z_C I_S \sinh \gamma z \qquad \ldots (12)$$

$$I(z) = -V_S \frac{\sinh \gamma z}{Z_C} + I_S \cosh \gamma z \qquad \ldots (13)$$

Similarly if the receiving end values V_T and I_T are known, then

$$V(l) = V_T \cosh \gamma l + Z_C I_T \sinh \gamma l \qquad \ldots (14)$$

$$I(l) = V_T \frac{\sinh \gamma l}{Z_C} + I_T \cosh \gamma l \qquad \ldots (15)$$

In this case l denotes the distance back along the line from the receiving end termination, as shown in fig. 3.3.

3. CHARACTERISTICS AND PARAMETERS OF TERMINATED TRANSMISSION LINES

(a) *Reflection and transmission coefficients*

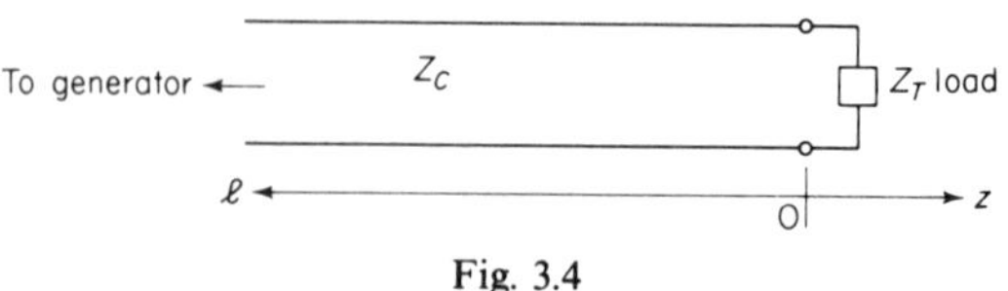

Fig. 3.4

Consider the conditions at the load end of line terminated in an impedance $Z_T \neq Z_C$. For convenience let the z axis origin coincide with the plane of this load. Then the general solutions

$$V(z) = V^+ e^{-\gamma z} + V^- e^{+\gamma z}, \quad I(z) = \frac{V^+}{Z_C} e^{-\gamma z} - \frac{V^-}{Z_C} e^{+\gamma z}$$

reduce to $V(0) = V^+ + V^-$, $I(0) = \dfrac{V^+ - V^-}{Z_C}$ at the load, $z = 0$;

but the ratio of voltage to current at the plane of the load equals the load impedance, thus

$$\frac{V(0)}{I(0)} = Z_T = Z_C \frac{V^+ + V^-}{V^+ - V^-} \qquad \ldots (16)$$

We define the ratio of the reflected wave voltage V^- to the incident wave voltage V^+ at the load as the reflection coefficient, $\rho_0 = |\rho_0| e^{j\phi_0}$

Thus from (16):
$$\frac{Z_T}{Z_C} = \frac{1 + \dfrac{V^-}{V^+}}{1 - \dfrac{V^-}{V^+}} = \frac{1 + \rho_0}{1 - \rho_0}$$

hence
$$\rho_0 = \frac{V^-}{V^+} = \frac{Z_T - Z_C}{Z_T + Z_C} \qquad \ldots (17)$$

is non-zero unless $Z_T = Z_C$.

The ratio of the voltage across the load to the incident wave voltage is defined as the transmission coefficient, τ:

$$\tau = \frac{V^+ + V^-}{V^+} = 1 + \rho_0 = \frac{2Z_T}{Z_T + Z_C}$$

The ratio of the reflected to incident voltage waves at any distance l from the load is

$$\rho(l) = \left[\frac{V^- e^{+\gamma z}}{V^+ e^{-\gamma z}}\right]_{z=-l} = \frac{V^-}{V^+} e^{-2\gamma l} = \rho_0\, e^{-2\gamma l}$$

$$= \left[\,|\rho_0|\, e^{-2\alpha l}\right] e^{j(\phi_0 - 2\beta l)} \qquad\qquad \dots (18)$$

(b) *The distribution of voltage and current on a terminated line: standing waves and the voltage standing wave ratio*

The voltage and current solutions on a mismatched line terminated in $Z_T \neq Z_C$ at a distance l from the load are (neglecting loss in the line, i.e. $\gamma = j\beta$):

$$V(l) = V^+ e^{j\beta l} + V^- e^{-j\beta l} = V^+(e^{j\beta l} + |\rho_0|\, e^{-j(\beta l - \phi_0)}) \quad \dots (19)$$

$$I(l) = \frac{V^+}{Z_C} e^{j\beta l} - \frac{V^-}{Z_C} e^{-j\beta l} = \frac{V^+}{Z_C}(e^{j\beta l} - |\rho_0|\, e^{-j(\beta l - \phi_0)}) \quad \dots (20)$$

where

$$\frac{V^-}{V^+} = \frac{Z_T - Z_C}{Z_T + Z_C} = \rho_0 = |\rho_0|\, e^{j\phi_0}$$

is the reflection coefficient at the load.

The resultant amplitude of line voltage and current may be found by the phasor addition of the incident and reflected wave components as shown in figs. 3.5(a) and (b) respectively:

$$|V(l)| = V^+[1 + |\rho_0|^2 + 2|\rho_0|\cos(2\beta l - \phi_0)]^{1/2} \qquad \dots (21)$$

$$|I(l)| = \frac{V^+}{Z_C}[1 + |\rho_0|^2 - 2|\rho_0|\cos(2\beta l - \phi_0)]^{1/2} \qquad \dots (22)$$

The voltage amplitude has a maximum amplitude V_{max} and the current a minimum amplitude I_{min} at positions on the line where

$$\cos(2\beta l - \phi_0) = +1,\ 2\beta l - \phi_0 = 2\pi m,\ m = 0, 1, 2, 3 \dots$$

i.e. when $\qquad l = \dfrac{1}{2\beta}(2\pi m + \phi_0) = \dfrac{1}{2}m\lambda + \dfrac{\lambda}{4\pi}\phi_0$ as $\beta = \dfrac{2\pi}{\lambda}\qquad \dots (23)$

and $\qquad V_{max} = V^+(1 + |\rho_0|),\ I_{min} = V^+(1 - |\rho_0|)/Z_C \qquad \dots (24)$

Also when $\cos(2\beta l - \phi_0) = -1,\ 2\beta l - \phi_0 = (2m+1)\pi$

$$l = \frac{1}{2}m\lambda + \frac{1}{4}\lambda + \frac{\lambda}{4\pi}\phi_0 \qquad\qquad \dots (25)$$

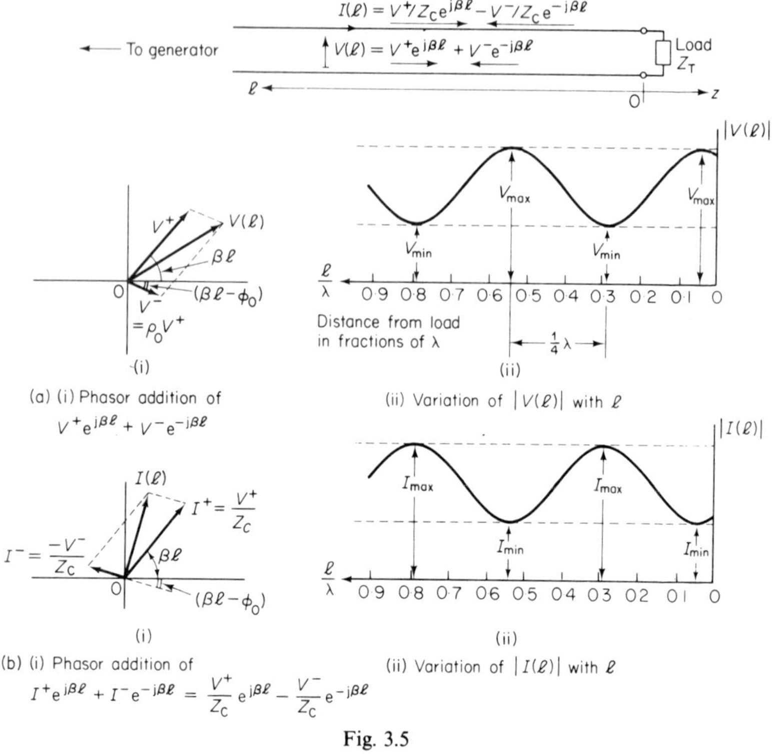

Fig. 3.5

the voltage and current amplitudes are respectively at their minimum and maximum values:

$$V_{min} = V^+(1 - |\rho_0|), \quad I_{max} = V^+(1 + |\rho_0|)/Z_C \qquad \ldots (26)$$

The variation of voltage and current amplitudes with distance l as defined by (21) and (22) give rise to what is known as a standing wave pattern on the line. The term standing wave qualifies the fact that the positions of voltage and current maximum and minimum always occur at the same positions on a loss-less line. A pure standing wave occurs on the line when $|\rho_0| = 1$, in which case $V_{min} = I_{min} = 0$. If $|\rho_0| \neq 1$ a partial standing wave is said to exist on the line. When $|\rho_0| = 0$, i.e. $Z_T = Z_C$, there are no standing waves set up since no reflection occurs at the load. In this case maximum power is transferred to the load and this is the condition we wish to achieve in practice.

A standing wave on a line may be quantified by 2 parameters. The first fixes the position of the standing wave pattern relative to the termination. This may be found experimentally by locating a voltage minimum, for example. The second parameter is used as a measure of the degree of mismatch produced by the load, and this parameter is known as the VOLTAGE STANDING WAVE RATIO, normally abbreviated to V.S.W.R. and denoted by the symbol S. It is defined as

$$S = \frac{V_{max}}{V_{min}} = \frac{I_{max}}{I_{min}} \qquad \ldots (27)$$

On substituting for V_{max} and V_{min} using (24) and (26) we have

$$S = \frac{1+|\rho_0|}{1-|\rho_0|} \qquad \ldots (28)$$

$$\text{and } |\rho_0| = \frac{S-1}{S+1} \qquad \ldots (29)$$

Also at positions on the line where the voltage is a maximum, the corresponding current is a minimum and we have

$$V = V^+(1+|\rho_0|e^{-j(2\beta l - \phi_0)})e^{j\beta l} = V^+(1+|\rho_0|)e^{j\beta l} = V_{max}\,e^{j\beta l}$$

$$I = \frac{V^+}{Z_C}(1-|\rho_0|e^{-j(2\beta l - \phi_0)})e^{j\beta l} = \frac{V^+}{Z_C}(1-|\rho_0|)e^{-j\beta l} = I_{min}\,e^{j\beta l}$$

since $(2\beta l - \phi_0) = 2m\pi$, $e^{-j(2\beta l - \phi_0)} = 1$ and therefore the effective input impedance looking in at these positions along the terminated line is

$$Z_{in} = \frac{V}{I} = \frac{V_{max}}{I_{min}} = Z_C\frac{1+|\rho_0|}{1-|\rho_0|} = Z_C S \qquad \ldots (30)$$

i.e. at positions of voltage maximum the input impedance is S times the characteristic impedance of the line.

Likewise at positions of voltage minimum, the current is at a maximum and the effective input impedance at these positions is

$$Z_{in} = \frac{V}{I} = \frac{V_{min}}{I_{max}} = Z_C\frac{1-|\rho_0|}{1+|\rho_0|} = \frac{Z_C}{S} \qquad \ldots (31)$$

(c) *The input impedance of a length l of terminated line*
The input impedance of a length l of line terminated in an impedance Z_T is defined as

$$Z_{in} = \frac{V(l)}{I(l)} \qquad \ldots (32)$$

where $\qquad V(l) = V^+(e^{\gamma l} + \rho_0 e^{-\gamma l}) \ldots$ the line voltage at l

$$I(l) = \frac{V^+}{Z_C}(e^{\gamma l} - \rho_0 e^{-\gamma l}) \ldots \text{the line current at } l$$

and $\qquad \rho_0 = \dfrac{Z_T - Z_C}{Z_T + Z_C}$ is the reflection coefficient at the load

On substituting the above expressions for $V(l)$ and $I(l)$ we obtain

$$Z_{in} = Z_C \frac{e^{\gamma l} + \rho_0 e^{-\gamma l}}{e^{\gamma l} - \rho_0 e^{-\gamma l}} \qquad \ldots \text{(33a)}$$

$$= Z_C \frac{1 + \rho_0 e^{-2\gamma l}}{1 - \rho_0 e^{-2\gamma l}} = Z_C \frac{1 + \rho(l)}{1 - \rho(l)} \qquad \ldots \text{(33b)}$$

where $\rho(l) = \rho_0 e^{-2\gamma l} \ldots$ the reflection coefficient (ratio of the reflected voltage wave to the incident voltage wave) at l from the termination.

To obtain a general formula for Z_{in} in terms of Z_T, γ, Z_C we substitute $\rho_0 = (Z_T - Z_C)/(Z_T + Z_C)$ into (33a), then

$$Z_{in} = Z_C \frac{(Z_T + Z_C)e^{\gamma l} + (Z_T - Z_C)e^{-\gamma l}}{(Z_T + Z_C)e^{\gamma l} - (Z_T - Z_C)e^{-\gamma l}}$$

$$= Z_C \frac{Z_T(e^{\gamma l} + e^{-\gamma l}) + Z_C(e^{\gamma l} - e^{-\gamma l})}{Z_T(e^{\gamma l} - e^{-\gamma l}) + Z_C(e^{\gamma l} + e^{-\gamma l})}$$

$$= Z_C \frac{Z_T \cosh \gamma l + Z_C \sinh \gamma l}{Z_T \sinh \gamma l + Z_C \cosh \gamma l} = Z_C \frac{Z_T + Z_C \tanh \gamma l}{Z_C + Z_T \tanh \gamma l} \qquad \ldots \text{(34)}$$

It is often common practice to specify impedances in transmission line problems as a fraction of Z_C, the characteristic impedance of the line. If this is done the impedances are known as normalized impedances and are written in small type. The normalized input impedance of a terminated line is then given by

$$z_{in} = \frac{Z_{in}}{Z_C} = \frac{\dfrac{Z_T}{Z_C} + \tanh \gamma l}{1 + \dfrac{Z_T}{Z_C}\tanh \gamma l} = \frac{z_T + \tanh \gamma l}{1 + z_T \tanh \gamma l} \qquad \ldots \text{(35)}$$

where $\quad z_T = \dfrac{Z_T}{Z_C}$ is the normalized load impedance.

Normalization of impedances and admittances has important practical advantages. For example it allows problems, regardless of different Z_C values, to be worked out graphically by use of the Smith chart.

For loss-less and low-loss for which αl is sufficiently small to be neglected, $\gamma = j\beta$ and $\tanh \gamma l = \tanh j\beta l = j \tan \beta l$. For these cases (34) reduces to

$$Z_{\text{in}} = Z_{\text{C}} \frac{Z_{\text{T}} + jZ_{\text{C}} \tan \beta l}{Z_{\text{C}} + jZ_{\text{T}} \tan \beta l} \qquad \dots (36)$$

Now since $\tan \beta l$ is a periodic function, Z_{in} is also periodic. That is

$$Z_{\text{in}}(l) = Z_{\text{in}}(l + \tfrac{1}{2}m\lambda), m = 1, 2, 3, 4 \dots$$

In particular if $l = \tfrac{1}{2}m\lambda$:

$$Z_{\text{in}}(\tfrac{1}{2}m\lambda) = Z_{\text{C}} \frac{Z_{\text{T}} + jZ_{\text{C}} \tan m\pi}{Z_{\text{C}} + jZ_{\text{T}} \tan m\pi} = Z_{\text{T}} \qquad \dots (37)$$

Thus a single or multiple half wavelength of line acts as a $1 : 1$ transformer. A single half wavelength of line is often referred to as a half wave transformer.

Also if $l = (2m + 1)\tfrac{1}{4}\lambda$, $m = 0, 1, 2, 3 \dots$,

$$\tan \beta l = \tan \frac{\pi}{4}(2m - 1) \rightarrow \infty$$

thus

$$Z_{\text{in}}\{(2m + 1)\tfrac{1}{4}\lambda\} = Z_{\text{C}} \left[\frac{Z_{\text{T}} + jZ_{\text{C}} \tan \beta l}{Z_{\text{C}} + jZ_{\text{T}} \tan \beta l} \right]_{l = (2m + 1)(\lambda_{\text{g}}/4)} = \frac{Z_{\text{C}}^2}{Z_{\text{T}}}$$
$$\dots (38)$$

Thus a quarter wavelength or an odd number of quarter wavelengths of line acts as an impedance 'inverter'. A single $\tfrac{1}{4}\lambda$ section of line is often referred to as a quarter wave transformer. The $\tfrac{1}{4}\lambda$ transformer finds important applications as an impedance matching device.

4. EQUIVALENT T AND Π NETWORKS FOR A LENGTH OF TRANSMISSION LINE

A length l of uniform transmission line may be represented, so far as its terminal behaviour is concerned, by an equivalent 2-port network. The values of the elements of the equivalent T and Π networks, shown in figs. 3.6(b) and (c), may be derived by equating the respective open and short-circuit immittances to those of transmission line under the same terminal conditions. For example, for the T network,

$$\text{the open-circuit input impedance,} \quad Z_1 + Z_2 = Z_{\text{C}} \coth \gamma l \qquad \dots (39)$$

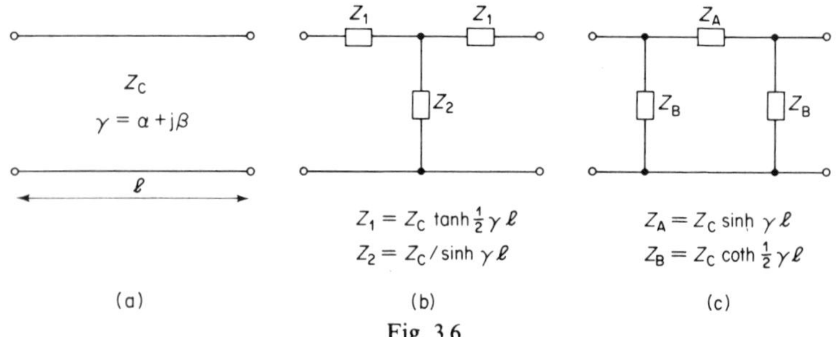

$$Z_1 = Z_C \tanh \tfrac{1}{2}\gamma \ell$$
$$Z_2 = Z_C / \sinh \gamma \ell$$

$$Z_A = Z_C \sinh \gamma \ell$$
$$Z_B = Z_C \coth \tfrac{1}{2}\gamma \ell$$

(a) (b) (c)

Fig. 3.6

the short-circuit input impedance, $Z_1 + \dfrac{Z_1 Z_2}{Z_1 + Z_2} = Z_C \tanh \gamma l \quad \ldots (40)$

and on eliminating Z_1 from (40) using (39) we obtain

$$Z_2 = \sqrt{[Z_C^2(\coth^2 \gamma l - 1)]} = Z_C \sqrt{(\mathrm{cosech}^2\, \gamma l)} = Z_C /\sinh \gamma l$$

and $Z_1 = Z_C \coth \gamma l - Z_2 = Z_C \tanh \tfrac{1}{2}\gamma l$

For the Π, the elements may be found by equating open and short circuit input admittances.

In the case of electrically short lines for which $\alpha l \ll 1$ and $\beta l \ll 1$, the equivalent T and Π circuits reduce to the forms shown in fig. 3.2(d) and (e) where l replaces δz. These forms are known as *nominal equivalent networks* for a transmission line and they may be used as a basis for the construction of models to investigate and simulate the behaviour of power and l.f. transmission lines.

5. THE SMITH CHART: APPLICATIONS TO IMMITANCE EVALUATION AND MATCHING

(a) *Form of the Smith chart*
The Smith chart is an extremely useful tool for calculating the normalized impedances and admittances of a terminated line by graphical means. The chart is shown in a simplified form in fig. 3.7.

(b) *Evaluation of the input admittance of a length l of terminated line*

(i) *Given z_T or y_T*
The normalized load, $Z_T/Z_C = z_T = r_T + jx_T$ (or $Y_T/Y_C = y_T = g_T + jb_T$) is first located on the chart. In fig. 3.8 we take this point to be A. The V.S.W.R. or S circle is then drawn by describing a circle centre $(1, 0)$, i.e. point M, and radius MA. If required, the value of V.S.W.R. on the line

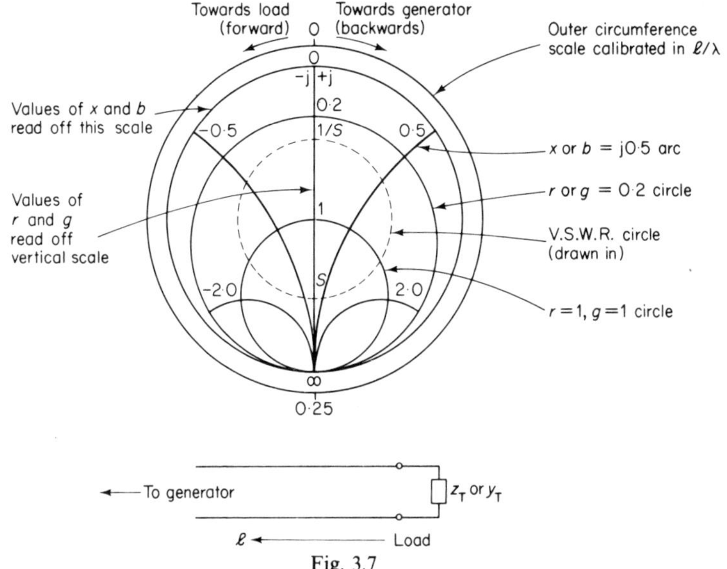

Fig. 3.7

may be read off where this circle cuts the vertical r scale, i.e. S corresponds to point B.

The input impedance, $z_{\text{in}}(l) = r + \mathrm{j}x$ may now be found as follows:

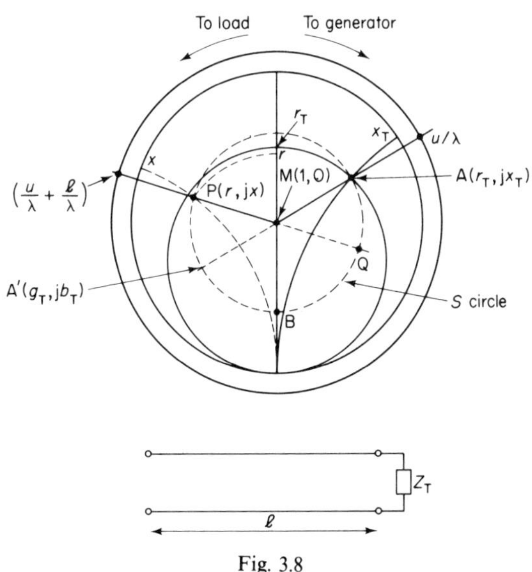

Fig. 3.8

(1) Draw a line through M passing through the load point A to outer l/λ scale. Read off the value, u/λ say, where this line cuts the l/λ scale.

(2) Add l/λ to this value and locate the new value $(u/\lambda + l/\lambda)$ in the clockwise (towards generator) direction. Join this point to the centre M.

(3) The normalized input impedance corresponds to the coordinates of point P where the line drawn in (2) cuts the S circle.

(4) Note that y_T corresponds to point A' and $y_{in}(l) = g + jb$ corresponds to point Q. In general normalized admittance and impedance points are always diametrically opposite points on the S circle ($\frac{1}{4}\lambda$ transformer principle).

(ii) *Given the V.S.W.R. S on the line and distance d of the first voltage minimum from the load*

In this case the S circle can be drawn immediately by locating S on the vertical r axis and describing a circle centre M(1,0) radius S as shown in fig. 3.9. Then referring to this figure

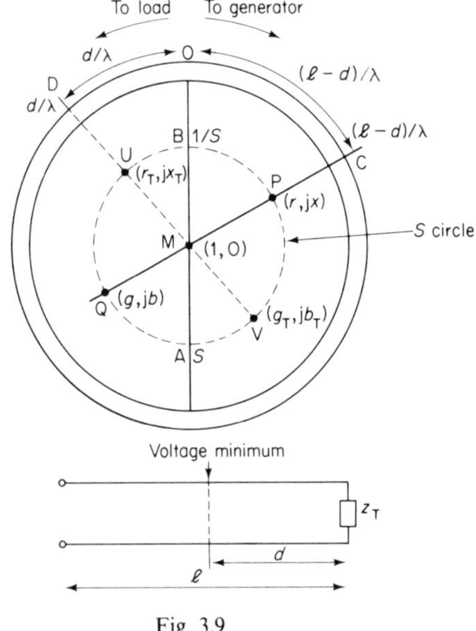

Fig. 3.9

(1) Point A $(S, 0)$ refers to positions on the line of voltage maximum where $z_{in} = S$, point B $(1/S, 0)$ refers to positions of voltage minimum where $z_{in} = 1/S$.

(2) To find $z_{in}(l)$ locate position $(l-d)/\lambda$ on outer wavelength scale (point C in fig. 3.9). Join MC. Co-ordinates of point P give $z_{in}(l) = r+jx$, point Q gives $y_{in}(l) = g+jb$.

(3) To find normalized load impedance z_T locate point D d/λ towards load with respect to B. Join DM. Point U gives (r_T, jx_T), point V gives normalized load admittance $y_T = (g_T, jb)$.

(iii) *For short- and open-circuited lines*
For these cases $S = \infty$ and the S circle corresponds to the outer circle $(r = \infty)$ on the Smith chart. The input immittances are purely imaginary and their evaluation is intimated in fig. 3.10.

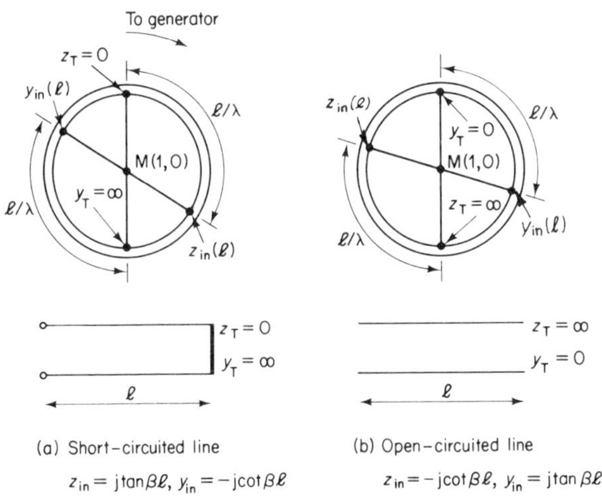

Fig. 3.10

(c) *Application to single stub matching*
Short and open-circuit lines (stubs) are extensively used as reactive-susceptive elements in the matching of transmission lines to a load termination.

(i) *Shunt-stub matching*
(1) Plot normalized load admittance y_T, point A in fig. 3.11, on Smith chart and draw in the S circle.
(2) Locate point (usually for practical considerations nearest the load to minimise the frequency sensitivity of the match) where $y_{in} = 1+jb$ i.e. point B, where the S circle cuts the $g=1$ circle, and note its distance u/λ from load.

(3) Find length of a short- (or open-) circuited stub which has an
admittance $y_{\text{stub}} = -jb$ and place this element in parallel with the
line at u/λ from load. Then at this point the combined input admittance $y_{\text{in}} + y_{\text{stub}} = 1$ and a match is achieved.

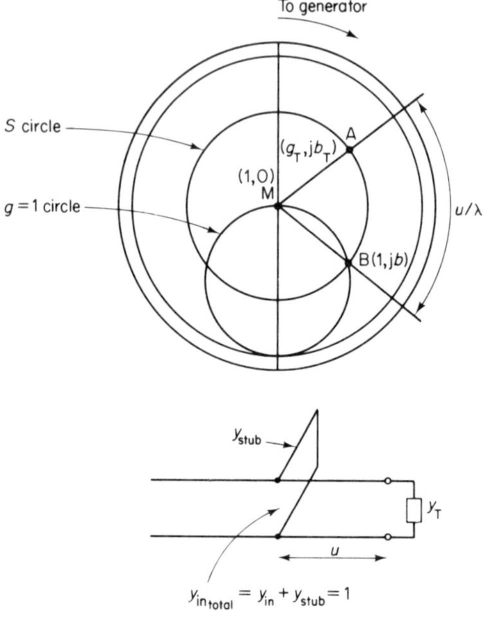

Fig. 3.11

(ii) *Series stub matching*
(1) Plot normalized load impedance z_T and draw S-circle.
(2) Locate point nearest load where S circle intersects $r = 1$ circle, say
v/λ from load. At this position $z_{\text{in}} = 1 + jx$.

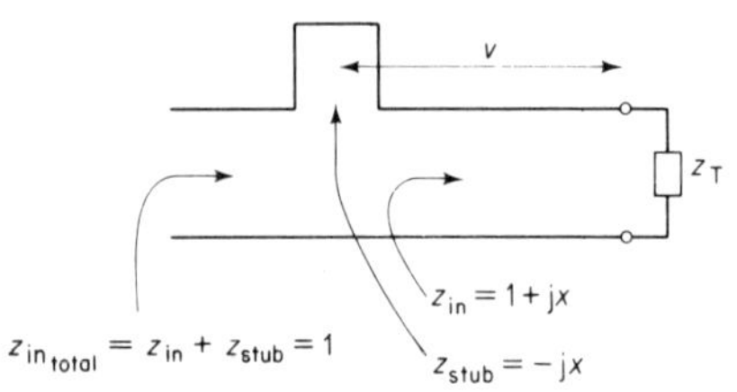

Fig. 3.12

(3) Find length of stub to cancel jx reactance, i.e. $z_{stub} = -jx$, and insert this stub in series with the line at v from termination, as shown in fig. 3.12.

3.2. Worked Problems

1. A transmission line cable is 120 km long and has the following distributed parameters:

$$R = 0.025\,\Omega/m,\ L = 0.620\,\mu H/m,\ C = 37\,pF/m,\ G = 0.$$

Calculate the characteristic impedance of the line at 10 kHz. If the line is terminated in this impedance, determine at 10 kHz (a) the wavelength; (b) the phase and group velocities; (c) the gain in dB of identical repeaters inserted at 15 km intervals between the sending and receiving ends so that the overall transmission loss on the line is zero.

Solution
In terms of its distributed parameters the characteristic impedance of the line,

$$Z_C = \left(\frac{R + j\omega L}{G + j\omega C}\right)^{1/2} = |Z_C|\,e^{j\phi}$$

where $|Z_C| = \left(\dfrac{R^2 + \omega^2 L^2}{G^2 + \omega^2 C^2}\right)^{1/4}$ and $\phi = \dfrac{1}{2}\left[\tan^{-1}\left(\dfrac{\omega L}{R}\right) - \tan^{-1}\left(\dfrac{\omega C}{G}\right)\right]$

On substituting $R = 0.025$, $L = 0.620 \times 10^{-6}$, $C = 37 \times 10^{-12}$, $G = 0$ and $\omega = 2\pi \times 10^4$ we have

$$|Z_C| = 141 \cdot 1\,\Omega,\ \phi = 16° \, 22'$$

The propagation constant γ is given by

$$\gamma = \alpha + j\beta = \sqrt{[(R + j\omega L)(G + j\omega C)]} \qquad \ldots (1)$$
$$= 0.328 \times 10^{-3} \angle 73° \, 38' = (0.0924 + j0.3147)10^{-3}$$

at 10 kHz. Thus $\alpha = 0.0924 \times 10^{-3}$ Np/m and $\beta = 0.3147 \times 10^{-3}$ rad/m, and so the wavelength,

$$\lambda = \frac{2\pi}{\beta} = \frac{2\pi}{0.3147 \times 10^{-3}} = 19\,965 \cdot 7\,m$$

and the phase velocity,

$$v_p = \frac{\omega}{\beta} = \frac{2\pi \times 10^4}{0.3147 \times 10^{-3}} = 1 \cdot 9966 \times 10^8\,m/s$$

On solving the complex equation (1) for β we obtain with $G = 0$,

$$\beta^2 = \tfrac{1}{2}[\omega^2 LC + \omega C(R^2 + \omega^2 L^2)^{1/2}]$$

and on differentiating this expression with respect to β, we have

$$2\beta\frac{\mathrm{d}\beta}{\mathrm{d}\omega} = \tfrac{1}{2}[2\omega LC + C(R^2 + \omega^2 L^2)^{1/2} + \tfrac{1}{2}\omega C(R^2 + \omega^2 L^2)^{-1/2}2\omega L^2]$$

so $$\frac{\mathrm{d}\omega}{\mathrm{d}\beta} = \frac{4\beta}{[2\omega LC + C(R^2 + \omega^2 L^2)^{1/2} + \omega^2 LC(R^2 + \omega^2 L^2)^{-1/2}]}$$
$$= v_g, \text{ the group velocity, which at } 10\,\mathrm{kHz} \text{ has the value}$$
$$v_g = 2{\cdot}1672 \times 10^8 \,\mathrm{m/s}$$

The overall line loss of the 120 km line, before insertion of repeater amplifiers, is

$$120\,000\alpha = 120\,000 \times 0{\cdot}0924 \times 10^{-3} = 11{\cdot}088 \text{ nepers}$$

$$= 11{\cdot}088 \times 8{\cdot}686 = 96{\cdot}31 \text{ decibels.}$$

In all 7 repeaters at 15 km intervals are required so gain of each to provide zero overall loss is $\tfrac{1}{7} \times 96{\cdot}31 = 13{\cdot}76\,\mathrm{dB}$.

2. A loss-less air-spaced transmission line of $600\,\Omega$ characteristic impedance connects a 150 MHz transmitter to an antenna of effective impedance of $75\,\Omega$. The antenna radiates a power of 6 W. Determine (a) the voltage reflection coefficient at the antenna's input and the V.S.W.R. on the line, (b) the position and values of maximum and minimum voltage and current on the line.

Solution
(a) The reflection coefficient at the load,

$$\rho = \frac{Z_T - Z_C}{Z_T + Z_C}$$

where Z_T = the load impedance = $75\,\Omega$, Z_C = the characteristic impedance of the line = $600\,\Omega$.

Thus $\rho = \dfrac{75 - 600}{75 + 600} = -0{\cdot}7778$

and the V.S.W.R. $S = \dfrac{1 + |\rho|}{1 - |\rho|} = \dfrac{1 + 0{\cdot}7778}{1 - 0{\cdot}7778} = 8$

(b) Since ρ is real and negative at the load, the incident and reflected voltage wave components are in antiphase at the load. Hence a voltage minimum occurs at the load. Voltage minima are also repeated at $\frac{1}{2}\lambda$ intervals along the line from the load. Since current maxima occur at positions of voltage minimum, current maxima also occur at $\frac{1}{2}\lambda$ intervals from the load. Since the line is air spaced, the wavelength (assuming TEM waves) is given by

$$\lambda = \frac{\text{velocity of light}}{\text{frequency}} = \frac{3 \times 10^8}{150 \times 10^6} = 2\,\text{m}$$

Thus voltage minima and current maxima occur at 0, 1 m, 2 m, 3 m . . . from the load. Let the peak amplitude of the voltage and current at the load be V_T, I_T. Then

$$\text{Power dissipated in load} = \frac{1}{2}\frac{V_T^2}{75} = \frac{1}{2} \times 75 \times I_T^2 = 6\,\text{W}$$

Hence $V_T = \sqrt{(2 \times 75 \times 6)} = 30\,\text{V}$ . . . amplitude of voltage minima, V_{min};

$$I_T = \sqrt{\left(\frac{2 \times 6}{75}\right)} = 0{\cdot}4\,\text{A} \ldots \text{amplitude of current maxima, } I_{max}.$$

Voltage maxima occur mid-way between the positions of voltage minimum. Likewise current minima. Thus V_{max} and I_{min} occur at $\frac{1}{4}\lambda$, $\frac{3}{4}\lambda$, $\frac{5}{4}\lambda$ . . . distances from load. Their amplitudes are

$$V_{max} = SV_{min} = 8 \times 30 = 240\,\text{V}$$

$$I_{min} = \frac{1}{S}I_{max} = \frac{0{\cdot}4}{8} = 0{\cdot}05\,\text{A}.$$

3. A loss-free line of characteristic impedance Z_C ohms is terminated in an admittance of $Y_T = (G_T + jB_T)$ siemens. If the voltage amplitude at the termination is V_T volts determine the amplitude of the voltage V at any distance l along the line from Y_T, and show that the condition for $|V|$ to be a maximum or minimum is

$$\tan 2\beta l = \frac{2Z_C B_T}{Z_C^2(G_T^2 + B_T^2) - 1}$$

If the line has a characteristic impedance of $50\,\Omega$, operates at 1 GHz, and is terminated in an admittance of $(17{\cdot}32 + j10)\,\text{mS}$ find the positions of minimum and maximum voltage and the V.S.W.R. on the line.

Solution
Using the general solutions of the transmission line equations we have for the voltage and current at any position l from the termination:

$$V = V^+ e^{j\beta l} + V^- e^{-j\beta l}$$

$$I = \frac{V^+}{Z_C} e^{j\beta l} - \frac{V^-}{Z_C} e^{-j\beta l}$$

At the termination, $l = 0$, we have

$$V(0) = V_T = V^+ + V^- \qquad\qquad \dots (1)$$

$$I(0) = Y_T V_T = \frac{1}{Z_C}(V^+ - V^-) \qquad\qquad \dots (2)$$

and solving (1) and (2) for V^+ and V^- we obtain

$$V^+ = \tfrac{1}{2}(1 + Y_T Z_C)V_T, \quad V^- = \tfrac{1}{2}(1 - Y_T Z_C)V_T$$

Hence
$$\begin{aligned}
V &= [\tfrac{1}{2}(1 + Y_T Z_C)e^{j\beta l} + \tfrac{1}{2}(1 - Y_T Z_C)e^{-j\beta l}]V_T \\
&= [\tfrac{1}{2}(e^{j\beta l} + e^{-j\beta l}) + \tfrac{1}{2}Y_T Z_C(e^{j\beta l} - e^{-j\beta l})]V_T \\
&= [\cos \beta l + j Y_T Z_C \sin \beta l]V_T \\
&= [\cos \beta l + j Z_C(G_T + j B_T)\sin \beta l]V_T \\
&= [(\cos \beta l - Z_C B_T \sin \beta l) + j Z_C G_T \sin \beta l]V_T
\end{aligned}$$

and therefore $|V|^2 = [(\cos \beta l - Z_C B_T \sin \beta l)^2 + Z_C^2 G_T^2 \sin^2 \beta l]V_T^2$

The condition for $|V|^2$ (and also $|V|$) to be a maximum or minimum is found from

$$\frac{\partial^2 |V|^2}{\partial l} = [2(\cos \beta l - Z_C B_T \sin \beta l)(-\beta \sin \beta l - \beta Z_C B_T \cos \beta l)$$

$$+ 2 Z_C^2 G_T^2 \beta \sin \beta l \cos \beta l]V_T^2 = 0$$

from which we obtain

$$\cos \beta l \sin \beta l[Z_C^2(B_T^2 + G_T^2) - 1] - Z_C B_T(\cos^2 \beta l - \sin^2 \beta l) = 0$$

i.e. $\quad \dfrac{\cos \beta l \sin \beta l}{\cos^2 \beta l - \sin^2 \beta l} = \dfrac{\tan \beta l}{1 - \tan^2 \beta l} = \tfrac{1}{2}\tan 2\beta l = \dfrac{Z_C B_T}{Z_C^2(B_T^2 + G_T^2) - 1}$

so the required condition is $\tan 2\beta l = \dfrac{2 Z_C B_T}{Z_C^2(B_T^2 + G_T^2) - 1}$

When $B_T = 10\,\text{mS}$, $G_T = 17\cdot32\,\text{mS}$, $Z_C = 50\,\Omega$, $Z_C^2(B_T^2 + G_T^2) = 1$

and therefore positions of voltage maxima and minima occur when $\tan 2\beta l \to \infty$, i.e. $2\beta l = n\pi + \tfrac{1}{2}\pi$, $n = 0, 1, 2, 3 \dots$ or $l = \tfrac{1}{8}\lambda + \tfrac{1}{4}n\lambda$

When $l = \tfrac{1}{8}\lambda$,

$$|V| = [\tfrac{1}{2}(1 - 50 \times 10^{-2})^2 + (50 \times 17\cdot32 \times 10^{-3} \times 1/\sqrt{2})^2]^{1/2}V_T = 0\cdot707 V_T$$

and when $l = \frac{3}{8}\lambda$,

$$|V| = [\tfrac{1}{2}(-1-0\cdot5)^2 + \tfrac{1}{2} \times 0\cdot75]^{1/2} V_T = 1\cdot225 \ V_T$$

Thus a voltage minimum occurs at a distance of $\frac{1}{8}\lambda$ from the load and a voltage maximum at $\frac{3}{8}\lambda$. Subsequent positions of voltage minima and maxima occur at $\frac{1}{8}\lambda + \frac{1}{2}n\lambda$ and $\frac{3}{8}\lambda + \frac{1}{2}n\lambda$, respectively.

The V.S.W.R. on the line is $S = \dfrac{V_{max}}{V_{min}} = \dfrac{1\cdot225}{0\cdot707} = 1\cdot73$

4. Find the power dissipated in the load $Z_L = \frac{1}{6}Z_C$ of the circuit shown in fig. 3.13, when $V_G = 5\,\text{V}$ peak and the line characteristic impedances Z_C are both $50\,\Omega$. Calculate also the reflection coefficient amplitudes and the V.S.W.R.s on both $\frac{1}{2}\lambda$ and $\frac{1}{4}\lambda$ line sections.

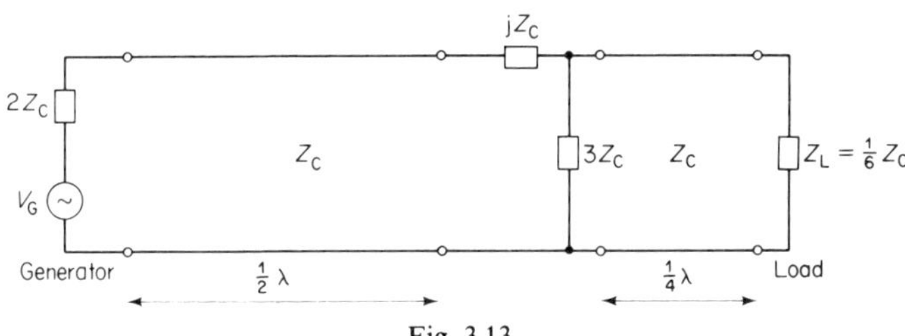

Fig. 3.13

Solution
The effective load impedance $\frac{1}{4}\lambda$ back towards the generator is

$$\frac{Z_C^2}{\frac{1}{6}Z_C} = 6Z_C$$

on applying the impedance transformation property of a $\frac{1}{4}\lambda$ line. Also since the $\frac{1}{2}\lambda$ section acts as a $1:1$ transformer, we may transfer the generator $\frac{1}{2}\lambda$ towards the load. The equivalent circuit which results and which we may use for the power calculation is shown in fig. 3.14.

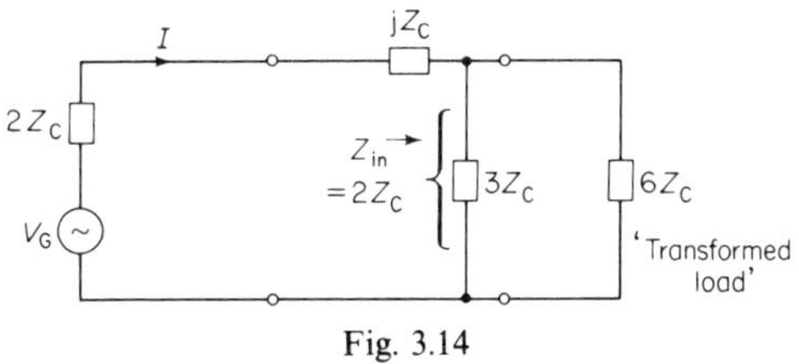

Fig. 3.14

85

Using this circuit, we have for the generator current,

$$I = \frac{V_G}{2Z_C + jZ_C + 2Z_C} = \frac{V_G}{Z_C(4+j)}$$

The current flowing in the $6Z_C$ impedance is therefore

$$I' = \frac{2Z_C I}{6Z_C} = \tfrac{1}{3}I$$

and the power dissipated in the $6Z_C$ impedance and therefore that dissipated in the load is

$$P = \frac{1}{2}|I'|^2 6Z_C = \frac{1}{2}\left|\frac{1}{3}\frac{V_G}{Z_C(4+j)}\right|^2 6Z_C$$

$$= \frac{V_G^2}{51Z_C} = \frac{25}{51 \times 50} = \frac{1}{102}\,\text{W}$$

The amplitude of the reflection coefficient on the $\tfrac{1}{2}\lambda$ line is

$$|\rho_1| = \left|\frac{Z_{L_1}-Z_C}{Z_{L_1}+Z_C}\right| \quad \text{where } Z_{L_1} = jZ_C + 3Z_C \parallel 6Z_C = jZ_C + 2Z_C$$

$$= \left|\frac{Z_C(1+j)}{Z_C(3+j)}\right| = \sqrt{\frac{2}{10}} = \frac{1}{\sqrt{5}} = 0.447$$

Hence the corresponding V.S.W.R.

$$S_1 = \frac{1+|\rho_1|}{1-|\rho_1|} = \frac{1+\dfrac{1}{\sqrt{5}}}{1-\dfrac{1}{\sqrt{5}}} = 2.62$$

The amplitude of the reflection coefficient on the $\tfrac{1}{4}\lambda$ line is

$$|\rho_2| = \left|\frac{\tfrac{1}{6}Z_C - Z_C}{\tfrac{1}{6}Z_C + Z_C}\right| = \frac{5}{7} \quad \text{and the V.S.W.R. } S_2 = \frac{1+|\rho_2|}{1-|\rho_2|} = 6$$

5. A quarter wavelength of low-loss short-circuited line may be represented as an equivalent lumped element parallel circuit as shown in fig. 3.15. Derive expressions for R, L and C in terms of the characteristic impedance Z_C, the attenuation coefficient α of the line, and the wavelength λ for the case of an air-spaced $\tfrac{1}{4}\lambda$ line, when the line losses are small but finite.

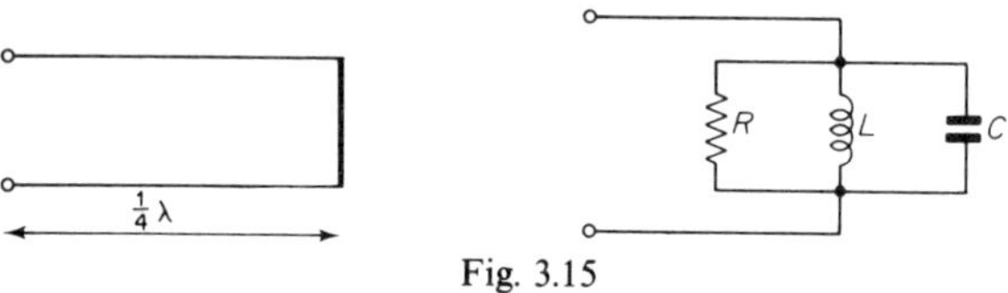

Fig. 3.15

Solution

To determine the equivalence between the two circuits we adopt the following procedures:

(i) To find R we equate the input impedance of the $\frac{1}{4}\lambda$ line to that of the parallel circuit at their resonant frequency, $\omega_0 = 2\pi f_0 = 2\pi c/\lambda$, $c = $ velocity of light in air.

The input impedance of the $\frac{1}{4}\lambda$ short-circuited line is

$$Z_{in} = Z_C \tanh \gamma l$$

$$= Z_C \frac{e^{\gamma l} - e^{-\gamma l}}{e^{\gamma l} + e^{-\gamma l}} = Z_C \frac{1 - e^{-2\gamma l}}{1 + e^{-2\gamma l}}$$

$$= Z_C \frac{1 - e^{-2\alpha l} e^{-j2\beta l}}{1 + e^{-2\alpha l} e^{-j2\beta l}} = \frac{1 + e^{-\alpha\lambda/2}}{1 - e^{-\alpha\lambda/2}}$$

since $2\gamma l = 2(\alpha + j\beta)\dfrac{\lambda}{4} = \alpha\dfrac{\lambda}{2} + j\beta\dfrac{\lambda}{2} = \alpha\dfrac{\lambda}{2} + j\pi$ as $\beta = \dfrac{2\pi}{\lambda}$.

Using the fact that losses are small, i.e.

$$\alpha\frac{\lambda}{2} \ll 1, e^{-\alpha\lambda/2} \approx 1 - \frac{\alpha\lambda}{2}$$

we have

$$Z_{in} \approx Z_C \frac{1 + \left(1 - \alpha\dfrac{\lambda}{2}\right)}{1 - \left(1 - \alpha\dfrac{\lambda}{2}\right)} = Z_C \frac{2 - \alpha\dfrac{\lambda}{2}}{\alpha\dfrac{\lambda}{2}} \approx \frac{4Z_C}{\alpha\lambda}$$

At resonance, $\omega_0 L = 1/(\omega_0 C)$, and therefore the impedance of the parallel circuit is R and so for equivalence we have

$$R = \frac{4Z_C}{\alpha\lambda} \text{ ohms}$$

(ii) To determine the two other parameters, L and C, we use the approximation that the slope of the susceptance curve of the $\frac{1}{4}\lambda$ line should equal the slope of the susceptance curve of the parallel R–L–C circuit at $\omega = \omega_0$.

The susceptance of the $\frac{1}{4}\lambda$ line,

$$B = \left[-\frac{\cot \beta l}{Z_C} \right]_{l = 1/4\lambda, \, \beta = \omega_0/C}$$

and its slope $\left(\dfrac{\mathrm{d}B}{\mathrm{d}\omega}\right)_{\omega=\omega_0} = \left[\dfrac{l}{Z_{\mathrm{C}}c}\,\mathrm{cosec}^2\,\beta l\right]_{l=1/4\lambda} = \dfrac{\lambda}{4Z_{\mathrm{C}}c} \qquad \dots (1)$

The susceptance of the parallel R–L–C circuit, $B' = \left(\omega C - \dfrac{1}{\omega L}\right)$

and its slope $\left(\dfrac{\mathrm{d}B'}{\mathrm{d}\omega}\right)_{\omega=\omega_0} = \left(C + \dfrac{1}{\omega^2 L}\right)_{\omega=\omega_0} = 2C = \dfrac{2}{\omega_0^2 L} \qquad \dots (2)$

Equating (1) and (2) we have

$$C = \dfrac{\lambda}{8Z_{\mathrm{C}}c}\ \text{farads},\quad L = \dfrac{8Z_{\mathrm{C}}c}{\omega_0^2 \lambda} = \dfrac{2Z_{\mathrm{C}}\lambda}{\pi^2 c}\ \text{henries}$$

6. Prove that the voltage at any distance z from the generator end on the terminated transmission line shown in fig. 3.16 is given by,

$$V = \dfrac{Z_{\mathrm{C}}V_{\mathrm{G}}}{Z_{\mathrm{C}}+Z_{\mathrm{G}}} \times \dfrac{1}{(1-\rho_{\mathrm{G}}\rho_{\mathrm{T}}\,\mathrm{e}^{-\mathrm{j}2\beta L})}\left[\mathrm{e}^{-\mathrm{j}\beta z} + \rho_{\mathrm{T}}\,\mathrm{e}^{\mathrm{j}\beta(z-2L)}\right]\ \text{volts}$$

where $\qquad \rho_{\mathrm{G}} = \dfrac{Z_{\mathrm{G}}-Z_{\mathrm{C}}}{Z_{\mathrm{G}}+Z_{\mathrm{C}}},\ \rho_{\mathrm{T}} = \dfrac{Z_{\mathrm{T}}-Z_{\mathrm{C}}}{Z_{\mathrm{T}}+Z_{\mathrm{C}}},$

$\qquad\qquad\quad \beta = $ the phase coefficient

and $\qquad\quad Z_{\mathrm{C}} = $ the characteristic impedance of the line

$\qquad\qquad\qquad$ (assumed to be loss-less)

Determine also the V.S.W.R. on the line and sketch the standing wave pattern, $|V|$ v z, for the case of a loss-free line when $L = 2\lambda$, $Z_{\mathrm{G}} = 2Z_{\mathrm{C}}$, $Z_{\mathrm{T}}=3Z_{\mathrm{C}}$. If $V_{\mathrm{G}}=100\,\text{V}$ peak, $Z_{\mathrm{C}}=50\,\Omega$, calculate also the power dissipated in the load.

Solution

The general solutions for the voltage and current on a transmission line are

$$V = V^+ \mathrm{e}^{-\mathrm{j}\beta z} + V^- \mathrm{e}^{\mathrm{j}\beta z} \qquad \dots (1)$$

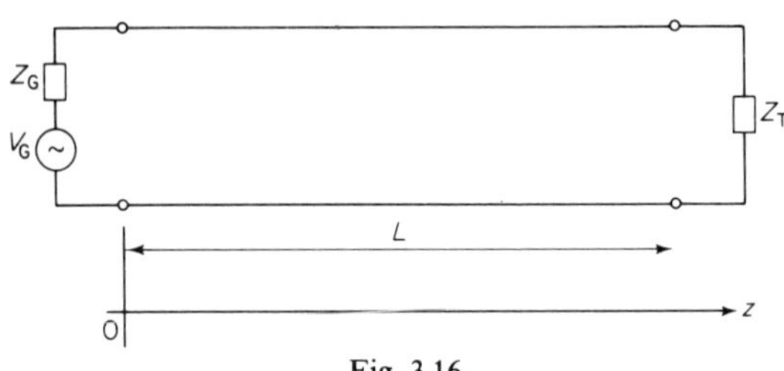

Fig. 3.16

$$Z_C I = V^+ e^{-j\beta z} - V^- e^{j\beta z} \qquad \dots (2)$$

In the above problem the constants V^+ and V^- may be uniquely determined by using the conditions at the generator and load. That is

at $z = 0$
$$V(0) = V^+ + V^- = V_G - Z_G I(0) = V_G - \frac{Z_G}{Z_C}(V^+ - V^-)$$

i.e.
$$(Z_C + Z_G)V^+ + (Z_C - Z_G)V^- = Z_C V_G \qquad \dots (3)$$

at $z = L$
$$V(L) = V^+ e^{-j\beta L} + V^- e^{j\beta L}$$

$$= Z_T I(L) = \frac{Z_T}{Z_C}(V^+ e^{-j\beta L} - V^- e^{j\beta L})$$

i.e.
$$(Z_C - Z_T)e^{-j\beta L}V^+ + (Z_C + Z_T)e^{j\beta L}V^- = 0 \qquad \dots (4)$$

On solving (3) and (4) we obtain

$$V^+ = \frac{1}{\Delta}[(Z_T + Z_C)Z_C \, e^{j\beta L}]V_G \qquad V^- = \frac{1}{\Delta}[(Z_T - Z_C)Z_C \, e^{-j\beta L}]V_G$$

where
$$\Delta = (Z_C + Z_G)(Z_C + Z_T)e^{j\beta L} - (Z_C - Z_G)(Z_C - Z_T)e^{-j\beta L}$$

$$= (Z_C + Z_G)(Z_C + Z_T)e^{+j\beta L}[1 - \rho_G \rho_T \, e^{-j2\beta L}]$$

On substituting into (1) for V^+ and V^- we obtain

$$V = \frac{[(Z_T + Z_C)Z_C \, e^{j\beta L}V_G]e^{-j\beta z} + [(Z_T - Z_C)Z_C \, e^{-j\beta L}V_G]e^{j\beta z}}{(Z_C + Z_G)(Z_C + Z_T)e^{j\beta L} - (Z_C - Z_G)(Z_C - Z_T)e^{-j\beta L}}$$

$$= \frac{Z_C V_G}{Z_C + Z_G} \times \frac{1}{(1 - \rho_G \rho_T \, e^{-j2\beta L})} [e^{-j\beta z} + \rho_T \, e^{j\beta(z - 2L)}] \text{ volts}$$

When
$$L = 2\lambda, \ Z_G = 2Z_C, \ Z_T = 3Z_C$$

$$V = \frac{V_G}{3} \times \frac{1}{(1 - \tfrac{1}{3} \times \tfrac{1}{2} \times e^{-j8\pi})}[e^{-j\beta z} + \tfrac{1}{2}e^{j\beta(z - 2\lambda)}]$$

$$= \tfrac{2}{5}V_G[e^{-j\beta z} + \tfrac{1}{2}e^{j\beta z}]$$

Thus
$$V_{\max} = \tfrac{2}{5}V_G(1 + \tfrac{1}{2}),$$

$$V_{\min} = \tfrac{2}{5}V_G(1 - \tfrac{1}{2}) \text{ and the V.S.W.R.,}$$

$$S = \frac{V_{\max}}{V_{\min}} = 3$$

Since $Z_T > Z_C$ and is purely resistive, a voltage maximum and a corresponding current minimum occur at the load. The power dissipated in the load,

$$P = \frac{1}{2}\left[\frac{V_{\max}}{Z_T}\right]^2 Z_T = \frac{1}{2}\frac{(\frac{3}{5}V_G)^2}{3Z_C} = \frac{3}{50}\frac{V_G^2}{Z_C}\ \text{watts}$$

When $V_G = 100\,\text{V}$, $Z_C = 50\,\Omega$, $P = 12\,\text{W}$.

A sketch of the standing wave pattern of the voltage on the line is shown in fig. 3.17.

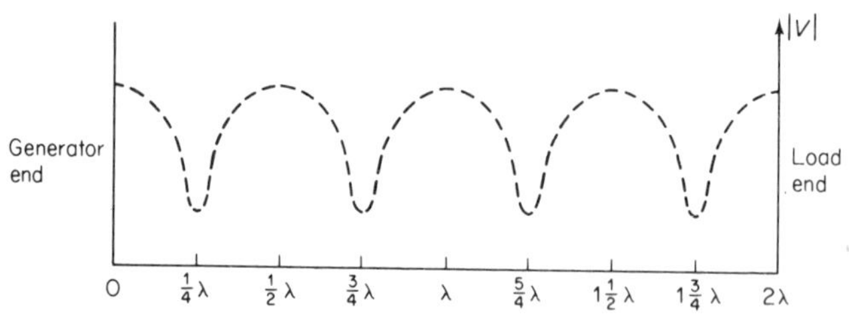

Fig. 3.17

3.3. Exercise Problems

(Note the qualification of the question numbers by the letters L, P, R and M is used to denote the bias of the question to the following areas: L low-frequency and telephone lines; P power; R radio-frequency and microwave, M matching.)

1. (L) A uniform telephone line of length $l = 16\,\text{km}$ has the following primary constants per loop kilometre, $R = 125\,\Omega/\text{km}$, $L = 4.4\,\text{mH/km}$, $C = 0.06\,\mu\text{F/km}$, $G = 0$. Determine the characteristic impedance Z_C, the attenuation constant α, and the phase constant β of the line at 5 kHz. If a voltage of 10 V at a frequency of 5 kHz is applied at the line input determine
   (a) the voltage amplitude at the output when the line is open-circuited,
   (b) the current amplitude at the output when the line is short-circuited.

2. (L) A low frequency line has a characteristic impedance of $Z_c = 331\ \angle -40.6°\,\Omega$, an attenuation constant $\alpha = 0.0526\,\text{Np/km}$, and a phase constant $\beta = 0.0609\,\text{rad/km}$ at a frequency of 1 kHz. Determine the distributed parameters of the line and construct a nominally equivalent Π network which may be used to simulate the

90

behaviour of a 1 km length of line. Derive also analytical values for the elements of an exactly equivalent Π.

3. (L) The distributed parameters of a coaxial cable used in a multi-channel telephone system were measured at 3 MHz with the following results: $R = 65\,\Omega/\text{km}$, $L = 0{\cdot}27\,\text{mH/km}$, $C = 0{\cdot}04\,\mu\text{F/km}$, $G = 0{\cdot}002\,\text{S/km}$. Estimate approximate values for the characteristic impedance and attenuation coefficient of the cable at 3 MHz. Calculate also the minimum gain in decibels of a repeater amplifier, inserted at the mid-point of a 20 km section, so as to provide an overall system loss of less than 40 dB between sending and receiving ends.

4. Derive approximate expressions for the characteristic impedance, phase and attenuation constants of a low-loss uniform line in terms of its distributed parameters: R, L, C and G, assuming that $\omega L \gg R$, $\omega C \gg G$. Assuming R, C and G are fixed but that L may be varied determine the condition that α should be a minimum and the corresponding values of α and β when this condition is satisfied.

5. (L) An open wire line of length 400 km has an attenuation coefficient α and a phase constant β at 1 kHz given by $\alpha = 0{\cdot}1375\,\text{dB/km}$, $\beta = 0{\cdot}0406\,\text{rad/km}$. Its characteristic impedance at 1 kHz is $Z_C = 900\ \angle -28°\ \Omega$. If a 1 kHz generator of e.m.f. 10 V is applied at the sending end of the line and the receiving end is terminated in a load equal to Z_C, determine
(a) the wavelength and phase velocity of waves at 1 kHz on the line;
(b) the voltage amplitude, the current amplitude, and the power at this load.

6. (L) Design a nominal T section to simulate the terminal behaviour of a length $l = 2\,\text{km}$ of line which has the following distributed parameters: $R = 27{\cdot}5\,\Omega/\text{km}$, $L = 2{\cdot}37\,\text{mH/km}$, $C = 0{\cdot}005\,\mu\text{F/km}$, $G = 3{\cdot}13\,\mu\text{S/km}$ which remain approximately constant up to a frequency of 10 kHz. What is the maximum length of line that could be represented by a nominal T network up to a frequency of 10 kHz, if the criterion for approximate equivalence is that $\alpha l < 0{\cdot}1$ and $\beta l < 0{\cdot}1$, where α and β are the attenuation and phase coefficients of the line?

7. (L) A telephone cable has a resistance of $20\,\Omega$ and a capacitance of $0{\cdot}04\,\mu\text{F}$ per loop kilometre; the inductance and leakage conductance are negligible. Determine the attenuation coefficient of the cable in dB/km at 5 kHz. To reduce the attenuation, loading coils of inductance 40 mH and resistance $3\,\Omega$ are added in cascade at 1 km intervals along the cable. Assuming that the effect of each coil may

be taken into account by considering its inductance and resistance to be uniformly distributed over the length of line between successive coils, determine the new value of attenuation coefficient.

8. A loss-less coaxial line, 20 km long, is terminated in a 600 Ω resistor. A leakage fault between the inner and outer conductors develops at an unknown point on the line. To locate it, a short rectangular pulse is applied at the input at $t = 0$ and a subsequent oscilloscope record of the transmitted pulse and the first two reflections from the fault and termination are shown in fig. 3.18. Determine

(a) the position of the fault,

(b) the characteristic impedance Z_C and the distributed parameters L and C of the line.

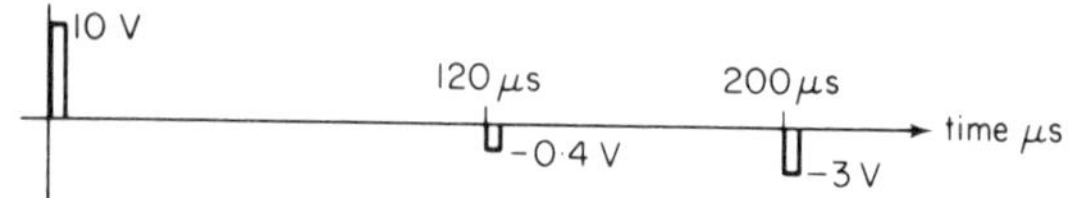

Fig. 3.18. For problem 8

9. (P) A 3-phase 330 kV 50 Hz overhead line has a phase inductance of 1·1 mH/km. Assuming that the line is loss-less, determine

(a) the phase capacitance per kilometre,

(b) the natural load (in ohms and in megawatts),

(c) the maximum length over which maximum power can be transmitted if the end-to-end phase angle of voltage and current is to be less than 20°.

10. (P) A 3-phase 150 km 50 Hz transmission line has the following distributed constants per phase: $R = 0\cdot1\,\Omega/\text{km}$, $L = 1\,\text{mH/km}$, $C = 0\cdot007\,\mu\text{F/km}$, $G = 0$. If the line is supplied at 140 kV, determine

(a) the natural load of the line,

(b) the receiving-end voltage on open circuit, using both the exact transmission line solution and a nominal T or Π equivalent circuit.

11. (P) A 3-phase transmission line of 100 km length has a series impedance $Z = (0\cdot1 + \text{j}0\cdot3)\,\Omega/\text{km}$ per phase and an admittance to neutral per conductor $Y = \text{j}2\cdot5\,\mu\text{S/km}$ per phase. The line supplies a load of 75 MW at 132 kV and 0·8 lagging power factor. Using a nominal Π circuit determine the percentage increase of the sending-end voltage with respect to the load voltage and the phase angle between them.

12. (P) A 3-phase 50 Hz power cable has the following distributed parameters per phase: $R = 0\cdot1\,\Omega/\text{km}$, $L = 0\cdot5\,\text{mH/km}$,

$C = 0.2\,\mu\text{F/km}$, $G = 0$. Thermal heating considerations limit the line current to 300 A. Construct a nominal T network to simulate a length l of cable. Hence estimate

(a) the length of cable for which the line current, when supplied at a line voltage of 120 kV, would equal the limiting current,

(b) the current amplitude and power at the receiving end at 120 kV and 0·94 lagging power factor that could be safely supplied via a 40 km length of the above cable.

13. (P) A 3-phase transmission line has total series impedance Z and shunt admittance Y per phase given by $Z = 200 \angle 80°\,\Omega$, $Y = 0.002 \angle 90°$ S. Draw a nominal Π network to represent the terminal behaviour of the line. A 60 MW load at unity power factor is connected across the mid-point of the line and the open-circuit voltage measured across the receiving end is 200 kV. Using the nominal Π estimate the amplitude of the sending end voltage.

14. (R) Evaluate the voltage standing wave ratio (V.S.W.R.) on the input line 1 for each of transmission line circuits shown in fig. 3.19.

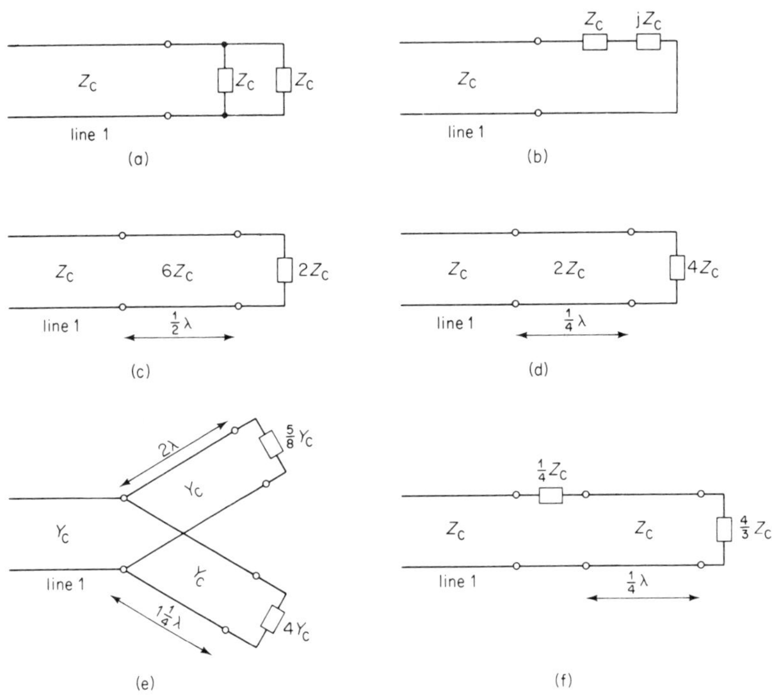

Fig. 3.19. For problem 14

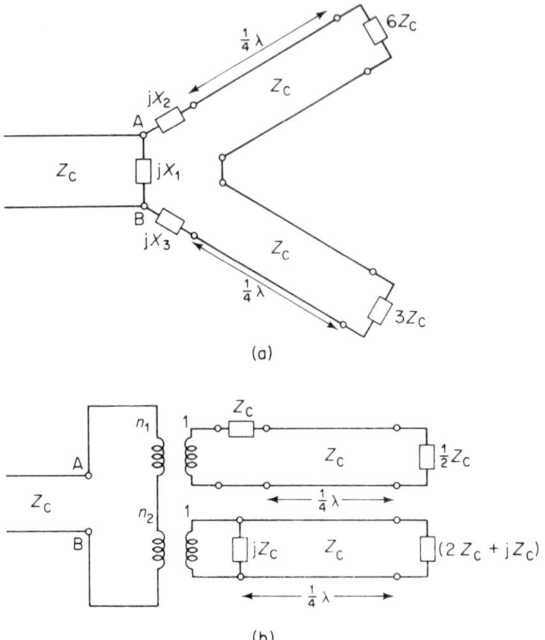

Fig. 3.20. For problem 15

15. (R) Determine the input impedance at terminals A B for both of the circuits shown in fig. 3.20.

16. (R) Determine the power dissipated in the load admittances Y_2 and Y_3 in the circuit shown in fig. 3.21, given that $I_G = 5\,\text{A}$ peak. $Y_C = 0.02\,\text{S}$. The characteristic admittances of lines 1, 2, 3 are Y_C, $4Y_C$ and $2Y_C$ respectively.

17. (R) In the circuit of fig. 3.22 a generator of $50\,\Omega$ internal impedance and operating at $1\,\text{GHz}$ feeds a $75\,\Omega$ load via a line of $50\,\Omega$ characteristic impedance. Determine,
(a) the reflection coefficient at the load and the V.S.W.R. on the line,
(b) the position of voltage and current minima on the line.
If the generator delivers $10\,\text{W}$ of power when operating into $50\,\Omega$, calculate the power supplied to the $75\,\Omega$ load and the values of maximum voltage and current amplitude on the line.

18. (R) Deduce expressions for the normalized resistive and reactive components of a load termination in terms of the V.S.W.R. S measured on the line and the distance d of the voltage minimum from the load.

94

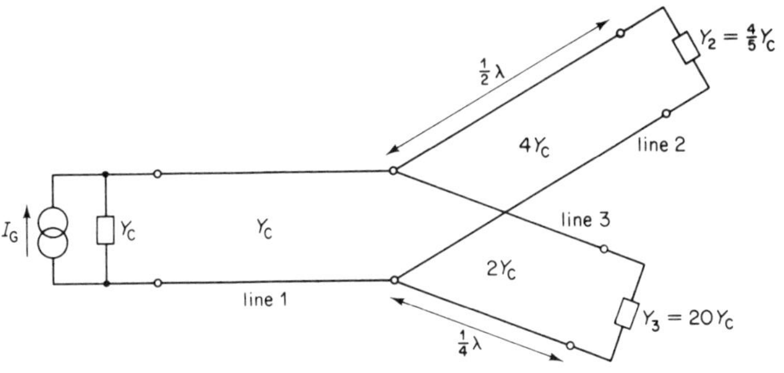

Fig. 3.21. For problem 16

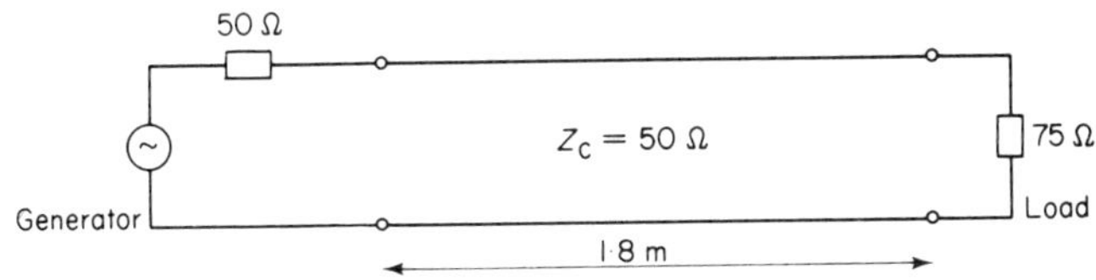

Fig. 3.22. For problem 17

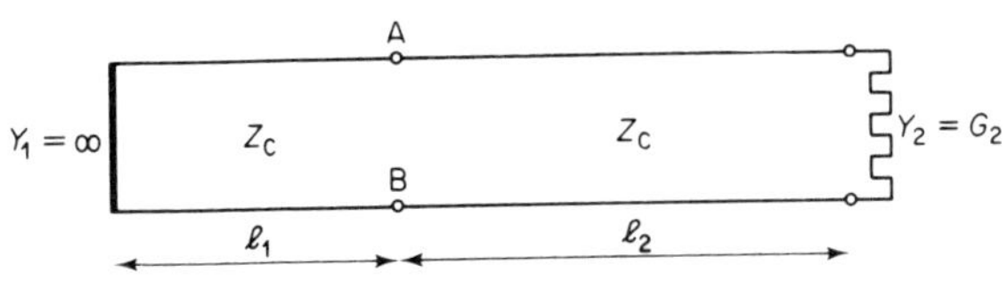

Fig. 3.23. For problem 19

Measurements made at a frequency of 1000 MHz on a 70 Ω loss-less air-spaced, TEM line terminated in an impedance Z_T gave a V.S.W.R. of 3 with a voltage minimum at 66 mm from the termination. Calculate the value of Z_T at this frequency.

19. (R) Calculate the input admittance looking into points AB for the doubly terminated loss-free transmission line shown in fig. 3.23. The line has a characteristic impedance of Z_C ohms and a phase constant of β radians per second.

20. (R) Show that the input admittance of a length l of low-loss transmission line short-circuited at its far end is given approximately by

$$Y_{in} = Y_C \left\{ \frac{\alpha l}{\sin^2 \beta l} - j \cot \beta l \right\} \text{ siemens}$$

assuming that $\sin \beta l \gg \alpha l$, where Y_C is the characteristic admittance of the line and α and β are the attenuation and phase constants.

It is found that, if a capacitor of C farads is connected in parallel with the input terminals of such a line, the combination acts as a pure conductance of G siemens at a radian frequency of ω_0 radians/second. Using these facts deduce expressions for α and the dielectric constant ε_r of the dielectric in the line.

21. (R) The distributed parameters of an air-spaced (low-loss) line are R ohms/metre, L henries/metre, C farads/metre with $G = 0$. Show that the input impedance of short-circuited sections of this line of lengths $\frac{1}{8}\lambda$ and $\frac{3}{8}\lambda$ may be represented, respectively, by R_1 in series with L_1 and R_2 in series with C_2, where

$$R_1 = \tfrac{1}{8}R\lambda,\; L_1 = \frac{\lambda}{2\pi} L \text{ and } R_2 = \tfrac{3}{8}R\lambda,\; C_2 = \frac{\lambda}{2\pi} C$$

22. (R) Find the magnitude of the reflection coefficients and V.S.W.R.'s on the l_1 and l_2 sections of line shown in the circuit of fig. 3.24. The one-way loss of the attenuator is 3 dB and the normalized load impedance is 4.

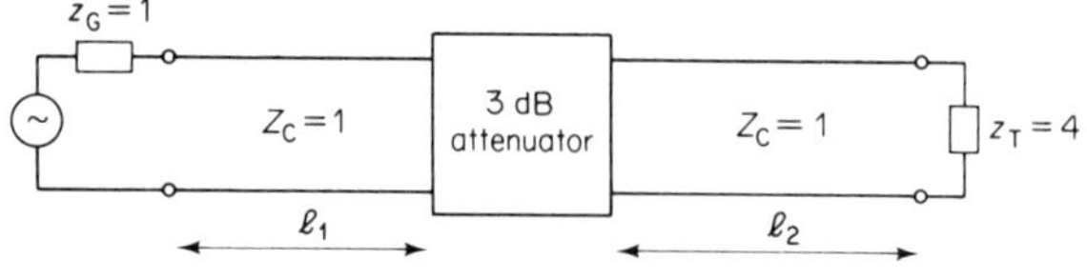

Fig. 3.24. For problem 22

23. (R) In fig. 3.25 a complex load admittance Y_L terminates a loss-less transmission line of characteristic admittance Y_C. The V.S.W.R. on the line is S and the first minimum of voltage occurs at a distance d from the load. Determine, in terms of S and λ, the wavelength on the line, the distance y from the voltage minimum and the value of the shunt susceptance jB that should be placed across the line at this point (AB in fig. 3.25) so that the input admittance looking in at AB equals Y_C.

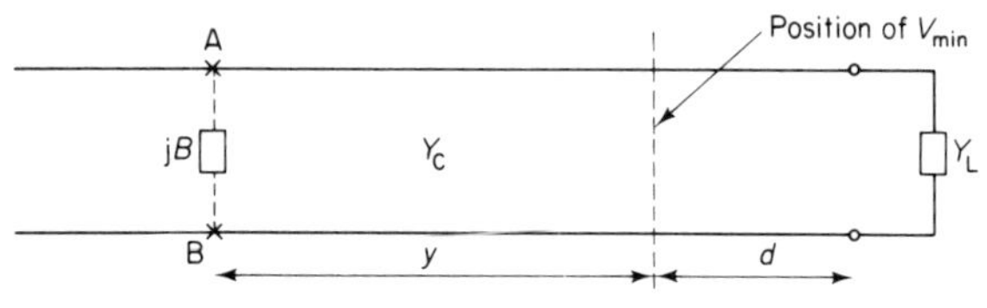

Fig. 3.25. For problem 23

24. (R) In fig. 3.26 the load admittance $Y_L = (Y_C - jB)$, $B > 0$, is supplied by a current generator of r.m.s. current amplitude I_G and internal conductance Y_C, via two sections of loss-less transmission line of characteristic admittance Y_C. Calculate:
   (a) the power delivered to the load,
   (b) the V.S.W.R. on both sections of line,
   (c) the power delivered to the load if the jB susceptance is removed.

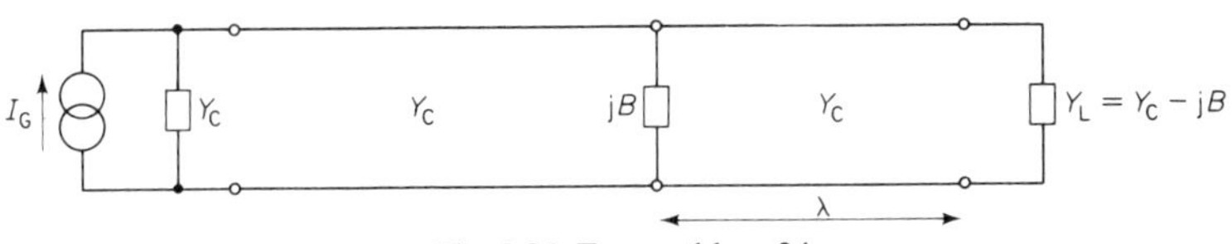

Fig. 3.26. For problem 24

25. (R) Determine the value of the normalized series reactance jx and the shortest line length y $(y>0)$ such that the normalized input impedance looking in at position AB, in the circuit of fig. 3.27, is unity.

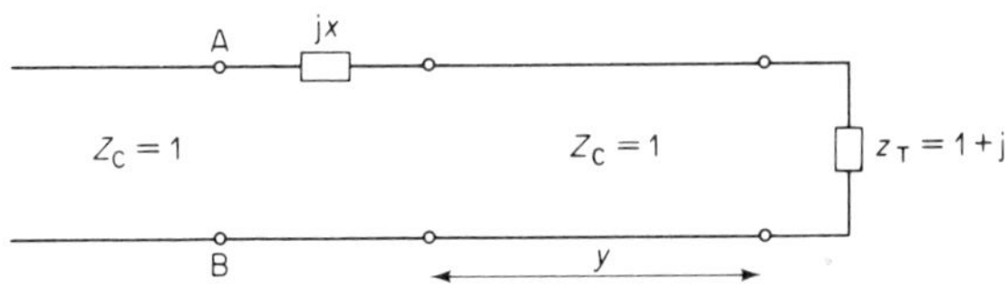

Fig. 3.27. For problem 25

26. (R) A load terminates a long length L metres of low-loss transmission line which has an attenuation constant of α nepers per metre. The V.S.W.R. measured at the line input is S'. Derive an expression for the V.S.W.R. S on the line immediately adjacent to the load in terms of S', α and L.

In an experiment made at 1 GHz to determine the attenuation coefficient of a coaxial line feeding a microwave aerial, the following results were obtained:

Position of	*Distance from load*	*Voltage magnitude*
first minimum	92·1 mm	8·89
first maximum	142·1 mm	20·00
second minimum	192·1 mm	8·89

At a distance of 40 m along the feeder back from the aerial the

V.S.W.R. was found to be 1·25. From these results and assuming TEM transmission, determine:
(a) the dielectric constant of the material between the inner and outer conductors of the coaxial line,
(b) the attenuation coefficient of the line.

27. (M) Determine the value of the admittance Y such that maximum power is dissipated in the load Z_L in the circuit shown in fig. 3.28.

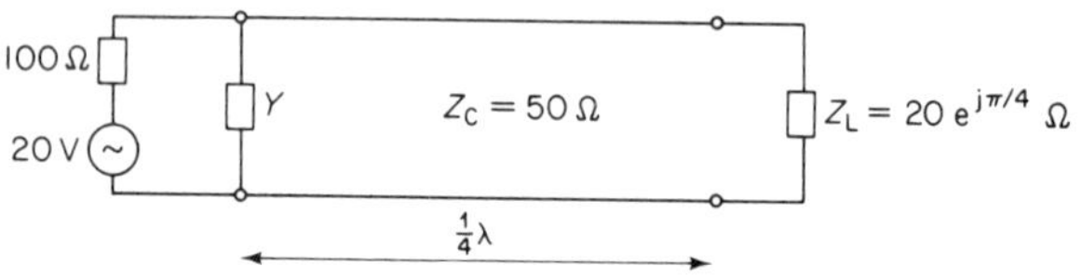

Fig. 3.28. For problem 27

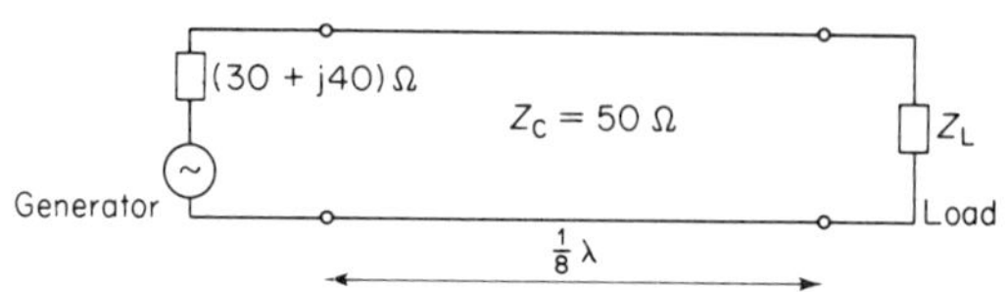

Fig. 3.29. For problem 28

28. (M) Determine the value of the load Z_L in the circuit shown in fig. 3.29 which will cause maximum power to be transferred from the generator to this load.

29. If the V.S.W.R. on a terminated transmission line of characteristic impedance Z_C ohms is S, determine the maximum possible power that may be transmitted on this line and dissipated in the termination, without causing voltage breakdown. The breakdown voltage is V_B volts.

30. (M) Fig. 3.30 shows a diagram of a 50 Ω transmission line supplying power to a 70 Ω load via a matching section consisting of a series

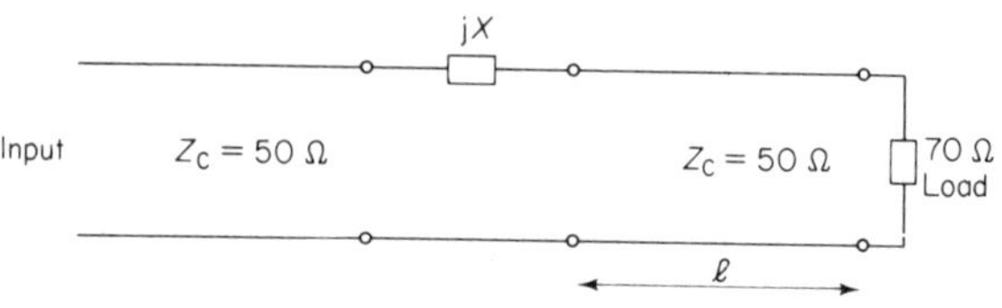

Fig. 3.30. For problem 30

reactance jX and a length l of $50\,\Omega$ line. Determine by analysis the smallest value of l and the corresponding value of X which will effect a match with the input line. Check your result using a Smith chart.

31. (M) Find the input impedance at terminals AB of the multi-quarter wave transmission line circuit of fig. 3.31.

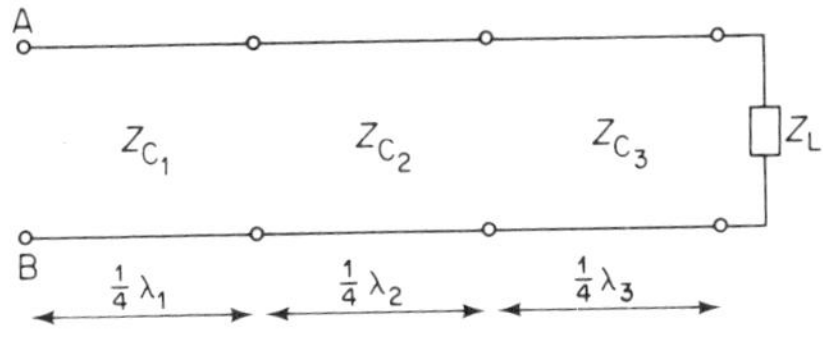

Fig. 3.31. For problem 31

32. (M) A generator of an internal impedance $800\,\Omega$ is to supply power to a load of $50\,\Omega$, situated many wavelengths away. Three types of low-loss line are available, having characteristic impedances of $50\,\Omega$, $100\,\Omega$ and $400\,\Omega$. Describe how these lines may be used to provide a connection between load and generator so that maximum power is delivered to the load.

33. (M) Determine the value of the susceptance jB and the characteristic impedance Z_{C2} of the right-hand quarter wave section in fig. 3.32, such that maximum power is transferred to the $100\,\Omega$ load. Calculate also the V.S.W.R.s on both sections of line.

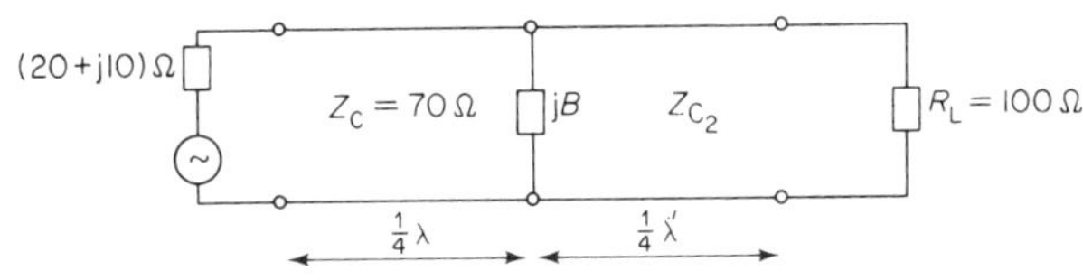

Fig. 3.32. For problem 33

34. (M) The V.S.W.R. on a terminated line is $2\cdot0$ and the position of the first voltage maximum occurs $\frac{3}{8}\lambda$ from the load. Determine the normalized load admittance.

   If a quarter wave section is used to match this load by inserting it in cascade with the line, as shown in fig. 3.33, determine
   (a) the minimum distance d between the load and this section,
   (b) the characteristic impedance of the section.

35. (M) The following measurements were taken on a $50\,\Omega$ transmission line terminated in a load impedance: V.S.W.R. $= 3$ and distance of the first voltage minimum from the load $= 0\cdot11\lambda$. Calculate the

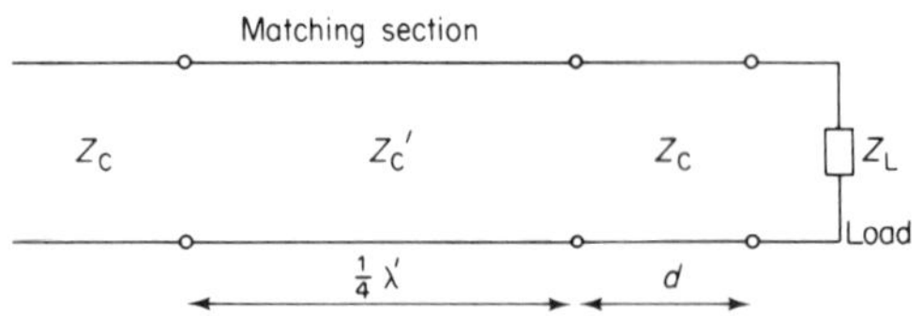

Fig. 3.33. For problem 34

value of the load impedance and determine the position nearest the load and the length of a single short-circuited stub, $Z_C = 50\,\Omega$, which will match the load to the line.

36. (M) A loss-less transmission line of $50\,\Omega$ characteristic impedance is to be matched to a load of normalized admittance $(0{\cdot}5+j0{\cdot}7)$S by means of a short-circuited stub line. Determine the stub position nearest the load and the stub length so that a match is obtained when the characteristic impedance of the stub line is (a) $50\,\Omega$, (b) $70\,\Omega$.

37. (M) Determine the characteristic impedance Z_C and the length l of the line shown in fig. 3.34 so that it effects a match between a generator of $50\,\Omega$ impedance and the load, $Z_L = (70+j30)\,\Omega$, with the provision that $l < \frac{1}{4}\lambda$.

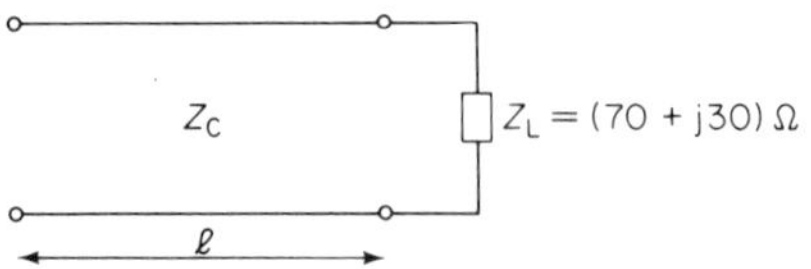

Fig. 3.34. For problem 37

38. (M) Determine, using a Smith chart, the input admittance at AB and the V.S.W.R. on the input line, in fig. 3.35. Calculate also the length and position nearest AB of a $50\,\Omega$ short-circuited, shunt stub that will improve the V.S.W.R. on the input line to unity.

39. (M) The impedance of a UHF antenna was measured at 3 frequencies and the following results were obtained:

f MHz	480	500	520
impedance, Ω	$35-j20$	$50+j0$	$70+j30$

Calculate the characteristic impedance of a $\frac{1}{4}\lambda$ transformer which will effect a match to a $70\,\Omega$ line at 500 MHz. Find also the V.S.W.R. on the feed line, with the $\frac{1}{4}\lambda$ transformer inserted, at 480 MHz and 520 MHz.

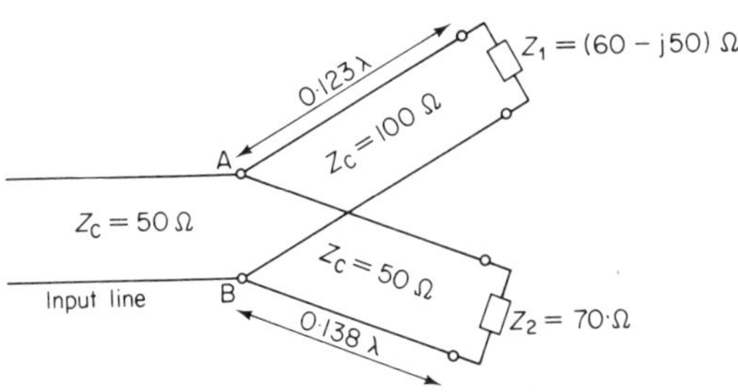

Fig. 3.35. For problem 38

40. The V.S.W.R. measured on a low-loss line at a position around the voltage minimum closest to the load is 2·80. At a position 20λ away, the V.S.W.R. measured is 1·9. Calculate the approximate attenuation constant of the line and, if the position of the closest voltage minimum is $0·08\lambda$ from the load, the location and length of an open-circuited shunt stub required to match the load.

41. (M) A double stub shunt matching unit, shown in fig. 3.36, is used to match a normalized load admittance, $y_L = (0·6 + j0·9)$S. Determine the required stub susceptances and the stub lengths, l_1 and l_2.

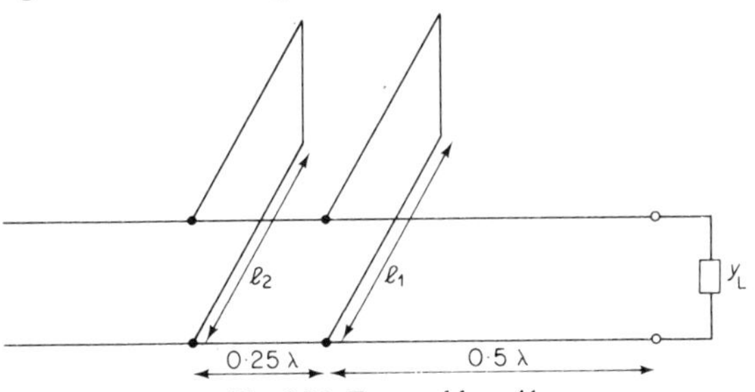

Fig. 3.36. For problem 41

42. (M) A low-loss coaxial feeder feeds an aerial of normalized impedance $(0·9 − j1·0)\,\Omega$. Provision is made for the insertion of a double stub tuner in the line at 2λ from the aerial. The shunt elements of the tuner are short-circuited lengths of line of the same characteristic impedance as the feeder and are separated by $0·375\lambda$. Calculate with the aid of a Smith chart,
(a) the V.S.W.R. on the feeder before insertion of the tuner,
(b) the lengths of the shunt elements of the tuner required to achieve matched conditions.

43. (M) Measurements made on a $50\,\Omega$ terminated line gave a V.S.W.R. equal to 3·0 with a voltage minimum $0·14\lambda$ from the load. Using this data, calculate with the aid of a Smith chart,

(a) the load impedance and admittance,
(b) the stub length and position nearest the load of a single short-circuited stub required to match the line to the load,
(c) whether it is possible to match the load with a double stub tuner, consisting of two short-circuited stubs spaced 0.25λ apart when the position of the first stub is 0, 0.125λ, 0.25λ from the load.

4 2-Port Networks

4.1. Theory Summary

Section 1: General Definitions, Parameters, and Properties

1. 2-PORT NETWORK PARAMETERS

The standard convention adopted in 2-port network analysis for the polarities of the terminal voltages V_1 and V_2 and for the directions of the terminal currents I_1 and I_2 is shown in fig. 4.1.

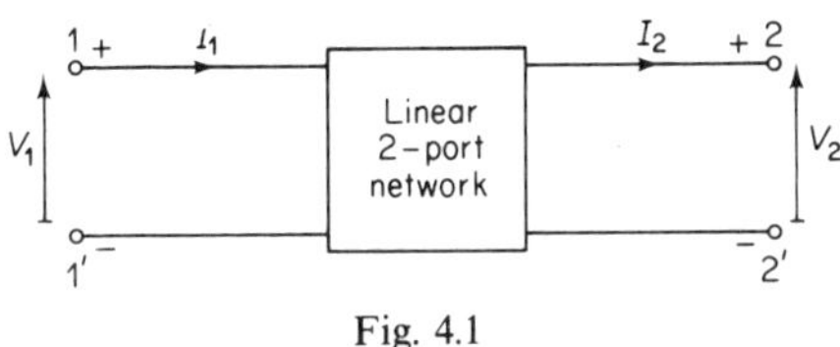

Fig. 4.1

The external behaviour of a linear 2-port network may be defined by relating the terminal currents and voltages by two simultaneous equations. This may be done in any one of six ways leading to the six sets of parameters summarized in the table of fig. 4.2.

The evaluation of the actual parameters themselves may be accomplished by analysis (if the 2-port network components are known) or by measurement. The parameters of some common 2-port passive networks are given in the table of fig. 4.3.

Each set of parameters defines individually the terminal behaviour of a given linear 2-port network. If one set is known then any other may be determined by rearrangement of the given pair of equations and solving them for the required dependent variables of another set. A summary of conversions from one set to another is given in the tables of fig. 4.4.

2. PARAMETERS OF A TERMINATED 2-PORT NETWORK

The six pairs of parameter equations are completely general in that they define the behaviour of a network without any assumption as to the nature of the external connections made at its terminals. Knowledge of the type of terminations supplies two further equations, thus allowing

Choice of independent variables	Choice of dependent variables	Defining equations of network at terminal ports	
V_1, V_2	I_1, I_2	$y_{11} V_1 + y_{12} V_2 = I_1$ $y_{21} V_1 + y_{22} V_2 = I_2$ y parameters (short–circuit driving point and transfer admittance parameters) $\Big\}\longrightarrow$	$y_{11} = \left(\dfrac{I_1}{V_1}\right)_{V_2=0}$ …short–circuit input admittance $y_{12} = \left(\dfrac{I_1}{V_2}\right)_{V_1=0}$ …s–c reverse transfer admittance $y_{21} = \left(\dfrac{I_2}{V_1}\right)_{V_2=0}$ …s–c forward transfer admittance $y_{22} = \left(\dfrac{I_2}{V_2}\right)_{V_1=0}$ …s–c output admittance
I_1, I_2	V_1, V_2	$z_{11} I_1 + z_{12} I_2 = V_1$ $z_{21} I_1 + z_{22} I_2 = V_2$ z parameters (open–circuit driving point and transfer impedance parameters) $\Big\}\longrightarrow$	$z_{11} = \left(\dfrac{V_1}{I_1}\right)_{I_2=0}$ …open–circuit input impedence $z_{12} = \left(\dfrac{V_1}{I_2}\right)_{I_1=0}$ …o–c reverse transfer impedence $z_{21} = \left(\dfrac{V_2}{I_1}\right)_{I_2=0}$ …o–c forward transfer impedence $z_{22} = \left(\dfrac{V_2}{I_2}\right)_{I_1=0}$ …o–c output impedence
I_1, V_2	V_1, I_2	$h_{11} I_1 + h_{12} V_2 = V_1$ $h_{21} I_1 + h_{22} V_2 = I_2$ h or hybrid parameters $\longrightarrow$	$h_{11} = \left(\dfrac{V_1}{I_1}\right)_{V_2=0}$ …s–c input impedence $h_{12} = \left(\dfrac{V_1}{V_2}\right)_{I_1=0}$ …reverse voltage transfer ratio $h_{21} = \left(\dfrac{I_2}{I_1}\right)_{V_2=0}$ …forward current transfer ratio $h_{22} = \left(\dfrac{I_2}{V_2}\right)_{I_1=0}$ …o–c output admittance
V_1, I_2	I_1, V_2	$g_{11} V_1 + g_{12} I_2 = I_1$ $g_{21} V_1 + g_{22} I_2 = V_2$	$[g] = [h]^{-1}$ g or inverse h parameters
V_2, I_2	V_1, I_1	$A V_2 - B I_2 = V_1$ $C V_2 - D I_2 = I_1$	$\begin{bmatrix} A & B \\ C & D \end{bmatrix}$ = voltage – current transfer or transmission matrix
V_1, I_1	V_2, I_2	$E V_1 - F I_1 = V_2$ $G V_1 - H I_1 = I_2$	$\begin{bmatrix} E & F \\ G & H \end{bmatrix} = \begin{bmatrix} A & B \\ C & D \end{bmatrix}^{-1}$

Fig. 4.2. Table of 6 sets of 2-port network parameters

	Transmission matrix	z parameter matrix	y parameter matrix
(2-port network with I_1, I_2, V_1, V_2)	$\begin{bmatrix} V_1 \\ I_1 \end{bmatrix} = \begin{bmatrix} A & B \\ C & D \end{bmatrix} \begin{bmatrix} V_2 \\ -I_2 \end{bmatrix}$	$\begin{bmatrix} V_1 \\ V_2 \end{bmatrix} = \begin{bmatrix} z_{11} & z_{12} \\ z_{21} & z_{22} \end{bmatrix} \begin{bmatrix} I_1 \\ I_2 \end{bmatrix}$	$\begin{bmatrix} I_1 \\ I_2 \end{bmatrix} = \begin{bmatrix} y_{11} & y_{12} \\ y_{21} & y_{22} \end{bmatrix} \begin{bmatrix} V_1 \\ V_2 \end{bmatrix}$
Series Z	$A = 1$ $B = Z$ $C = 0$ $D = 1$	$z_{11} = z_{12} = z_{21} = z_{22} = \infty$	$y_{11} = \dfrac{1}{Z}$ $y_{12} = y_{21} = \dfrac{-1}{Z}$ $y_{22} = \dfrac{1}{Z}$
Shunt Y	$A = 1$ $B = 0$ $C = Y$ $D = 1$	$z_{11} = \dfrac{1}{Y}$ $z_{12} = z_{21} = \dfrac{1}{Y}$ $z_{22} = \dfrac{1}{Y}$	$y_{11} = y_{12} = y_{21} = y_{22} = \infty$
L-section (Z series, Y shunt)	$A = 1 + ZY$ $B = Z$ $C = Y$ $D = 1$	$z_{11} = Z + \dfrac{1}{Y}$ $z_{12} = z_{21} = \dfrac{1}{Y}$ $z_{22} = \dfrac{1}{Y}$	$y_{11} = \dfrac{1}{Z}$ $y_{12} = \dfrac{-1}{Z} = y_{21}$ $y_{22} = \dfrac{1}{Z} + Y$
T-section (Z_1, Z_3 series, Z_2 shunt)	$A = 1 + \dfrac{Z_1}{Z_2}$ $B = Z_1 + Z_3 + \dfrac{Z_1 Z_3}{Z_2}$ $C = \dfrac{1}{Z_2}, \; D = 1 + \dfrac{Z_3}{Z_2}$	$z_{11} = Z_1 + Z_2$ $z_{12} = z_{21} = Z_2$ $z_{22} = Z_2 + Z_3$	$y_{11} = \dfrac{1}{\Delta_z}(Z_2 + Z_3)$ $y_{12} = y_{21} = \dfrac{-Z_2}{\Delta_z}$ $y_{22} = \dfrac{1}{\Delta_z}(Z_1 + Z_2)$ $\Delta_z = Z_1 Z_2 + Z_2 Z_3 + Z_3 Z_1$
π-section (Y_A, Y_B, Y_C)	$A = 1 + \dfrac{Y_C}{Y_B}, \; B = \dfrac{1}{Y_B}$ $C = Y_A + Y_C + \dfrac{Y_A Y_C}{Y_B}$ $D = 1 + \dfrac{Y_A}{Y_B}$	$z_{11} = \dfrac{1}{\Delta_y}(Y_B + Y_C)$ $z_{12} = z_{21} = \dfrac{1}{\Delta_y} Y_B$ $z_{22} = \dfrac{1}{\Delta_y}(Y_A + Y_B)$ $\Delta_y = Y_A Y_B + Y_B Y_C + Y_C Y_A$	$y_{11} = Y_A + Y_B$ $y_{12} = y_{21} = -Y_B$ $y_{22} = Y_B + Y_C$
γ, Z_C Transmission line (γ = propagation constant Z_C = characteristic impedance)	$A = \cosh \gamma \ell$ $B = Z_C \sinh \gamma \ell$ $C = \dfrac{1}{Z_C} \sinh \gamma \ell$ $D = \cosh \gamma \ell$	$z_{11} = Z_C \coth \gamma \ell$ $z_{12} = z_{21} = Z_C / \sinh \gamma \ell$ $z_{22} = Z_C \coth \gamma \ell$	$y_{11} = \dfrac{1}{Z_C} \coth \gamma \ell$ $y_{12} = y_{21} = \dfrac{-1}{Z_C \sinh \gamma \ell}$ $y_{22} = \dfrac{1}{Z_C} \coth \gamma \ell$
$a : 1$ Ideal tranformer	$A = a$ $B = 0$ $C = 0$ $D = \dfrac{1}{a}$	$z_{11} = z_{12} = z_{21} = z_{22} = \infty$	$y_{11} = y_{12} = y_{21} = y_{22} = \infty$
b Ideal gyrator	$A = 0$ $B = b$ $C = \dfrac{1}{b}$ $D = 0$	$z_{11} = 0$ $z_{12} = -b$ $z_{21} = b$ $z_{22} = 0$	$y_{11} = 0$ $y_{12} = +\dfrac{1}{b}$ $y_{21} = -\dfrac{1}{b}$ $y_{22} = 0$

Fig. 4.3. Table of some parameters of some common 2-port networks

	Parameter interrelations in terms of					
	y	z	h	g	$ABCD$	$EFGH$
y_{11}		$\dfrac{z_{22}}{\Delta^z}$	$\dfrac{1}{h_{11}}$	$\dfrac{\Delta^g}{g_{22}}$	$\dfrac{D}{B}$	$\dfrac{E}{F}$
y_{12}		$\dfrac{-z_{12}}{\Delta^z}$	$\dfrac{-h_{12}}{h_{11}}$	$\dfrac{g_{12}}{g_{22}}$	$\dfrac{-\Delta^A}{B}$	$\dfrac{-1}{F}$
y_{21}		$\dfrac{-z_{21}}{\Delta^z}$	$\dfrac{h_{21}}{h_{11}}$	$\dfrac{-g_{21}}{g_{22}}$	$\dfrac{-1}{B}$	$\dfrac{-\Delta^E}{F}$
y_{22}		$\dfrac{z_{11}}{\Delta^z}$	$\dfrac{\Delta^h}{h_{11}}$	$\dfrac{1}{g_{22}}$	$\dfrac{A}{B}$	$\dfrac{H}{F}$
z_{11}	$\dfrac{y_{22}}{\Delta^y}$		$\dfrac{\Delta^h}{h_{22}}$	$\dfrac{1}{g_{11}}$	$\dfrac{A}{C}$	$\dfrac{H}{G}$
z_{12}	$\dfrac{-y_{12}}{\Delta^y}$		$\dfrac{h_{12}}{h_{22}}$	$\dfrac{-g_{12}}{g_{11}}$	$\dfrac{\Delta^A}{C}$	$\dfrac{1}{G}$
z_{21}	$\dfrac{-y_{21}}{\Delta^y}$		$\dfrac{-h_{12}}{h_{22}}$	$\dfrac{g_{21}}{g_{11}}$	$\dfrac{1}{C}$	$\dfrac{\Delta^E}{G}$
z_{22}	$\dfrac{y_{11}}{\Delta^y}$		$\dfrac{1}{h_{22}}$	$\dfrac{\Delta^g}{g_{11}}$	$\dfrac{D}{C}$	$\dfrac{E}{G}$
h_{11}	$\dfrac{1}{y_{11}}$	$\dfrac{\Delta^z}{z_{22}}$		$\dfrac{g_{22}}{\Delta^g}$	$\dfrac{B}{D}$	$\dfrac{F}{E}$
h_{12}	$\dfrac{-y_{12}}{y_{11}}$	$\dfrac{z_{12}}{z_{22}}$		$\dfrac{-g_{12}}{\Delta^g}$	$\dfrac{\Delta^A}{D}$	$\dfrac{1}{E}$
h_{21}	$\dfrac{y_{21}}{y_{11}}$	$\dfrac{-z_{21}}{z_{22}}$		$\dfrac{-g_{21}}{\Delta^g}$	$\dfrac{-1}{D}$	$\dfrac{-\Delta^E}{E}$
h_{22}	$\dfrac{\Delta^y}{y_{11}}$	$\dfrac{1}{z_{22}}$		$\dfrac{g_{11}}{\Delta^g}$	$\dfrac{C}{D}$	$\dfrac{G}{E}$
g_{11}	$\dfrac{\Delta^y}{y_{12}}$	$\dfrac{1}{z_{11}}$	$\dfrac{h_{22}}{\Delta^h}$		$\dfrac{C}{A}$	$\dfrac{G}{H}$
g_{12}	$\dfrac{y_{12}}{y_{22}}$	$\dfrac{-z_{12}}{z_{11}}$	$\dfrac{-h_{12}}{\Delta^h}$		$\dfrac{-\Delta^A}{A}$	$\dfrac{-1}{H}$
g_{21}	$\dfrac{-y_{21}}{y_{22}}$	$\dfrac{z_{21}}{z_{11}}$	$\dfrac{-h_{21}}{\Delta^h}$		$\dfrac{1}{A}$	$\dfrac{\Delta^E}{H}$
g_{22}	$\dfrac{1}{y_{22}}$	$\dfrac{\Delta^z}{z_{11}}$	$\dfrac{h_{11}}{\Delta^h}$		$\dfrac{B}{A}$	$\dfrac{F}{H}$
A	$\dfrac{-y_{22}}{y_{21}}$	$\dfrac{z_{11}}{z_{21}}$	$\dfrac{-\Delta^h}{h_{21}}$	$\dfrac{1}{g_{21}}$		$\dfrac{H}{\Delta^E}$
B	$\dfrac{-1}{y_{21}}$	$\dfrac{\Delta^z}{z_{21}}$	$\dfrac{-h_{11}}{h_{21}}$	$\dfrac{g_{22}}{g_{21}}$		$\dfrac{F}{\Delta^E}$
C	$\dfrac{-\Delta^y}{y_{21}}$	$\dfrac{1}{z_{21}}$	$\dfrac{-h_{22}}{h_{21}}$	$\dfrac{g_{11}}{g_{21}}$		$\dfrac{G}{\Delta^E}$
D	$\dfrac{-y_{11}}{y_{21}}$	$\dfrac{z_{22}}{z_{21}}$	$\dfrac{-1}{h_{21}}$	$\dfrac{\Delta^g}{g_{21}}$		$\dfrac{E}{\Delta^E}$
E	$\dfrac{-y_{11}}{y_{12}}$	$\dfrac{z_{22}}{z_{12}}$	$\dfrac{1}{h_{12}}$	$\dfrac{-\Delta^g}{g_{12}}$	$\dfrac{D}{\Delta^A}$	
F	$\dfrac{-1}{y_{12}}$	$\dfrac{\Delta^z}{z_{12}}$	$\dfrac{h_{11}}{h_{12}}$	$\dfrac{-g_{22}}{g_{12}}$	$\dfrac{B}{\Delta^A}$	
G	$\dfrac{-\Delta^y}{y_{12}}$	$\dfrac{1}{z_{12}}$	$\dfrac{h_{22}}{h_{12}}$	$\dfrac{-g_{11}}{g_{12}}$	$\dfrac{C}{\Delta^A}$	
H	$\dfrac{-y_{22}}{y_{12}}$	$\dfrac{z_{11}}{z_{12}}$	$\dfrac{\Delta^h}{h_{12}}$	$\dfrac{-1}{g_{12}}$	$\dfrac{A}{\Delta^A}$	

Fig. 4.4(a). Parameter conversion table

	Determinant interrelations in terms of					
	y	z	h	g	$ABCD$	$EFGH$
$\Delta^y = \begin{vmatrix} y_{11} & y_{12} \\ y_{21} & y_{22} \end{vmatrix} = y_{11}y_{22} - y_{12}y_{21}$		$\dfrac{1}{\Delta^z}$	$\dfrac{h_{22}}{h_{11}}$	$\dfrac{g_{11}}{g_{22}}$	$\dfrac{C}{B}$	$\dfrac{G}{F}$
$\Delta^z = \begin{vmatrix} z_{11} & z_{12} \\ z_{21} & z_{22} \end{vmatrix} = z_{11}z_{22} - z_{12}z_{21}$	$\dfrac{1}{\Delta^y}$		$\dfrac{h_{11}}{h_{22}}$	$\dfrac{g_{22}}{g_{11}}$	$\dfrac{B}{C}$	$\dfrac{F}{G}$
$\Delta^h = \begin{vmatrix} h_{11} & h_{12} \\ h_{21} & h_{22} \end{vmatrix} = h_{11}h_{22} - h_{12}h_{21}$	$\dfrac{y_{22}}{y_{11}}$	$\dfrac{z_{11}}{z_{22}}$		$\dfrac{1}{\Delta^g}$	$\dfrac{A}{D}$	$\dfrac{H}{E}$
$\Delta^g = \begin{vmatrix} g_{11} & g_{12} \\ g_{21} & g_{22} \end{vmatrix} = g_{11}g_{22} - g_{12}g_{21}$	$\dfrac{y_{11}}{y_{22}}$	$\dfrac{z_{22}}{z_{11}}$	$\dfrac{1}{\Delta^h}$		$\dfrac{D}{A}$	$\dfrac{E}{H}$
$\Delta^A = \begin{vmatrix} A & B \\ C & D \end{vmatrix} = AD - BC$	$\dfrac{y_{12}}{y_{21}}$	$\dfrac{z_{12}}{z_{21}}$	$\dfrac{-h_{12}}{h_{21}}$	$\dfrac{-g_{12}}{g_{21}}$		$\dfrac{1}{\Delta^E}$
$\Delta^E = \begin{vmatrix} E & F \\ G & H \end{vmatrix} = EH - FG$	$\dfrac{y_{21}}{y_{12}}$	$\dfrac{z_{21}}{z_{12}}$	$\dfrac{-h_{21}}{h_{12}}$	$\dfrac{-g_{21}}{g_{12}}$	$\dfrac{1}{\Delta^A}$	

Fig. 4.4(b). Parameter determinant conversion table

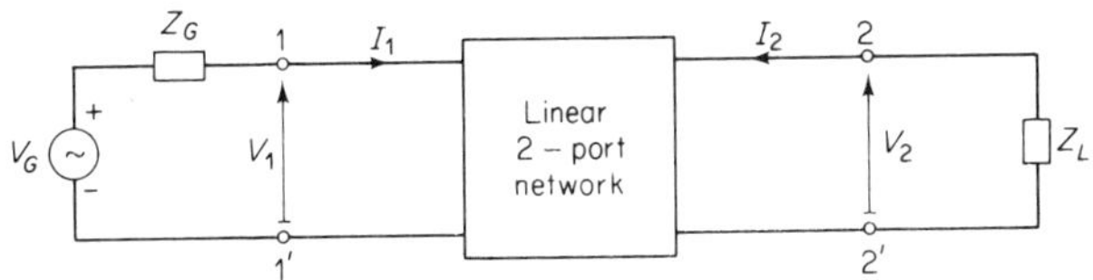

Fig. 4.5

all four terminal quantities to be uniquely determined. Consider fig. 4.5 in which a 2-port network is driven at port 11′ by a generator of e.m.f. V_G volts and impedance Z_G and terminated at port 22′ by a load Z_L. The terminal equations are:

$$V_1 = V_G - Z_G I_1 \quad \text{and} \quad V_2 = -Z_L I_2$$

and these two relations can then be combined with a given pair of parametric equations to determine I_1, I_2, V_1, V_2. Summarized in the table of fig. 4.6 are the important parameters: input and output impedance, voltage and current gain of a 2-port network, in terms of Z_G, Z_L and the six sets of parameters.

107

	Quantities in terms of					
	y parameters	z parameters	h parameters	g parameters	A parameters	E parameters
Input impedance Z_{in} $Z_{\text{in}} = \dfrac{V_1}{I_1}$	$\dfrac{1 + y_{22}\,Z_{\text{L}}}{y_{11} + \Delta^y Z_{\text{L}}}$	$\dfrac{\Delta^z + z_{11}\,Z_{\text{L}}}{z_{22} + Z_{\text{L}}}$	$\dfrac{h_{11} + \Delta^h Z_{\text{L}}}{1 + h_{22}\,Z_{\text{L}}}$	$\dfrac{g_{22} + Z_{\text{L}}}{\Delta^g + g_{11}\,Z_{\text{L}}}$	$\dfrac{B + A Z_{\text{L}}}{D + C Z_{\text{L}}}$	$\dfrac{F + H Z_{\text{L}}}{E + G Z_{\text{L}}}$
Output impedance Z_{out} $Z_{\text{out}} = \left(\dfrac{V_2}{I_2}\right)_{V_{\text{G}}=0}$	$\dfrac{1 + y_{11}\,Z_{\text{G}}}{y_{22} + \Delta^y Z_{\text{G}}}$	$\dfrac{\Delta^z + z_{22}\,Z_{\text{G}}}{z_{11} + Z_{\text{G}}}$	$\dfrac{h_{11} + Z_{\text{G}}}{\Delta^h + h_{22}\,Z_{\text{G}}}$	$\dfrac{g_{22} + \Delta^g Z_{\text{G}}}{1 + g_{11}\,Z_{\text{G}}}$	$\dfrac{B + D Z_{\text{G}}}{A + C Z_{\text{G}}}$	$\dfrac{F + E Z_{\text{G}}}{H + G Z_{\text{G}}}$
Current gain $\dfrac{I_2}{I_1}$	$\dfrac{y_{21}}{y_{11} + \Delta^y Z_{\text{L}}}$	$\dfrac{-z_{21}}{z_{22} + Z_{\text{L}}}$	$\dfrac{h_{21}}{1 + h_{22}\,Z_{\text{L}}}$	$\dfrac{-g_{21}}{\Delta^g + g_{11}\,Z_{\text{L}}}$	$\dfrac{-1}{D + C Z_{\text{L}}}$	$\dfrac{-\Delta^E}{E + F Z_{\text{L}}}$
Voltage gain $\dfrac{V_2}{V_1}$	$\dfrac{-y_{21}\,Z_{\text{L}}}{1 + y_{22}\,Z_{\text{L}}}$	$\dfrac{z_{21}\,Z_{\text{L}}}{\Delta^z + z_{11}\,Z_{\text{L}}}$	$\dfrac{-h_{21}\,Z_{\text{L}}}{h_{11} + \Delta^h Z_{\text{L}}}$	$\dfrac{g_{21}\,Z_{\text{L}}}{g_{22} + Z_{\text{L}}}$	$\dfrac{Z_{\text{L}}}{B + A Z_{\text{L}}}$	$\dfrac{\Delta^E Z_{\text{L}}}{F + H Z_{\text{L}}}$

Fig. 4.6. Properties of a terminated 2-port network

3. INSERTION LOSS OR GAIN OF A 2-PORT NETWORK

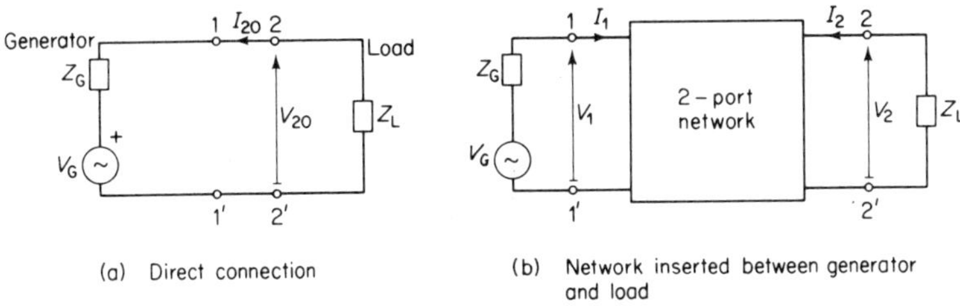

Fig. 4.7

The insertion loss (or gain) of a 2-port network refers to the loss (or gain) resulting from the insertion of the network in a given system. Thus on referring to fig. 4.7, where the generator and load may be regarded as the equivalent circuits representing the respective input and output sections of a given transmission system, we define formally:

The *insertion loss ratio*, $F = \dfrac{V_{20}}{V_2} = \dfrac{I_{20}}{I_2}$

where V_{20} = voltage across load when a direct connection between the generator and load is made

V_2 = voltage across load when the 2-port network is inserted between generator and load.

The *power insertion loss ratio* $= \dfrac{P_{20}}{P_2}$

where P_{20} = power delivered to the load on direct connection of load to generator

P_2 = power delivered to load when 2-port network is inserted between generator and load.

$$\frac{P_{20}}{P_2} = \frac{|I_{20}|^2 \operatorname{Re}(Z_L)}{|I_2|^2 \operatorname{Re}(Z_L)} = \left|\frac{I_{20}}{I_2}\right|^2 = |F|^2$$

When this ratio is expressed in decibels (or nepers) it is known as the

$$\textit{insertion loss} = 10\log_{10}\frac{P_{20}}{P_2} = 20\log_{10} F \text{ dB (or } \log_e F \text{ nepers).}$$

Obviously if the network in question is an amplifier the term insertion gain is used and is then normally defined as $10\log_{10}(P_2/P_{20})$ dB.

The use of the term insertion loss has a disadvantage in that maximum power transfer from a source via a 2-port network to a load may correspond to an arbitrary value of the ratio P_{20}/P_2. For this reason the network power loss ratio defined as

$$\text{network power loss ratio} = \frac{P_{\text{available}}}{P_2}$$

$$\text{where } P_{\text{available}} = \frac{|V_G|^2}{4\,\text{Re}\,(Z_G)}$$

is the maximum power available from the source, (i.e. when the source sees a matched load, $Z_L = $ complex conjugate of Z_G) is used. In this case complete power transfer always corresponds to $P_{\text{available}}/P_2 = 1$ regardless of the relative values of Z_G and Z_L.

4. INTERCONNECTION OF 2-PORT NETWORKS

2-port networks may be interconnected in five basic ways as illustrated in fig. 4.8. The individual matrices of the interconnected networks may be combined to form a single matrix to represent the resultant behaviour of the combination as follows (two 2-port networks being considered):

for a series combination: $[z] = [z' + z'']$

for a parallel combination: $[y] = [y' + y'']$

for a series connection at the input and a parallel connection at the output: $[h] = [h' + h'']$

for a parallel connection at the input and a series connection at the output: $[g] = [g' + g'']$

for a cascade connection: $[A] = [A'][A'']$

provided the connections do not destroy the condition that at any port of an individual network the current entering and leaving this port are equal. This condition, however, is automatically satisfied in a cascade connection.

Section 2: Image Parameters of 2-Port Networks

The image parameters considered below form the basis of the image method of filter and impedance matching network design.

110

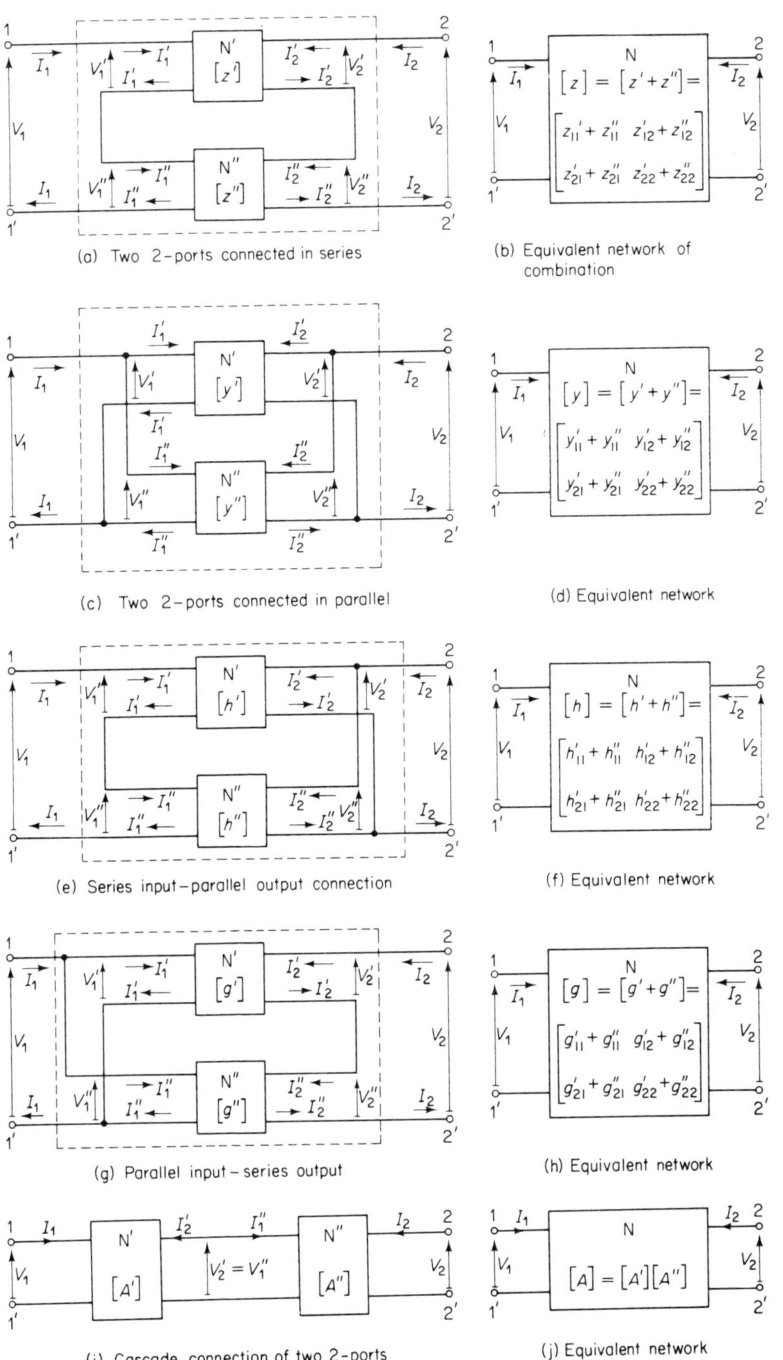

(a) Two 2-ports connected in series

(b) Equivalent network of combination

(c) Two 2-ports connected in parallel

(d) Equivalent network

(e) Series input–parallel output connection

(f) Equivalent network

(g) Parallel input–series output

(h) Equivalent network

(i) Cascade connection of two 2-ports

(j) Equivalent network

Fig. 4.8. Interconnections of 2-port networks

5. THE IMAGE IMPEDANCES AND IMAGE PROPAGATION FUNCTION

The image method for the analysis of 2-port networks is directly similar to the wave viewpoint used in the analysis of uniform transmission lines. In fact, the characteristic impedance of a uniform transmission line is also its image impedance and the propagation constant γ is related to the image propagation function θ by $\theta = \gamma l$, l being the line length.

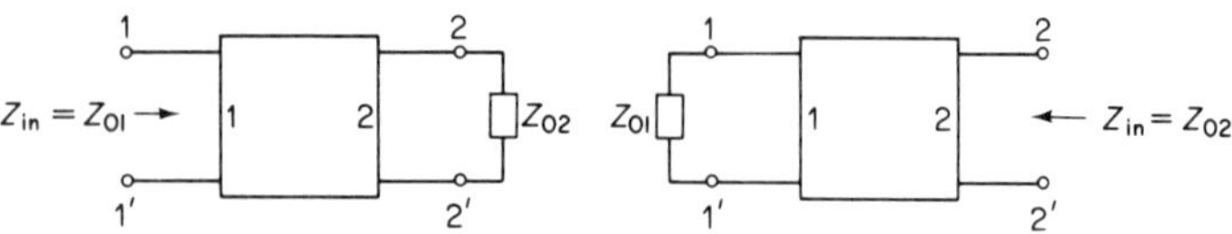

Fig. 4.9. The image impedance of a 2-port network

(a) The *image impedances*, Z_{01} and Z_{02}, of a 2-port network are those impedances such that if one of them, say Z_{02}, is connected across terminals 22', the impedance looking in at terminals 11' is equal to the other Z_{01}. Likewise if Z_{01} terminates 11' the impedance looking in at 22' is Z_{02}. The formal definition is:

> The two terminating impedances which are such that, when they are simultaneously connected to the appropriate terminals of a network, each terminating impedance is equal to the impedance presented to it.

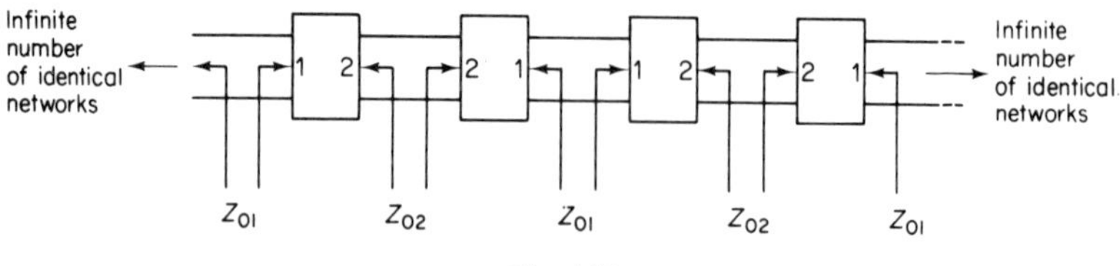

Fig. 4.10

Note also the meaning of Z_{01} and Z_{02} in the infinite chain of symmetrical networks shown in fig. 4.10. The same impedance Z_{01} is seen looking to the left and right of a junction connecting two end 1's, whilst at any junction connecting two end 2's Z_{02} will be seen looking in both left and right directions.

(b) The *image propagation* function θ (also known as the image transfer function) is defined by the following relation:

$$\frac{V_1 I_1}{V_2 I_2'} = e^{2\theta}, \text{ i.e. } \theta = \tfrac{1}{2}\log_e\left(\frac{V_1 I_1}{V_2 I_2'}\right)$$

where $V_1 I_1$ is the product of input voltage and current, $V_2 I_2'$ the product of output voltage and current, the network being terminated in its image impedances as shown in fig. 4.11. Note that as $I_1 = (V_1/Z_{01})$, $I_2' = (V_2/Z_{02})$,

$$\frac{V_1}{V_2}\sqrt{\left(\frac{Z_{02}}{Z_{01}}\right)} = \frac{I_1}{I_2'}\sqrt{\left(\frac{Z_{01}}{Z_{02}}\right)} = e^{\theta}$$

For a reciprocal network the image propagation function is equal for both directions of transmission through the network.

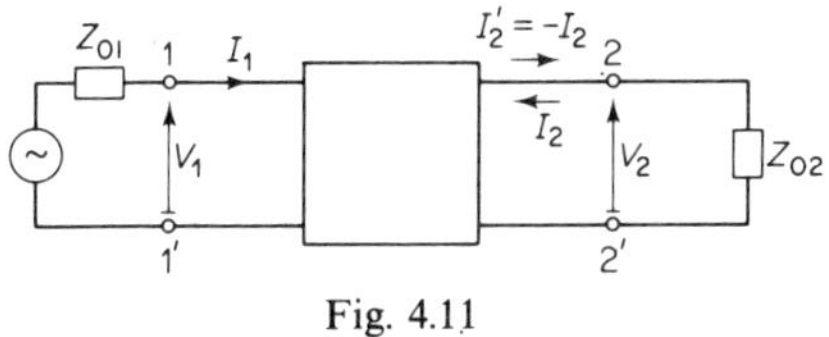

Fig. 4.11

6. THE SYMMETRICAL 2-PORT NETWORK: CHARACTERISTIC IMPEDANCE AND PROPAGATION CONSTANT

If a 2-port network is symmetrical, i.e. there is no change in the electrical behaviour external to the network when an interchange is made in the role of the terminal ports, then the image impedances are equal:

$$Z_{01} = Z_{02} = Z_C$$

where Z_C is known as the characteristic impedance of the network.

The ratio of input to output voltage equals the ratio of input to output current when a symmetrical network is terminated in its characteristic impedance. These ratios define the propagation constant γ of a symmetrical network as:

$$\frac{V_1}{V_2} = \frac{I_1}{I_2} = e^{\gamma}$$

7. THE ITERATIVE IMPEDANCES OF A 2-PORT NETWORK

As well as the image impedances, there are also another pair of impedances—the iterative impedances—which are of importance in 2-port network theory. The *iterative impedance* of a 2-port network is defined as the value of the impedance measured at one pair of terminals when the other pair of terminals is terminated with an impedance of the same value.

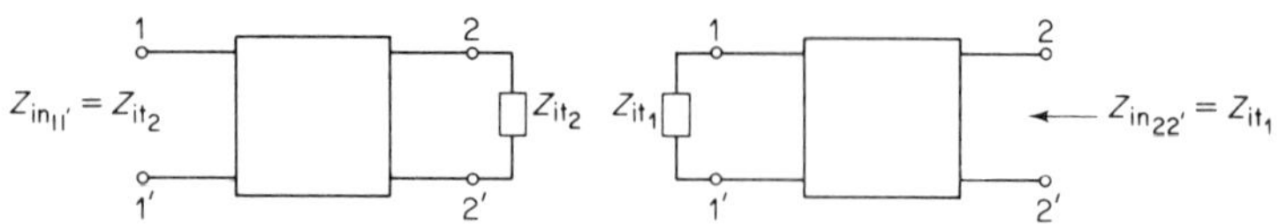

Fig. 4.12. The iterative impedances Z_{it_1} and Z_{it_2} of a 2-port network

8. IMAGE PARAMETERS OF SOME IMPORTANT NETWORKS

	Image impedances	θ or γ
Symmetrical T	$Z_{01} = Z_{02} = Z_C = \sqrt{(\tfrac{1}{4}Z_1^2 + Z_1 Z_2)}$	$\gamma = \tanh^{-1}\left[Z_C/(\tfrac{1}{2}Z_1 + Z_2)\right]^{\frac{1}{2}}$ $= \cosh^{-1}\left[1 + Z_1/(2Z_2)\right]$ $= \ln\left[1 + Z_1/(2Z_2) + Z_C/Z_2\right]$
Symmetrical Π	$Z_{01} = Z_{02} = Z_C = \dfrac{(Z_1 Z_2)}{\sqrt{(\tfrac{1}{4}Z_1^2 + Z_1 Z_2)}}$	$\gamma = \cosh^{-1}\left[1 + Z_1/(2Z_2)\right]$ $= 2\cosh^{-1}\sqrt{\left[1 + Z_1/(4Z_2)\right]}$ $= 2\coth^{-1}\sqrt{(1 + 4Z_2/Z_1)}$ Note γ of T and Π are equal
Symmetrical lattice	$Z_{01} = Z_{02} = Z_C = \sqrt{(Z_A Z_B)}$	$\gamma = 2\tanh^{-1}\sqrt{(Z_A/Z_B)}$ $= \ln\left(\dfrac{\sqrt{Z_A} + \sqrt{Z_B}}{\sqrt{Z_A} - \sqrt{Z_B}}\right)$
Symmetrical bridged T	$Z_{01} = Z_{02} = Z_C = \sqrt{\left[\dfrac{Z_1 Z_3(\tfrac{1}{4}Z_1 + Z_2)}{Z_1 + Z_3}\right]}$	$\gamma = \tanh^{-1}\dfrac{\sqrt{[Z_1 Z_3(\tfrac{1}{4}Z_1 + Z_2)(Z_1 + Z_3)]}}{Z_2(Z_1 + Z_3) + \tfrac{1}{2}Z_1(\tfrac{1}{2}Z_1 + Z_3)}$ $\gamma = \ln\left(1 + \dfrac{Z_3}{Z_C}\right)$ if $\tfrac{1}{2}Z_1 = Z_C$
Half T section	$Z_{01} = \sqrt{(\tfrac{1}{4}Z_1^2 + Z_1 Z_2)}$ $Z_{02} = \dfrac{Z_1 Z_2}{\sqrt{(\tfrac{1}{4}Z_1^2 + Z_1 Z_2)}}$	$\theta = \cosh^{-1}\sqrt{\left(1 + \dfrac{Z_1}{2Z_2}\right)}$ $= \sinh^{-1}\sqrt{\left(\dfrac{Z_1}{2Z_2}\right)}$ $= \coth^{-1}\sqrt{\left(1 + \dfrac{2Z_2}{Z_1}\right)}$

Fig. 4.13

9. EQUIVALENT T AND Π NETWORKS IN TERMS OF THE IMAGE PARAMETERS

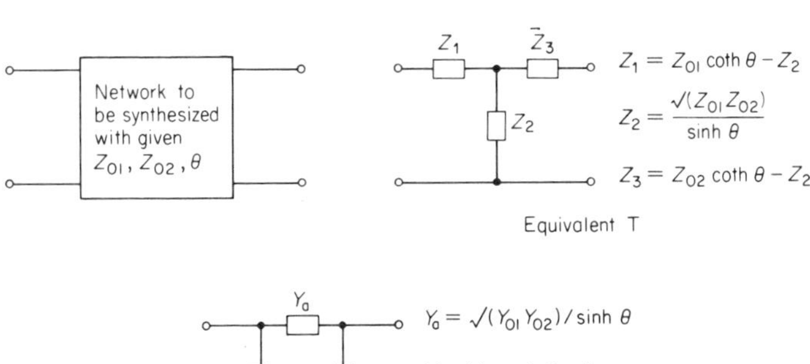

Fig. 4.14

A 2-port network with given values of Z_{01}, Z_{02} and θ may be synthesized in the form of an equivalent T and Π network as shown in fig. 4.14, although the impedance and admittance elements may not necessarily be physically realizable.

In the case of a symmetrical network for which $Z_{01} = Z_{02} = Z_c, \theta = \gamma$ the T elements are:

$$Z_1 = Z_3 = Z_c \tanh \tfrac{1}{2}\gamma, \quad Z_2 = Z_c/\sinh \gamma$$

whilst the elements of the equivalent Π are:

$$Y_b = Y_c = Y_c \tanh \tfrac{1}{2}\gamma, \quad Y_a = Y_c/\sinh \gamma$$

10. INTERRELATIONSHIPS BETWEEN THE IMAGE AND THE GENERAL 2-PORT NETWORK PARAMETERS

A summary of the relationships between the image impedances and propagation function and the *ABCD*, z and y parameters are given below in fig. 4.15.

Image parameters	In terms of			
	$ABCD$ parameters	z open-circuit impedance parameters	y short-circuit admittance parameters	driving point immittances $Z_{oc_1} = z_{11}, Z_{sc_1} = \frac{1}{y_{11}}, Z_{oc_2} = z_{22},$ $Z_{sc\,2} = 1/y_{22}$
Z_{01}	$\sqrt{\left(\dfrac{AB}{CD}\right)}$	$\sqrt{\left(\dfrac{z_{11}\Delta^z}{z_{22}}\right)}$	$\sqrt{\left(\dfrac{y_{22}}{y_{11}\Delta^y}\right)}$	$\sqrt{(Z_{oc_1}Z_{sc_1})} = \sqrt{\left(\dfrac{z_{11}}{y_{11}}\right)}$
Z_{02}	$\sqrt{\left(\dfrac{BD}{AC}\right)}$	$\sqrt{\left(\dfrac{z_{22}\Delta^z}{z_{11}}\right)}$	$\sqrt{\left(\dfrac{y_{11}y_{22}}{\Delta^y}\right)}$	$\sqrt{(Z_{oc_2}Z_{sc_2})} = \sqrt{\left(\dfrac{z_{22}}{y_{22}}\right)}$
$\theta = \theta_r + j\theta_i$	$\cosh^{-1}\sqrt{(AD)}$	$\cosh^{-1}\left(\dfrac{\sqrt{[z_{11}z_{22}]}}{z_{21}}\right)$	$\cosh^{-1}\left(\dfrac{\sqrt{[y_{11}y_{22}]}}{y_{21}}\right)$	$\cosh^{-1}\sqrt{\left(\dfrac{Z_{oc_1}}{Z_{oc_1}-Z_{sc_1}}\right)}$
$\theta = \theta_r + j\theta_i$	$\tanh^{-1}\sqrt{\left(\dfrac{BC}{AD}\right)}$	$\tanh^{-1}\sqrt{\left(\dfrac{\Delta^z}{z_{11}z_{22}}\right)}$	$\tanh^{-1}\sqrt{\left(\dfrac{\Delta^y}{y_{11}y_{22}}\right)}$	$\tanh^{-1}\sqrt{\left(\dfrac{Z_{sc_1}}{Z_{oc_1}}\right)}$
$\theta = \theta_r + j\theta_i$	$\sinh^{-1}\sqrt{(BC)}$	$\sinh^{-1}\left(\dfrac{\sqrt{\Delta^z}}{z_{21}}\right)$	$\sinh^{-1}\left(\dfrac{\sqrt{\Delta^y}}{y_{21}}\right)$	

(a) Image parameters in terms of $ABCD$, z, y and driving point immittances

$$V_1 = AV_2 - BI_2 \qquad A = \sqrt{\left(\frac{Z_{01}}{Z_{02}}\right)}\cosh\theta\ ,\quad B = \sqrt{(Z_{01}Z_{02})}\sinh\theta$$

$$I_1 = CV_2 - DI_2 \qquad C = \frac{1}{\sqrt{Z_{01}Z_{02}}}\sinh\theta\ ,\quad D = \sqrt{\left(\frac{Z_{02}}{Z_{01}}\right)}\cosh\theta$$

$$V_1 = z_{11}I_1 + z_{12}I_2 \qquad z_{11} = Z_{01}\coth\theta\ ,\quad z_{12} = \frac{\sqrt{(Z_{01}Z_{02})}}{\sinh\theta}$$

$$V_2 = z_{21}I_1 + z_{22}I_2 \qquad z_{12} = z_{21}\ ,\quad z_{22} = Z_{02}\coth\theta$$

$$I_1 = y_{11}V_1 + y_{12}V_2 \qquad y_{11} = Y_{01}\coth\theta\ ,\quad y_{12} = -\frac{\sqrt{(Y_{01}Y_{02})}}{\sinh\theta}$$

$$I_2 = y_{21}V_1 + y_{22}V_2 \qquad y_{12} = y_{21}\ ,\quad y_{22} = Y_{02}\coth\theta$$

(b) $ABCD$, z and y parameters in terms of image parameters.
$Y_{01} = (1/Z_{01})$, $Y_{02} = (1/Z_{02})$ and networks are assumed reciprocal

Fig. 4.15

11. LADDER NETWORKS

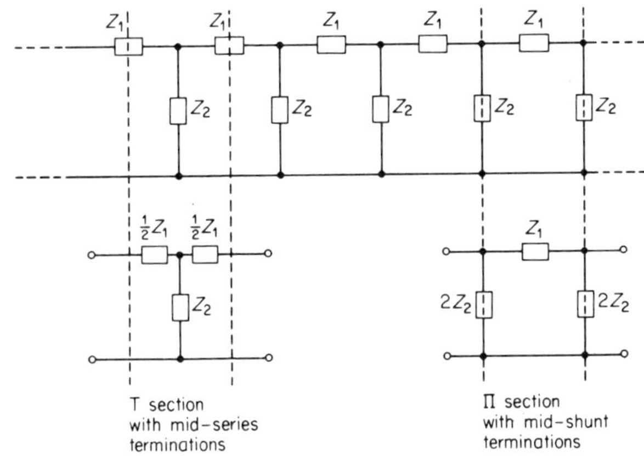

Fig. 4.16

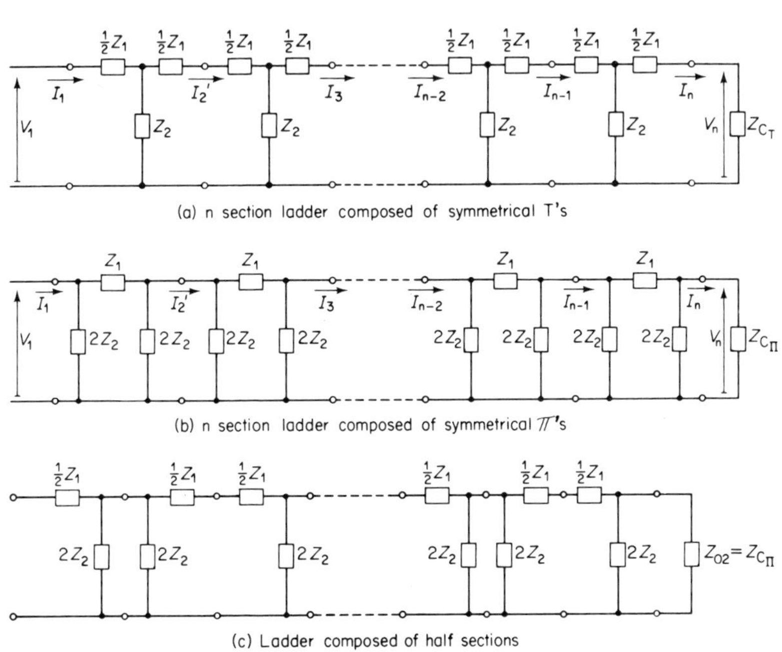

(a) n section ladder composed of symmetrical T's

(b) n section ladder composed of symmetrical π's

(c) Ladder composed of half sections

Fig. 4.17

117

An important type of recurrent network structure is the uniform ladder, shown in fig. 4.16. From the analysis viewpoint it is convenient to divide the ladder into a series of identical symmetrical T or Π network sections, or half T or Π sections as shown in fig. 4.17.

T sections are formed by dividing the ladder at points half way along the Z_1 series impedance elements on either side of the shunt impedances, Z_2. Hence the constituent T sections are said to have mid-series terminations. The Π sections are formed by dividing the shunt impedance Z_2 on either side of Z_1 into two equal shunt impedances of $2Z_2$. The individual Π sections are said to have mid-shunt terminations.

If the 'T' ladder of fig. 4.17(a) is terminated in the image or rather characteristic impedance $Z_{C_T} = \sqrt{(\frac{1}{4}Z_1^2 + Z_1 Z_2)}$ of the individual T section then this impedance is repeated at all points half way along the series branches of the ladder. Thus Z_{C_T} is often referred to as the *mid-series iterative impedance*. Likewise if the 'Π' ladder of fig. 4.17(b) is terminated in the characteristic impedance $Z_{C_\pi} = Z_1 Z_2 / \sqrt{(\frac{1}{4}Z_1^2 + Z_1 Z_2)}$ of the individual Π sections, then this impedance is repeated at all mid-shunt points of Z_2. Thus Z_{C_π} is referred to as the *mid-shunt iterative impedance*.

For the T and Π ladder networks of fig. 4.17 we have:

$$\frac{I_1}{I_2'} = \frac{I_2'}{I_3} = \cdots = \frac{I_{n-2}}{I_{n-1}} = \frac{I_{n-1}}{I_n} = e^\gamma,$$

so $$\frac{I_1}{I_n} = \frac{I_1}{I_2'} \times \frac{I_2'}{I_3} \cdots \times \frac{I_{n-2}}{I_{n-1}} \times \frac{I_{n-1}}{I_n} = e^{n\gamma},$$

hence $I_n = I_1 e^{-n\gamma}$ and also $V_n = V_1 e^{-n\gamma}$ where $\gamma = \cosh^{-1}(1 + Z_1/2Z_2)$.

4.2. Worked Problems

1. Write down the mesh equations for the network shown in fig. 4.18 and hence determine its z and y parameters, referred to ports 11' and 22'.

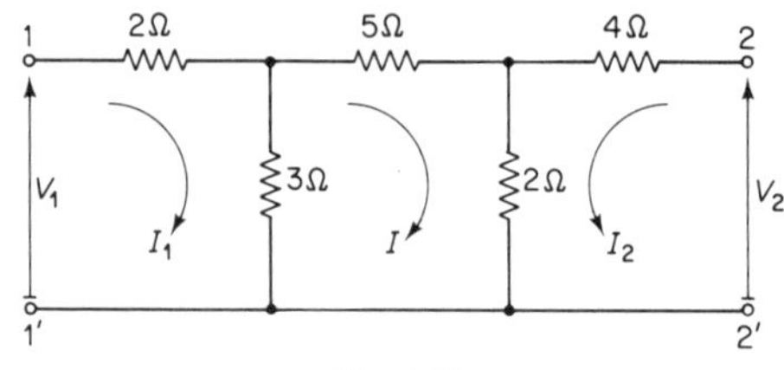

Fig. 4.18

118

Solution

The mesh equations for the network of fig. 4.18 are

$$5I_1 - 3I \qquad = V_1$$

$$-3I_1 + 10I + 2I_2 = 0$$

$$2I + 6I_2 = V_2$$

and on eliminating I from the set we obtain

$$4{\cdot}1I_1 + 0{\cdot}6I_2 = V_1$$

$$0{\cdot}6I_1 + 5{\cdot}6I_2 = V_2 \qquad \qquad \ldots (1)$$

Hence $z_{11} = \left(\dfrac{V_1}{I_1}\right)_{I_2=0} = 4{\cdot}1\,\Omega$, $z_{12} = \left(\dfrac{V_1}{I_2}\right)_{I_1=0} = 0{\cdot}6\,\Omega$

$z_{21} = \left(\dfrac{V_2}{I_1}\right)_{I_2=0} = 0{\cdot}6\,\Omega$, $z_{22} = \left(\dfrac{V_2}{I_2}\right)_{I_1=0} = 5{\cdot}6\,\Omega$

Equations (1) may be written in matrix form as

$$\begin{bmatrix} z_{11} & z_{12} \\ z_{21} & z_{21} \end{bmatrix} \begin{bmatrix} I_1 \\ I_2 \end{bmatrix} = \begin{bmatrix} V_1 \\ V_2 \end{bmatrix}$$

and on inverting the z-parameter matrix, i.e. solving for I_1 and I_2:

$$\begin{bmatrix} I_1 \\ I_2 \end{bmatrix} = \begin{bmatrix} z_{11} & z_{12} \\ z_{21} & z_{22} \end{bmatrix}^{-1} \begin{bmatrix} V_1 \\ V_2 \end{bmatrix} = \frac{1}{\Delta^z} \begin{bmatrix} z_{22} & -z_{12} \\ -z_{21} & z_{11} \end{bmatrix} \begin{bmatrix} V_1 \\ V_2 \end{bmatrix}$$

$$\equiv \begin{bmatrix} y_{11} & y_{12} \\ y_{21} & y_{22} \end{bmatrix} \begin{bmatrix} V_1 \\ V_2 \end{bmatrix}$$

where $\Delta^z = z_{11} z_{22} - z_{12} z_{21} = 4{\cdot}1 \times 5{\cdot}6 - 0{\cdot}6 \times 0{\cdot}6 = 22{\cdot}6$
and the y-parameters

$$y_{11} = \left(\frac{I_1}{V_1}\right)_{V_2=0} = \frac{z_{22}}{\Delta^z} = \frac{5{\cdot}6}{22{\cdot}6} = 0{\cdot}248\,\text{S}$$

$$y_{12} = \left(\frac{I_1}{V_2}\right)_{V_1=0} = \frac{-z_{12}}{\Delta^z} = \frac{-0{\cdot}6}{22{\cdot}6} = -0{\cdot}0265\,\text{S}$$

$$y_{21} = \left(\frac{I_2}{V_1}\right)_{V_2=0} = \frac{-z_{21}}{\Delta^z} = y_{12} = -0{\cdot}0265\,\text{S}$$

$$y_{22} = \left(\frac{I_2}{V_2}\right)_{V_1=0} = \frac{z_{11}}{\Delta^z} = \frac{4{\cdot}1}{22{\cdot}6} = 0{\cdot}181\,\text{S}$$

2. Find the *ABCD* transmission parameters of a 2-port network in terms of its z parameters and hence show that (a) if the network is reciprocal $AD - BC = 1$, (b) if the network is loss—less (i.e. consists of pure reactances), A and D are purely real, whilst B and C are purely imaginary.

Solution

From the z-parameter equations:

$$V_1 = z_{11}I_1 + z_{12}I_2 \qquad \ldots (1)$$

$$V_2 = z_{21}I_1 + z_{22}I_2 \qquad \ldots (2)$$

we obtain directly from (2):

$$I_1 = \frac{1}{z_{21}} V_2 - \frac{z_{22}}{z_{21}} I_2 \qquad \ldots (3)$$

and indirectly by substituting for I_1 in (1) using (3):

$$V_1 = \frac{z_{11}}{z_{21}} V_2 - \left(\frac{z_{11} z_{22}}{z_{21}} - z_{12} \right) I_2 \qquad \ldots (4)$$

Thus (4) and (3) are in the identical form to the transmission parameter equations,

$$V_1 = AV_2 - BI_2$$

$$I_1 = CV_2 - DI_2$$

and by direct comparison,

$$A = \frac{z_{11}}{z_{21}}, \; B = \frac{z_{11} z_{22}}{z_{21}} - z_{12}, \; C = \frac{1}{z_{21}}, \; D = \frac{z_{22}}{z_{21}}$$

(a) If the network is reciprocal $z_{12} = z_{21}$ so

$$AD - BC = \frac{z_{11} z_{22}}{z_{21}^2} - \left(\frac{z_{11} z_{22}}{z_{21}^2} - \frac{z_{12}}{z_{21}} \right) = \frac{z_{12}}{z_{21}} = 1$$

If the network is loss-less its z-parameters must be purely imaginary. For example a loss-less 2-port network may be represented by an equivalent T consisting of pure reactance such as that shown in fig. 4.19. For this network

$$z_{11} = j(X_1 + X_2), \; z_{12} = z_{21} = jX_2, \; z_{22} = j(X_3 + X_2)$$

120

(b)

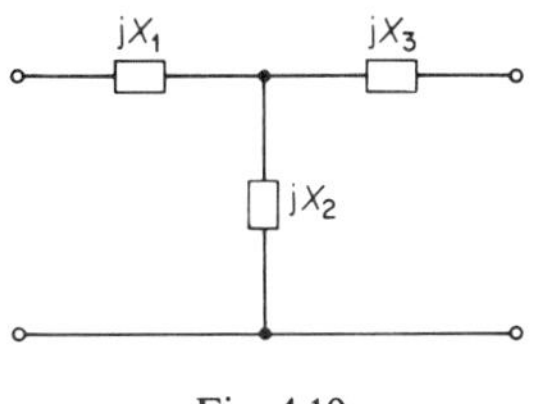

Fig. 4.19

Thus since A and D are the ratios of two imaginary quantities, they are real. Whilst,

$$B = \frac{z_{11}z_{22} - z_{12}z_{21}}{z_{21}} = \frac{-(X_1 + X_2)(X_3 + X_2) + X_2^2}{jX_2} \quad \text{and} \quad C = \frac{1}{z_{21}} = \frac{1}{jX_2}$$

are purely imaginary.

3. Fig. 4.20 shows the small signal equivalent circuit of a common-emitter transistor amplifier. The h-parameters of the transistor are:

$$h_{ie} = 1 \cdot 6 \, k\Omega, \ h_{fe} = 22, \ h_{re} = 6 \cdot 0 \times 10^{-4}, \ h_{oe} = 80 \, \mu S$$

If the amplifier input is taken from a sinusoidal signal generator which has an output resistance of 75 Ω and the amplifier output load is 20 kΩ, find the r.m.s. value of e.m.f. required from the generator if a power of 5 mW is to be dissipated in this load. Evaluate also the output resistance of the amplifier under the above conditions.

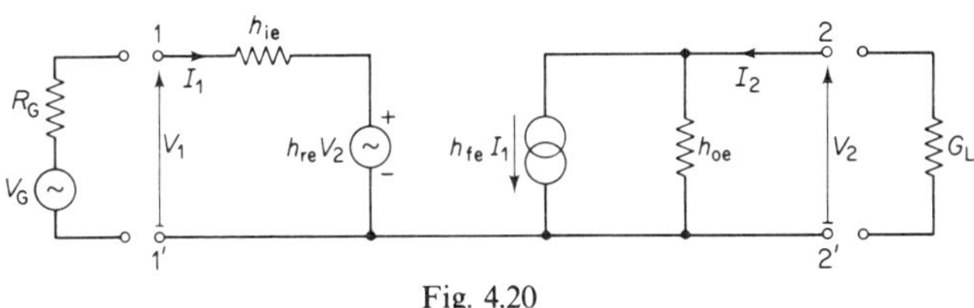

Fig. 4.20

Solution
Let the required r.m.s. e.m.f. of the generator be V_G volts, then on applying Kirchhoff's voltage law to the input mesh and the current law at the output (node 2), we obtain:

$$V_G = (R_G + h_{ie})I_1 + h_{re}V_2 \qquad \ldots (1)$$

$$h_{fe}I_1 + (h_{oe} + G_L)V_2 = 0 \qquad \ldots (2)$$

121

On using (2) to eliminate I_1 from (1) we have

$$V_G = \frac{-(R_G + h_{ie})(h_{oe} + G_L)}{h_{fe}} V_2 + h_{re} V_2$$

$$= \left[\frac{h_{re} h_{fe} - (R_G + h_{ie})(h_{oe} + G_L)}{h_{fe}} \right] V_2$$

and if a power P is to be dissipated in the load conductance G_L,

$$P = (G_L V_2)^2 \times \frac{1}{G_L}, \text{ i.e. } V_2 = \left(\frac{P}{G_L} \right)^{1/2}$$

Hence

$$V_G = \left[\frac{h_{re} h_{fe} - (R_G + h_{ie})(h_{oe} + G_L)}{h_{fe}} \right] \left(\frac{P}{G_L} \right)^{1/2}$$

$$= \left[\frac{6 \cdot 0 \times 10^{-4} \times 22 - (1675)(80 + 50)10^{-6}}{22} \right] \left(\frac{5 \times 10^{-3}}{50 \times 10^{-6}} \right)^{1/2} = 93 \, \text{mV}$$

On substituting for the h-parameters and $P = 5 \, \text{mW}$, $G_L = (20 \times 10^3)^{-1} = 50 \, \mu\text{S}$, $R_G = 75 \, \Omega$.

To find the output resistance we remove G_L, apply a generator V_2 at 22' and set $V_G = 0$, keeping R_G in the circuit. Then at node 2,

$$I_2 = h_{oe} V_2 + h_{fe} I_1 \qquad \qquad \ldots (3)$$

whilst from equation (1) with $V_G = 0$, we obtain

$$I_1 = \frac{-h_{re}}{R_G + h_{ie}} V_2$$

and on substituting for I_1 in (3):

$$I_2 = \left(h_{oe} - \frac{h_{re} h_{fe}}{R_G + h_{ie}} \right) V_2 = \left[\frac{R_G h_{oe} + h_{ie} h_{oe} - h_{re} h_{fe}}{(R_G + h_{ie})} \right] V_2$$

Hence the output resistance,

$$\frac{V_2}{I_2} = \frac{R_G + h_{ie}}{R_G h_{oe} + h_{ie} h_{oe} - h_{re} h_{fe}} = 13 \cdot 87 \, \text{k}\Omega$$

4. Find the transmission matrix of the network of fig. 4.21 and use the result to determine the $ABCD$ parameters of the cascaded $C-R$ network shown in fig. 4.22. If the output terminals 22' of this network are open circuited determine the frequency at which V_2 is in antiphase with V_1. Calculate also the ratio V_1/V_2 at this frequency.

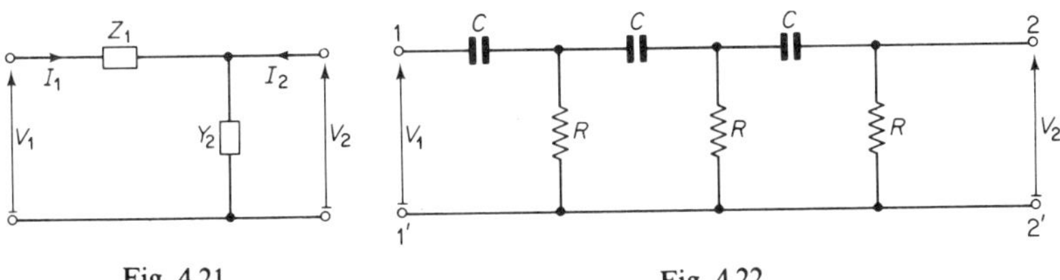

Solution
On applying the voltage law to the input mesh and the current law at the output node in the network of fig. 4.21 we obtain:

$$V_1 = Z_1 I_1 + \frac{1}{Y_2}(I_1 + I_2) \qquad \ldots (1)$$

$$I_1 = Y_2 V_2 - I_2 \qquad \ldots (2)$$

and substituting for I_1 in (1) using (2) we have

$$V_1 = Z_1(Y_2 V_2 - I_2) + \frac{1}{Y_2}(Y_2 V_2 - I_2 + I_2) = (Z_1 Y_2 + 1)V_2 - Z_1 I_2$$

$$\ldots (3)$$

Equations (3) and (2) are in the form of the *ABCD* equations, i.e.

$$\begin{bmatrix} V_1 \\ I_1 \end{bmatrix} = \begin{bmatrix} Z_1 Y_2 + 1 & Z_1 \\ Y_2 & 1 \end{bmatrix} \begin{bmatrix} V_2 \\ -I_2 \end{bmatrix} \equiv \begin{bmatrix} A & B \\ C & D \end{bmatrix} \begin{bmatrix} V_2 \\ -I_2 \end{bmatrix}$$

Hence $A = Z_1 Y_2 + 1$, $B = Z_1$, $C = Y_2$, $D = 1$.

The network of fig. 4.22 can be considered as composed of three C–R networks of the form of fig. 4.21 connected in cascade. Therefore substituting $Z_1 = 1/j\omega C$, $Y_2 = 1/R$, the overall transmission equation for the C–R ladder is:

$$\begin{bmatrix} V_1 \\ I_1 \end{bmatrix} = \begin{bmatrix} Z_1 Y_2 + 1 & Z_1 \\ Y_2 & 1 \end{bmatrix} \begin{bmatrix} Z_1 Y_2 + 1 & Z_1 \\ Y_2 & 1 \end{bmatrix} \begin{bmatrix} Z_1 Y_2 + 1 & Z_1 \\ Y_2 & 1 \end{bmatrix} \begin{bmatrix} V_2 \\ -I_2 \end{bmatrix}$$

$$= \begin{bmatrix} (Z_1 Y_2 + 1)^2 + Z_1 Y_2 & Z_1(Z_1 Y_2 + 2) \\ Y_2(Z_1 Y_2 + 2) & Z_1 Y_2 + 1 \end{bmatrix} \begin{bmatrix} Z_1 Y_2 + 1 & Z_1 \\ Y_2 & 1 \end{bmatrix} \begin{bmatrix} V_2 \\ -I_2 \end{bmatrix}$$

$$= \begin{bmatrix} 1 + 6Z_1 Y_2 + 5Z_1^2 Y_2^2 + Z_1^3 Y_2^3 & 3Z_1 + 4Z_1^2 Y_2 + Z_1^3 Y_2^2 \\ 3Y_2 + 4Z_1 Y_2^2 + Z_1^2 Y_2^3 & 1 + 3Z_1 Y_2 + Z_1^2 Y_2^2 \end{bmatrix} \begin{bmatrix} V_2 \\ -I_2 \end{bmatrix}$$

$$\equiv \begin{bmatrix} A & B \\ C & D \end{bmatrix} \begin{bmatrix} V_2 \\ -I_2 \end{bmatrix} \qquad \ldots (4)$$

Now $V_1 = AV_2 - BI_2 = AV_2$ when $I_2 = 0$ (i.e. 22' open-circuited).

123

Therefore if V_2 is to be in antiphase with V_1, A must be real and negative.

i.e. $\quad A = 1 + 6Z_1 Y_2 + 5Z_1^2 Y_2^2 + Z_1^3 Y_2^3$

$$= 1 - \frac{5}{(\omega CR)^2} + j\left(\frac{1}{(\omega CR)^3} - \frac{6}{\omega CR}\right)$$

so A is real when

$$\text{Im}(A) = \left(\frac{1}{(\omega CR)^3} - \frac{6}{\omega CR}\right) = 0$$

$$\omega = \frac{1}{\sqrt{6}CR}, f = \frac{\omega}{2\pi} = \frac{1}{2\pi\sqrt{6}CR} \text{ hertz}$$

and at this frequency,

$$\left(\frac{V_1}{V_2}\right)_{I_2=0} = A = 1 - \frac{5}{(1/\sqrt{6})^2} = -29$$

5. Show that the image impedances, Z_{01} and Z_{02}, and the image transfer function θ of a 2-port network are given by:

$$Z_{01} = \sqrt{\left(\frac{AB}{CD}\right)} = \sqrt{(Z_{SC_1} Z_{OC_1})}$$

$$Z_{02} = \sqrt{\left(\frac{BD}{AC}\right)}$$

$$\theta = \ln\left[\sqrt{(AD)} + \sqrt{(BC)}\right]$$

where A, B, C, D are the transfer parameters of the 2-port, and Z_{SC_1}, Z_{OC_1} are the input impedances at port $11'$ when port $22'$ is respectively short and open-circuited.

Find also expressions for the A, B, C, D parameters in terms of the image parameters and show that the input impedance at the port $11'$ when port $22'$ is terminated in Z_L is given by

$$Z_{in} = Z_{01} \frac{Z_L + Z_{02} \tanh \theta}{Z_{02} + Z_L \tanh \theta}$$

and that $\qquad \theta = \tanh^{-1} \sqrt{\left(\frac{Z_{SC_1}}{Z_{OC_1}}\right)}.$

The network is taken as reciprocal so that the relation $AD - BC = 1$ holds.

Solution

From the *ABCD* parameter equations we have:

$$\frac{V_1}{I_1} = \frac{AV_2 - BI_2}{CV_2 - DI_2} \qquad \ldots (1)$$

Thus if port 22' is terminated in Z_{02}, we have from (1)

$$\frac{V_1}{I_1} = Z_{01} = \frac{A(V_2/-I_2) + B}{C(V_2/-I_2) + D}$$

$$= \frac{AZ_{02} + B}{CZ_{02} + D} \text{ since } Z_{02} = V_2/-I_2,$$

from which

$$CZ_{01}Z_{02} + DZ_{01} - AZ_{02} - B = 0 \qquad \ldots (2)$$

Likewise if port 11' is terminated in Z_{01} (i.e. $V_1/-I_1 = Z_{01}$), then (1) gives

$$\frac{V_1}{I_1} = -Z_{01} = \frac{A(V_2/I_2) - B}{C(V_2/I_2) - D}$$

$$= \frac{AZ_{02} - B}{CZ_{02} - D} \text{ since } Z_{02} = V_2/I_2,$$

from which

$$-CZ_{01}Z_{02} + DZ_{01} - AZ_{02} + B = 0 \qquad \ldots (3)$$

On addition and subtraction of equations we obtain, respectively,

$$Z_{01}/Z_{02} = A/D, \quad Z_{01}Z_{02} = B/C$$

and hence
$$Z_{01} = \sqrt{\left(\frac{AB}{CD}\right)} \text{ and } Z_{02} = \sqrt{\left(\frac{BD}{AC}\right)}$$

Also from (1) we may find the short-circuit and open-circuit input impedance at port 11' by substituting respectively $V_2 = 0$ and $I_2 = 0$,

i.e. $\qquad Z_{SC_1} = B/D, \quad Z_{OC_1} = A/C$

so $\quad Z_{SC_1}Z_{OC_1} = (AB)/(CD) = Z_{01}^2$, hence $Z_{01} = \sqrt{(Z_{SC_1}Z_{OC_1})}$

When the 2-port is terminated in its image impedances the *ABCD* equations reduce to

$$\frac{V_1}{V_2} = A + B\left(\frac{-I_2}{V_2}\right) = A + \frac{B}{Z_{02}} = A + \sqrt{\left(\frac{ACB}{D}\right)}$$

$$= \sqrt{\left(\frac{A}{D}\right)}\{\sqrt{(AD)}+\sqrt{(BC)}\}$$

$$\frac{I_1}{-I_2}=C\left(\frac{V_2}{-I_2}\right)+D=CZ_{02}+D=\sqrt{\left(\frac{BDC}{A}\right)}+D$$

$$= \sqrt{\left(\frac{D}{A}\right)}\{\sqrt{(AD)}+\sqrt{(BC)}\}$$

and hence

$$\frac{V_1 I_1}{V_2(-I_2)}=\{\sqrt{(AD)}+\sqrt{(BC)}\}^2=e^{2\theta}\,(\text{by definition})$$

so $\quad e^{\theta}=\{\sqrt{(AD)}+\sqrt{(BC)}\}$ and $\theta=ln\{\sqrt{(AD)}+\sqrt{(BC)}\}$

To find the A, B, C, D parameters in terms of the image parameters we may proceed as follows:

$$e^{\theta}=\cosh\theta+\sinh\theta$$
$$= \sqrt{(AD)}+\sqrt{(BC)}=\sqrt{(AD)}+\sqrt{\{AD-1\}} \qquad \ldots (4)$$

since the network is reciprocal and therefore $AD-BC=1$
Equation (4) suggests we may identify

$$\sqrt{(AD)}=\cosh\theta \qquad\qquad\qquad \ldots (5)$$

$$\sqrt{\{AD-1\}}=\sqrt{\{\cosh^2\theta-1\}}=\sinh\theta=\sqrt{(BC)} \qquad \ldots (6)$$

Also as $Z_{01}/Z_{02}=A/D \ldots (7)$ $\qquad\qquad Z_{01}Z_{02}=B/C \ldots (8)$

$(5)\times\sqrt{(7)}$ gives $A=\sqrt{\left(\dfrac{Z_{01}}{Z_{02}}\right)}\cosh\theta$

$(5)\div\sqrt{(7)}$ gives $D=\sqrt{\left(\dfrac{Z_{02}}{Z_{01}}\right)}\cosh\theta$

$(6)\times\sqrt{(8)}$ gives $B=\sqrt{(Z_{01}Z_{02})}\sinh\theta$

$(6)\div\sqrt{(8)}$ gives $C=\dfrac{1}{\sqrt{(Z_{01}Z_{02})}}\sinh\theta$

and on substituting these values in (1) when 22' is terminated in Z_{L}, we have

$$Z_{\text{in}}=\frac{V_1}{I_1}=\frac{AZ_{\mathrm{L}}+B}{CZ_{\mathrm{L}}+D}\;(\text{since } Z_{\mathrm{L}}=V_2/-I_2)$$

126

$$= \frac{Z_L\sqrt{(Z_{01}/Z_{02})}\cosh\theta + \sqrt{(Z_{01}Z_{02})}\sinh\theta}{Z_L\{1/\sqrt{(Z_{01}Z_{02})}\}\sinh\theta + \sqrt{(Z_{02}/Z_{01})}\cosh\theta}$$

$$= Z_{01}\frac{Z_L\cosh\theta + Z_{02}\sinh\theta}{Z_{02}\cosh\theta + Z_L\sinh\theta} = Z_{01}\frac{Z_L + Z_{02}\tanh\theta}{Z_{02} + Z_L\tanh\theta} \qquad \cdots (9)$$

Further on substituting $Z_L = 0$ and $Z_L \to \infty$ in (9) we obtain the short and open-circuit input impedances at $11'$:

$$Z_{sc_1} = Z_{01}\tanh\theta, \quad Z_{OC_1} = Z_{01}/\tanh\theta$$

and hence $\tanh^2\theta = Z_{sc_1}/Z_{OC_1}$, $\theta = \tanh^{-1}\sqrt{(Z_{SC_1}/Z_{OC_1})}$.

6. Determine the characteristic impedance of the symmetrical lattice network of fig. 4.23(a) where $Z_A Z_B = R_0^2$ and show that the attenuation constant of the network is given by

$$\alpha = \ln\left|\frac{Z_A + R_0}{Z_A - R_0}\right| \text{ nepers}$$

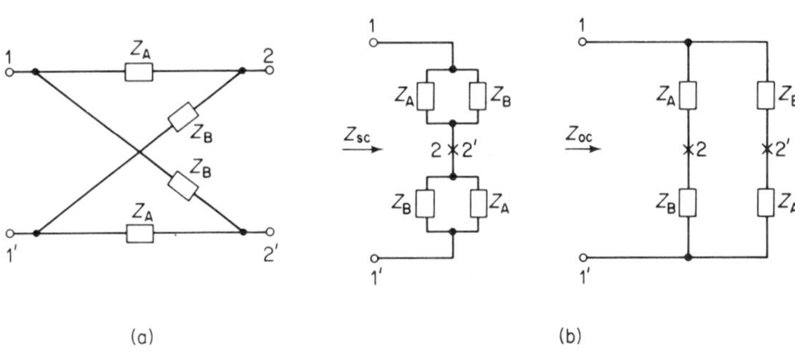

Fig. 4.23

Solution
The characteristic impedance of the symmetrical lattice may be found using the relation,

$$Z_C = \sqrt{(Z_{SC}Z_{OC})}$$

where the short-circuit and open-circuit input impedances at $11'$ (see fig. 4.23(b)) are

$$Z_{SC} = \frac{2Z_A Z_B}{Z_A + Z_B} \qquad Z_{OC} = \tfrac{1}{2}(Z_A + Z_B)$$

Thus
$$Z_C = \sqrt{(Z_A Z_B)} = R_0$$

127

The propagation constant γ may be found from the relation

$$\tanh \gamma = \sqrt{\left(\frac{Z_{SC}}{Z_{OC}}\right)} = \sqrt{\left[\frac{4Z_A Z_B}{(Z_A + Z_B)^2}\right]} = \frac{2R_0}{Z_A + Z_B}$$

$$= \frac{e^\gamma - e^{-\gamma}}{e^\gamma + e^{-\gamma}} = \frac{e^{2\gamma} - 1}{e^{2\gamma} + 1}$$

i.e. $\quad e^{2\gamma} = \dfrac{Z_A + Z_B + 2R_0}{Z_A + Z_B - 2R_0} = \dfrac{Z_A + R_0^2/Z_A + 2R_0}{Z_A + R_0^2/Z_A - 2R_0}$

$$= \frac{Z_A^2 + 2R_0 Z_A + R_0^2}{Z_A^2 - 2R_0 Z_A + R_0^2} = \left(\frac{Z_A + R_0}{Z_A - R_0}\right)^2$$

Hence

$$e^\gamma = e^\alpha e^{j\beta} = \frac{Z_A + R_0}{Z_A - R_0}$$

and

$$|e^\gamma| = e^\alpha = \left|\frac{Z_A + R_0}{Z_A - R_0}\right|$$

so

$$\alpha = \ln\left|\frac{Z_A + R_0}{Z_A - R_0}\right| \text{ nepers.}$$

7. Determine the insertion loss ratio of the symmetrical T, shown in fig. 4.24(a), when it is inserted between a generator of internal resistance R_0 ohms and a load of the same value. Hence show that the network of fig. 4.24(b) has an insertion loss of $10\log_{10}(1 + \omega^6 L^3 C^3/64)\,\text{dB}$ when inserted between $R_0 = \sqrt{(L/C)}$ ohm terminations.

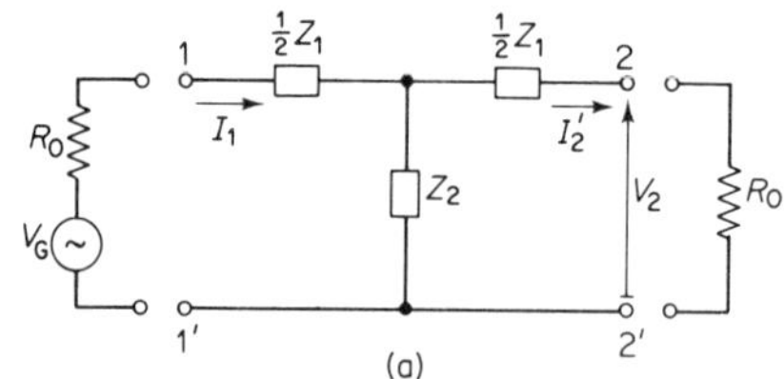

Fig. 4.24.

Solution
The insertion loss ratio,

$$F = \frac{V_{20}}{V_2}$$

128

where $V_{20} = (V_G/2R_0)R_0 = \frac{1}{2}V_G$ is the voltage across 22' when a direct connection is made between generator and load,

and $\quad V_2 = $ voltage across 22' when the network is inserted between the generator and load.

V_2 is determined as follows from fig. 4.24(a):

$$V_2 = R_0 I_2'$$

where $I_2' = \dfrac{Z_2}{Z_2 + \frac{1}{2}Z_1 + R_0} I_1$

$$I_1 = \frac{V_G}{R_0 + \frac{1}{2}Z_1 + Z_2(\frac{1}{2}Z_1 + R_0)/(Z_2 + \frac{1}{2}Z_1 + R_0)}$$

so $\quad V_2 = \dfrac{R_0 Z_2 V_G}{(R_0 + \frac{1}{2}Z_1)(Z_2 + \frac{1}{2}Z_1 + R_0) + Z_2(\frac{1}{2}Z_1 + R_0)}$

and $\quad F = \dfrac{(R_0 + \frac{1}{2}Z_1)(Z_2 + \frac{1}{2}Z_1 + R_0) + Z_2(\frac{1}{2}Z_1 + R_0)}{2R_0 Z_2}$

$$= 1 + \frac{Z_1}{2Z_2} + \frac{\frac{1}{4}Z_1^2 + Z_1 Z_2 + R_0^2}{2R_0 Z_2} \qquad \dots (1)$$

The insertion loss of the network of fig. 4.24(b) is found from (1) by substituting for $Z_1 = j\omega L$, $Z_2 = 1/(j\omega C)$ and $R_0 = \sqrt{(L/C)}$, i.e.

insertion loss $= 20 \log_{10} |F|$

$$= 20 \log_{10} |\{1 - \tfrac{1}{2}LC\omega^2\} + j\{\sqrt{(LC)}\omega - \tfrac{1}{8}(LC)^{3/2}\omega^3\}|$$

$$= 10 \log_{10} [(1 - \tfrac{1}{2}LC\omega^2)^2 + \{\sqrt{(LC)}\omega - \tfrac{1}{8}(LC)^{3/2}\omega^3\}^2]$$

$$= 10 \log_{10} [1 + \omega^6 L^3 C^3/64] \, \text{dB}$$

8. Fig. 4.25 shows a diagram of a length l of loss-less transmission line of phase constant β and characteristic impedance $Z_C = 1\,\Omega$, shunted at each end by elements of susceptance jb. Find the $ABCD$ transmission matrix of this network and show that its insertion loss in a $1\,\Omega$ system is given by

$$10 \log_{10} [1 + \tfrac{1}{4}b^2(2\cos\beta l - b\sin\beta l)^2] \, \text{dB}$$

Solution

The individual transmission matrices of the shunt jb and transmission line elements are, respectively:

$$\begin{bmatrix} 1 & 0 \\ jb & 1 \end{bmatrix} \begin{bmatrix} \cos\beta l & j\sin\beta l \\ j\sin\beta l & \cos\beta l \end{bmatrix}$$

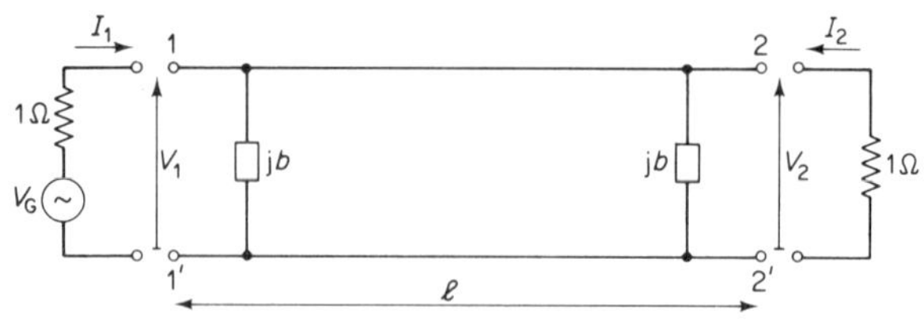

Fig. 4.25

and thus the overall transmission matrix of the 3 elements in cascade is

$$\begin{bmatrix} A & B \\ C & D \end{bmatrix} = \begin{bmatrix} 1 & 0 \\ jb & 1 \end{bmatrix} \begin{bmatrix} \cos \beta l & j \sin \beta l \\ j \sin \beta l & \cos \beta l \end{bmatrix} \begin{bmatrix} 1 & 0 \\ jb & 1 \end{bmatrix}$$

$$= \begin{bmatrix} \cos \beta l - b \sin \beta l & j \sin \beta l \\ j[2b \cos \beta l + (1 - b^2) \sin \beta l] & \cos \beta l - b \sin \beta l \end{bmatrix}$$

The insertion loss for the network is a $1\,\Omega$ system, in terms of its *ABCD* parameters, may be determined as follows:

$$V_{20} = \tfrac{1}{2} V_G = \text{voltage across load in direct connection,}$$

and $\quad V_2 = \text{voltage across load when network inserted}$

is found from the transmission equations

$$V_1 = AV_2 - BI_2 \quad I_1 = CV_2 - BI_2$$

using the terminal conditions $V_2/-I_2 = 1\,\Omega$, $V_1 = V_G - 1 \times I_1$.

On solving for V_2 we have $V_2 = V_G/(A + B + C + D)$ and so the insertion loss

$$= 20 \log_{10} \left| \frac{V_{20}}{V_2} \right|$$

$$= 20 \log_{10} |\tfrac{1}{2}(A + B + C + D)|$$

$$= 20 \log_{10} |\tfrac{1}{2}[\{2(\cos \beta l - b \sin \beta l)\} + j\{\sin \beta l + 2b \cos \beta l + (1 - b^2) \sin \beta l]|$$

$$= 10 \log_{10} \tfrac{1}{4}[\{2(\cos \beta l - b \sin \beta l)\}^2 + \{2 \sin \beta l + 2b \cos \beta l - b^2 \sin \beta l\}^2]$$

$$= 10 \log_{10} [1 + \tfrac{1}{4}b^2(2 \cos \beta l - b \sin \beta l)^2]\,\text{dB}.$$

4.3. Exercise Problems

Note Questions 1 to 24 refer to section 1, whilst 25 to 36 refer to section 2 of theory summary.

1. Determine the y parameters of the Π and T networks of fig. 4.26. Hence show that at $\omega = 1/\sqrt{(LC)}$ radians per second the networks are equivalent provided $G = \sqrt{(C/L)}$.

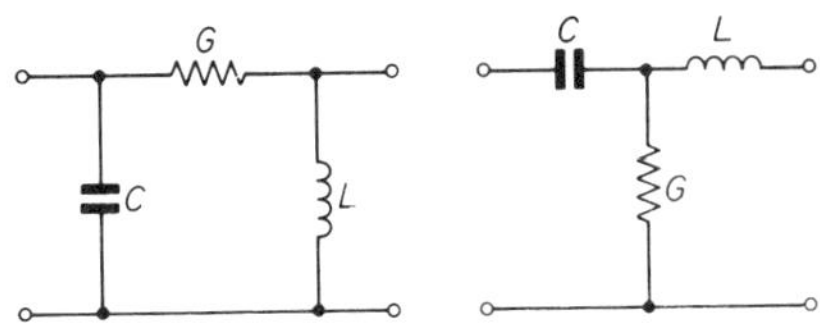

Fig. 4.26. For problem 1

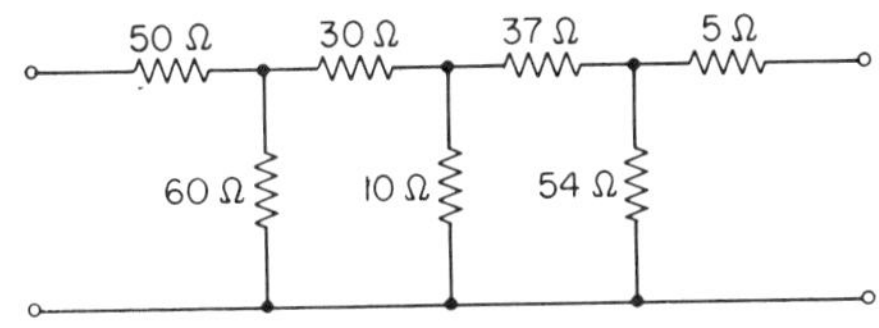

Fig. 4.27. For problem 2

2. Evaluate the z and $ABCD$ parameters for the network of fig. 4.27.

3. The results of open and short-circuit tests on a T network of resistors are given below:

the resistance measured at the input terminals when the output terminals are open-circuited is $100\,\Omega$,

the resistance measured at the input terminals when the output terminals are short-circuited is $55\,\Omega$,

the resistance measured at the output terminals when the input terminals are open-circuited is $80\,\Omega$.

From these results, determine the $ABCD$ parameters.

4. (a) Show that the input impedance of a 2-port network which is terminated in a load impedance Z_L is given by

$$Z_{in} = \frac{AZ_L + B}{CZ_L + D}$$

where A, B, C and D are the transmission parameters of the network.

(b) A 2-port network having parameters $A=D=0{\cdot}5$, $B=-j7{\cdot}5\,\Omega$, $C=-j0{\cdot}105$ is terminated in a resistive-capacitive load Z_L which can be varied between 0 and $5-j5\,\Omega$ and ganged in such a way that the resistive and reactive components are always equal. Plot the locus diagram of the input impedance as the load Z_L is varied over its range. From the diagram, or otherwise, determine the value of Z_L required to make the input impedance purely resistive.

5. Write down the transfer matrices ($ABCD$ parameters) for the networks shown in fig. 4.28(a) and (b). Hence determine those of the networks of fig. 4.28(c) and (d).

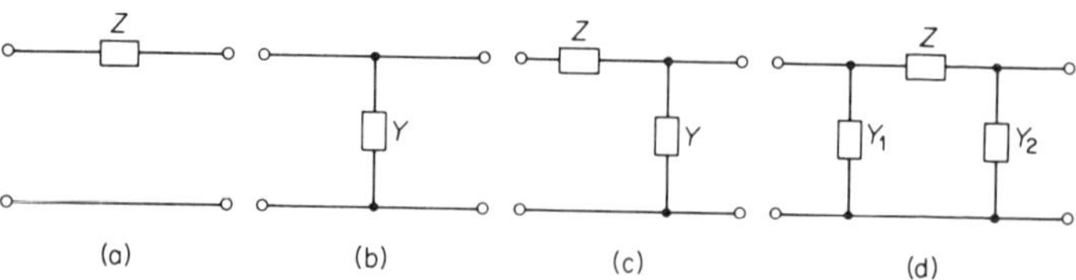

(a)   (b)   (c)   (d)

Fig. 4.28. For problem 5

6. Determine the z parameter matrix for the Π network of fig. 4.29(a). Hence, by considering the bridged T network of fig. 4.29(b) as a series input–series output connection of a Π and a shunt impedance, find its z parameter matrix.

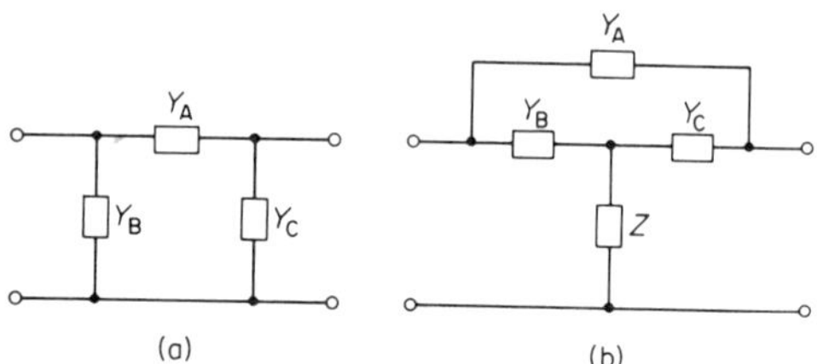

(a)   (b)

Fig. 4.29. For problem 6

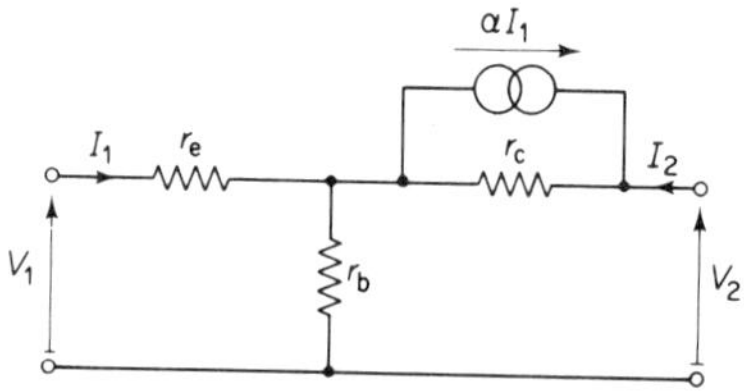

Fig. 4.30. For problem 8

7. A 2-port network with known y parameters is supplied by a sinusoidal current generator of internal conductance G_G siemens. A conductance of G_L siemens terminates the output port. Determine an expression for the current transfer function I_G/I_2 where I_G and I_2 are the phasor generator and load currents, respectively, and determine also I_G if $I_2 = 5(1/\sqrt{2} + j1/\sqrt{2})$, $G_G = G_L = G$, $y_{11} = jG$, $y_{12} = -j\sqrt{2}B$ and the 2-port network is symmetrical and passive.

8. Determine the hybrid parameters of the low frequency equivalent T circuit of a transistor, which is shown in fig. 4.30.

9. Fig. 4.31 shows the small signal hybrid equivalent circuit of a transistor operated in the common-emitter configuration. Determine the input resistance at terminals 11′ and the current gain I_2/I_1 when a load of G_L siemens is placed across 22′.

10. Using the equivalent circuit of fig. 4.31, draw the small signal a.c. equivalent circuit of the common emitter amplifier of fig. 4.32. Calculate the input impedance and the voltage gain at a frequency of 159 Hz, given that $h_{fe} = 60$, $h_{oe} = 40\,\mu S$, $h_{ie} = 1\,k\Omega$ and $h_{re} = 10^{-4}$.

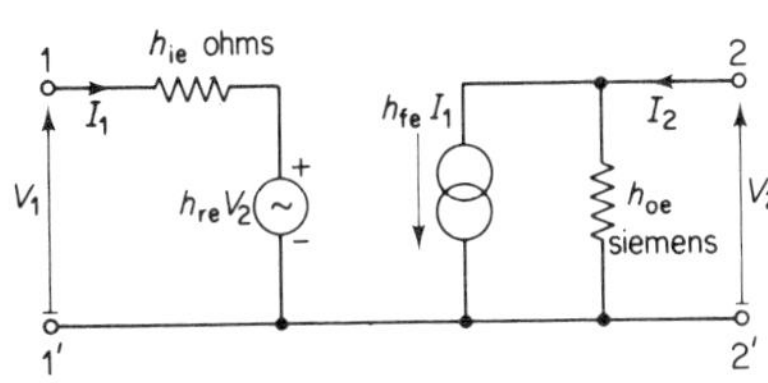

Fig. 4.31. For problems 9 and 10

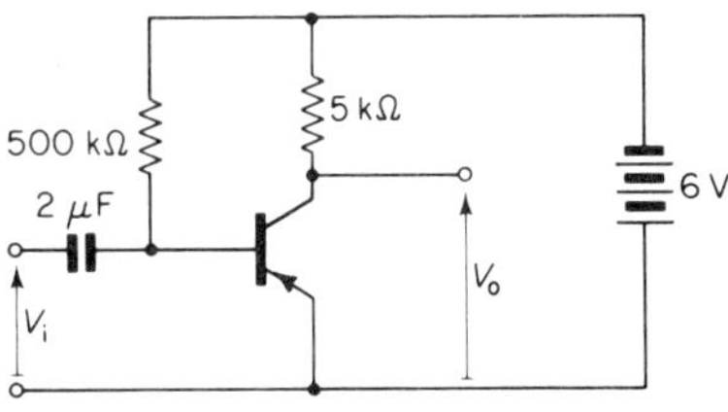

Fig. 4.32. For problem 10

11. The h-parameters of an n–p–n transistor at a particular d.c. operating point are: $h_{ie} = 580\,\Omega$, $h_{re} = 7 \times 10^{-4}$, $h_{fe} = 16$, $h_{oe} = 20\,\mu S$.

Calculate the power output when this transistor is used in a common-emitter amplifier with a load resistor $R_L = 20\,k\Omega$ and driven from a source of $250\,\Omega$ resistance, and an e.m.f. of $0.67\,V$ r.m.s.

12. Derive expressions for the voltage gain V_2/V_1, and the input and output impedances for the small signal equivalent amplifier circuit shown in fig. 4.33.

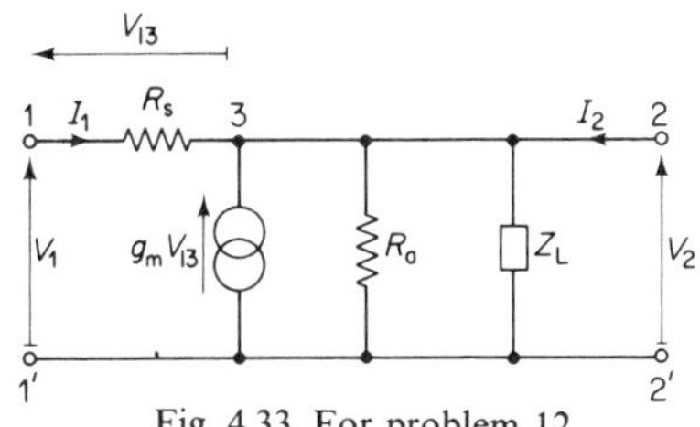

Fig. 4.33. For problem 12

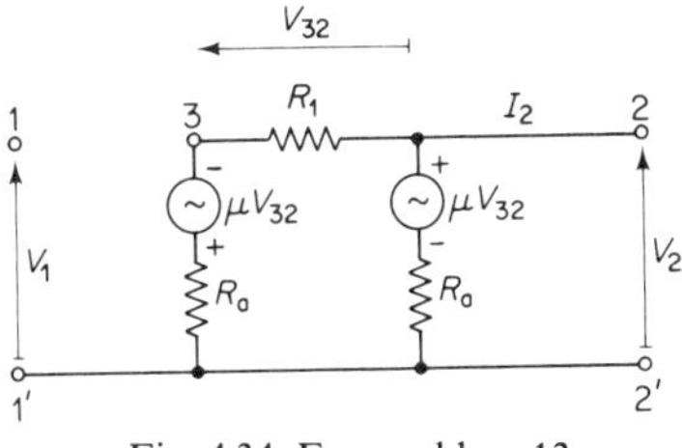

Fig. 4.34. For problem 13

13. Derive expressions for the voltage gain and output impedance of the small signal equivalent circuit shown in fig. 4.34. Evaluate this values when $R_a = 6\,k\Omega$, $R_1 = 1\,k\Omega$ and $\mu = 25$.

14. A 2-port active network is defined by the z-parameter equations:
$$z_{11}I_1 + z_{12}I_2 = V_1$$
$$z_{21}I_1 + z_{22}I_2 = V_2$$
(a) Using these equations form an equivalent circuit containing two current controlled generators, one controlled by I_2 at the input side and the other controlled by I_1 at the output.
(b) Find in terms of the z-parameters the values of Z_1, Z_2, Z_3 and α

in the equivalent T circuit of fig. 4.35 which may also be used to represent the terminal behaviour of the network.

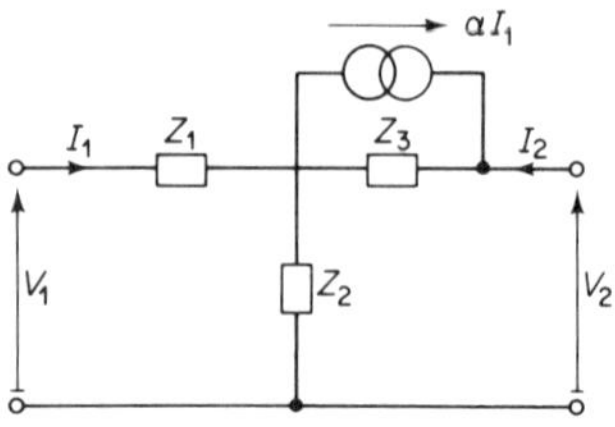

Fig. 4.35. For problem 14

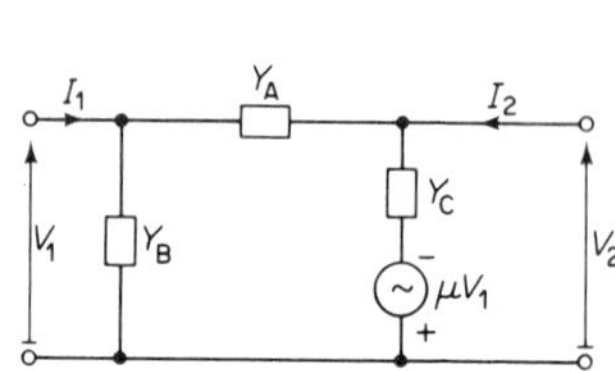

Fig. 4.36. For problem 15

15. Find the values of the admittances Y_A, Y_B, Y_C and μ in the equivalent Π circuit (shown in fig. 4.36) of an active 2-port network in terms of its y parameters.

16. Given the common base small signal hybrid equations of a transistor:

$$V_{eb} = h_{ib} I_e + h_{rb} V_{cb}$$
$$I_c = h_{fb} I_e + h_{ob} V_{cb}$$

determine the common emitter (see fig. 4.37(b)) and the common collector (see fig. 4.37(c)) hybrid parameters in terms of the common base parameters.

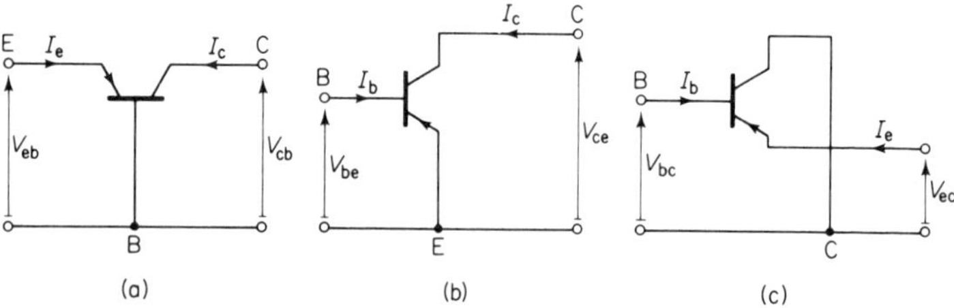

Fig. 4.37. For problem 16

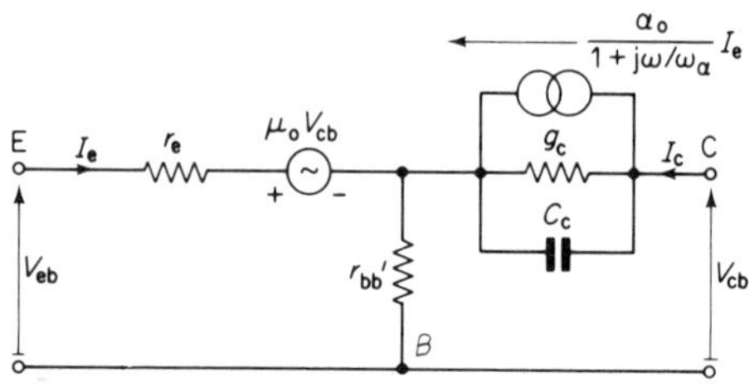

Fig. 4.38. For problem 17

17. Fig. 4.38 shows an approximate equivalent circuit model of a transistor based on physical principles. The circuit is valid up to frequencies of $\omega \sim \omega_\alpha$ and it may be assumed that $r_{bb'}g_c \ll 1$, $\omega Cr_{bb'} \ll 1$. Derive approximate values for the common base and common collector hybrid parameters of the transistor in terms of its physical parameters.

18. A 3-phase transmission circuit has the parameters $A = D = 0.95 \angle 0°$, $B = 35 \angle 90°$ ohms per phase. If the input end line voltage is 275 kV, evaluate (a) the open-circuit output end voltage, (b) the input end current when capacitors each of $-j8\,\Omega$ are placed in series with each phase at the input end, and the output end is short-circuited.

19. Fig. 4.39 shows the 'one phase' equivalent circuit of a 132 kV 3-phase transmission line supplying, a 40 MVA 132/33 kV transformer. The parameters of the line are $A = 0.94 \angle 0°$, $B = 62 \angle 75° \,\Omega/$ phase, and the transformer has a leakage reactance $Z_1 = j40\,\Omega$ and a magnetizing reactance Z_m of $j6000\,\Omega$/phase referred to the 132 kV side. Write down an algebraic expression for the resultant transmission matrix between ports 11′ and 22′ and determine the phase voltage amplitude at the input end of the line to supply 40 MVA, 33 kV, and 0.9 lagging power factor at the output of the transformer.

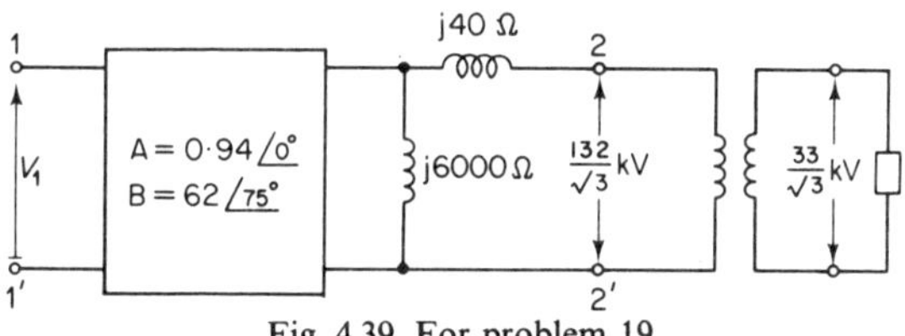

Fig. 4.39. For problem 19

20. A 3-phase transmission has the parameters $A_1 = D_1 = 0.96 \angle 0.6°$, $B_1 = 60 \angle 75°$ ohms per phase. Determine the parameter C_1 and the sending end current amplitude when a power of 50 MW at 132 kV and 0.9 lagging power factor is delivered at the receiving end. If a second line having the parameters $A_2 = 0.96 \angle 0.2°$ and $B_2 = 50 \angle 70°$ ohms per phase is connected in parallel with the first, determine the y parameters of the combination and hence construct an equivalent Π for the combination.

21. Determine (a) the insertion loss ratio, (b) the network power loss ratio, of a 2-port network in terms of its $ABCD$ parameters when inserted between a generator and a load of resistive impedances R_G and R_L respectively.

22. Determine the insertion loss in decibels of a length l metres of low-loss transmission line of attenuation constant α nepers per metre and resistive characteristic impedance Z_C when used in a system working between a generator and a load of resistive impedances R_G and R_L ohms when

(a) $R_G = R_L = Z_C$

(b) $R_G = Z_C, R_L = 2Z_C$

and l is an integral number of wavelengths.

23. Determine the insertion loss of the pad attenuator networks shown in fig. 4.40(a), (b), (c) when working between generator and load impedances of $600\,\Omega$.

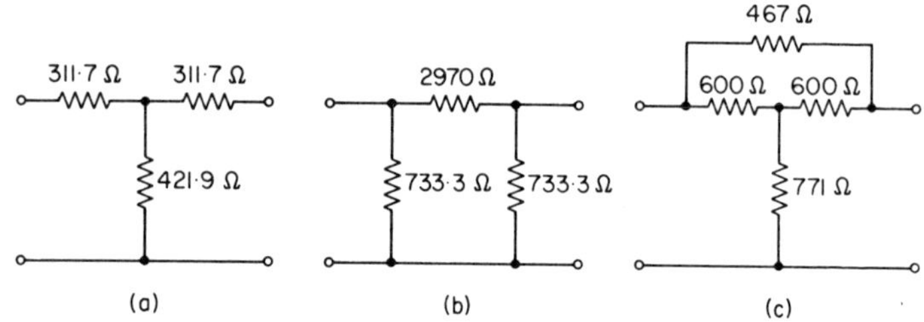

Fig. 4.40. For problem 23

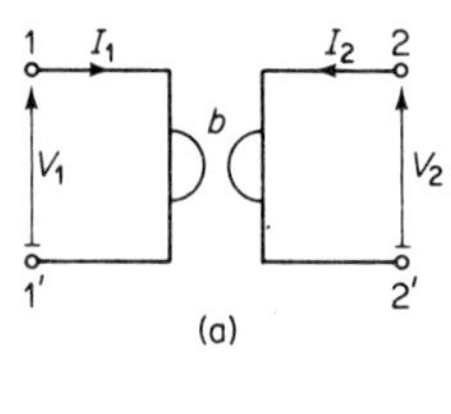
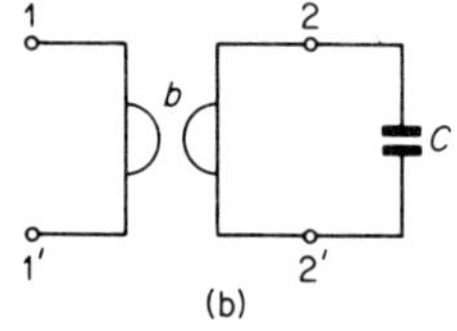

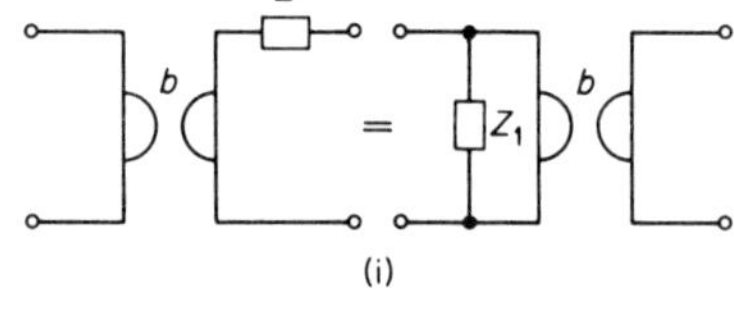
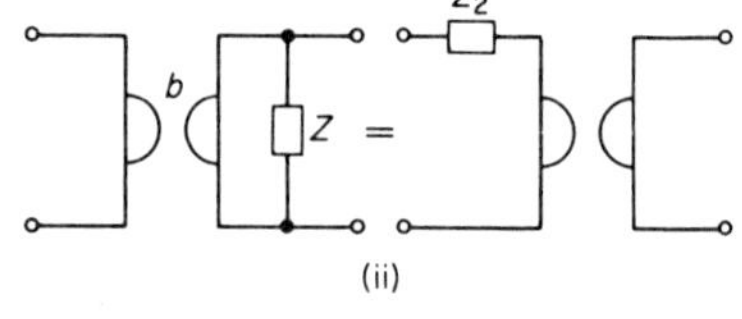

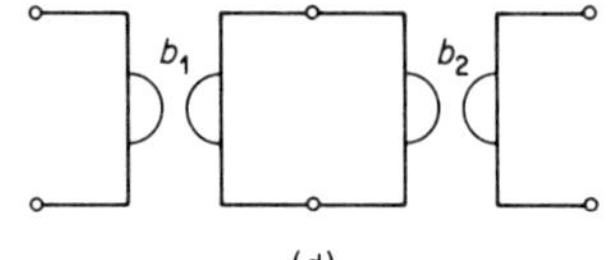
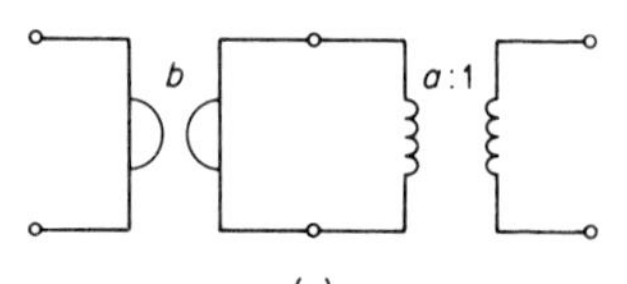

Fig. 4.41. For problem 24

24. The ideal gyrator (circuit symbol shown in fig. 4.41(a)) is a linear passive loss-less 2-port network element defined by the terminal relations:

$$V_1 = -bI_2, \quad V_2 = bI_1$$

(a) Determine its z and transmission matrices.
(b) Determine the input impedance at port 11' when the gyrator is terminated in a capacitor of C farads as shown in fig. 4.41(b).
(c) Determine the values of the impedance elements Z_1 and Z_2 such that the circuits of fig. 4.41(c) (i) and (ii) are equivalent
(d) The resultant transmission matrix of two gyrators connected in cascade as shown in fig. 4.41(d)
(e) The transmission matrix of the cascade connection of a gyrator and an ideal transformer (turns ratio a) as shown in fig. 4.41(e)

25. Determine the characteristic impedance and the propagation constant of the symmetrical T and Π networks shown in fig. 4.42.

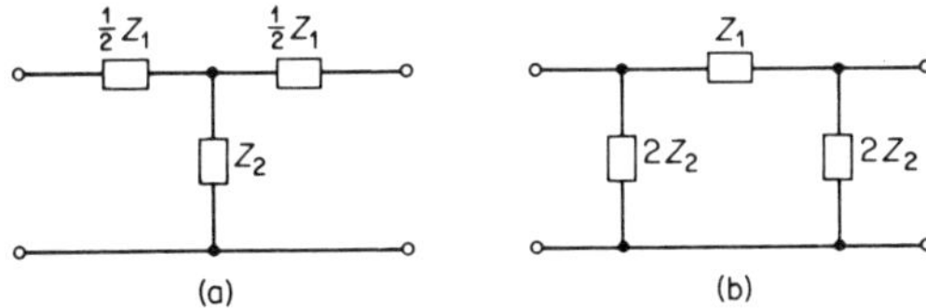

(a) (b)

Fig. 4.42. For problem 25

26. Determine the characteristic impedance and propagation constant of the symmetrical lattice network shown in fig. 4.43. Under what conditions does the network present infinite transmission loss between ports 1 and 2.

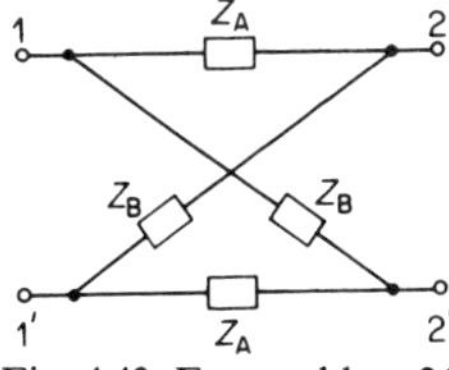

Fig. 4.43. For problem 26

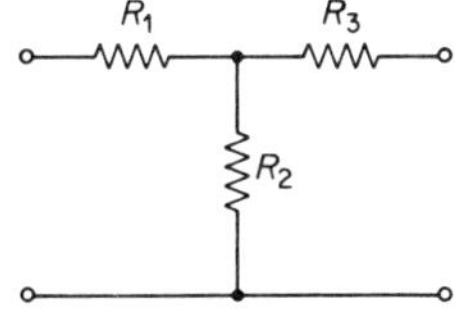

Fig. 4.44. For problem 27

27. Determine the values R_1, R_2 and R_3 of the T network shown in fig. 4.44 which effects a match between a generator of $50\,\Omega$ impedance and a load of 600Ω, and also provides an attenuation of $20\,\text{dB}$.

28. The ladder network of fig. 4.45(a) is composed of five identical Π

137

networks, the individual Π being shown in fig. 4.45(b). Show that the characteristic impedance of the latter is $600\,\Omega$ and determine the attenuation produced by the Π section when working between $600\,\Omega$ terminations. If a 300 Vr.m.s, $600\,\Omega$ generator is applied at the ladder input and the output is also terminated in a $600\,\Omega$ load, calculate the input power supplied to the ladder and the power dissipated in the load.

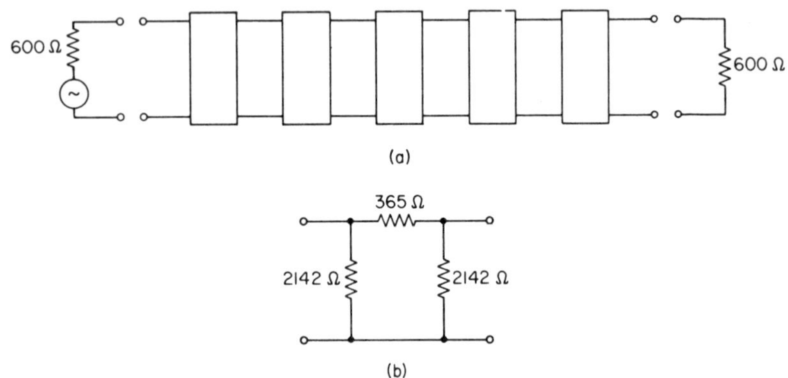

Fig. 4.45. For problem 28

29. Derive an expression for the input impedance of a 2-port network, terminated by an impedance Z_L at the other port, in terms of the *ABCD* transfer parameters of the network. Find also in terms of the *ABCD* parameters the image impedances Z_{01} and Z_{02} and the image propagation function θ. Finally, show that

$$Z_{01} = \sqrt{(Z_{SC_1} Z_{OC_1})}, \quad Z_{02} = \sqrt{(Z_{SC_2} Z_{OC_2})}$$

and $\quad \theta = \tanh^{-1}\sqrt{(Z_{SC_1}/Z_{OC_1})} = \tanh^{-1}\sqrt{(Z_{SC_2}/Z_{OC_2})}$

where Z_{SC}, Z_{OC} are the respective short-circuit and open-circuit driving point impedances at the respective ports.

30. Determine the characteristic impedance Z_C of the symmetrical bridged T network shown in fig. 4.46. If $Z_1 = 2\sqrt{(Z_2 Z_3)}$ show that $Z_C = \tfrac{1}{2}Z_1$ and the propagation constant $\gamma = \log_e[1 + \sqrt{(Z_3/Z_2)}]$. Hence design a bridged T network to work between $50\,\Omega$ terminations and provide an attenuation of $\log_e(1+x)$ nepers.

31. Determine the values of the reactances X_1 and X_2 in the L section of fig. 4.47 such that the section effects a match between a generator of impedance $R_1\,\Omega$ and a load of impedance R_2 ohms, $(R_2 > R_1)$.

32. Determine the elements of an L section which will effect a match

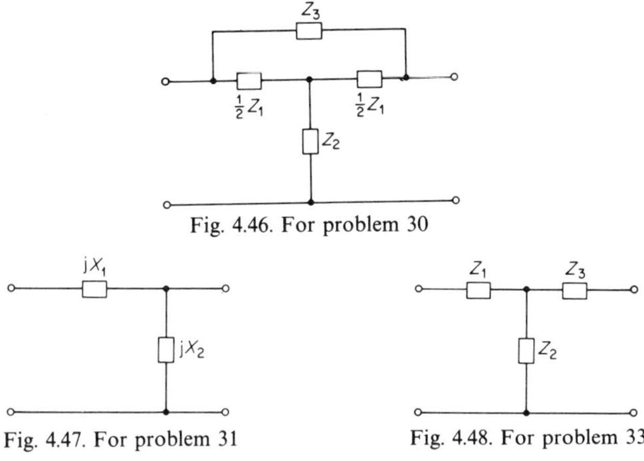

Fig. 4.46. For problem 30

Fig. 4.47. For problem 31

Fig. 4.48. For problem 33

between a generator of $70\,\Omega$ impedance and a load of $600\,\Omega$ at a frequency of $200\,\text{kHz}$. What is the phase shift produced by the section at this frequency?

33. Determine the image impedances of the asymmetrical T section shown in fig. 4.48. If the section is composed of pure reactances show that the section may effect a match between a generator of R_1 ohms and a load of R_2 ohms, provided $X_2^2 \geqslant (R_1 R_2)$. If $X_2 = \sqrt{(R_1 R_2)}$ determine the values of the X_1 and X_3 elements.

34. Find the iterative impedances of
    (a) a 2-port network in terms of its $ABCD$ transfer parameters,
    (b) the half section shown in fig. 4.49.

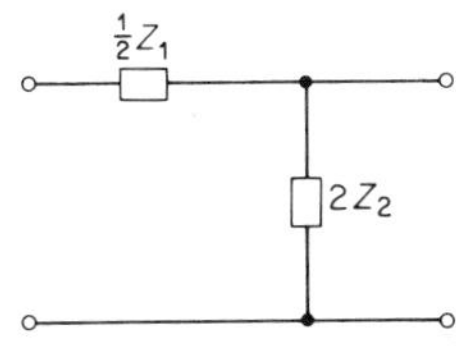

Fig. 4.49. For problem 34(b)

35. Determine the image impedances, Z_{01} and Z_{02}, of the half section shown in fig. 4.50(a). Show that if the section is terminated in its image impedances that $|I_1| = |I_2'|$, i.e. the insertion loss of the network is zero, provided $X_{\text{SC}_1} = \frac{1}{2}X_1$ and $X_{\text{OC}_1} = \frac{1}{2}X_1 + 2X_2$ are of opposite sign. Hence determine the range of frequencies for which the insertion loss of the sections shown in figs. 4.50(b) and (c) are zero (assuming the sections are terminated in their image impedances.

139

36. Show that the image impedances of the half m-derived section shown in fig. 4.51 are:

$$Z_{01} = R_0(1-x^2)^{1/2}, \quad Z_{02} = R_0 \frac{[1-(1-m^2)x^2]}{(1-x^2)^{1/2}}$$

where Z_1 and Z_2 are related by $Z_1 Z_2 = R_0^2$ and $-Z_1/4Z_2 = x^2$, R_0 and $x\,(x<1)$ being real positive quantities.

Sketch graphs of Z_{02} versus x for the range $0 \leqslant x \leqslant 1$ for values of $m = 0\cdot4,\ 0\cdot5,\ 0\cdot6,\ 0\cdot7,\ 0\cdot8$. Hence deduce that if $m = 0\cdot6$ the section may be used to effect an approximate match between a symmetrical T of total series and shunt impedances Z_1 and Z_2, and a load of R_0 ohms over the range $0 \leqslant x < 0\cdot9$.

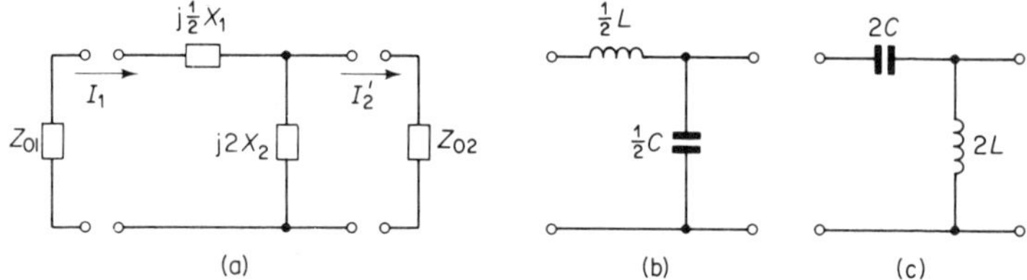

(a) (b) (c)

Fig. 4.50. For problem 35

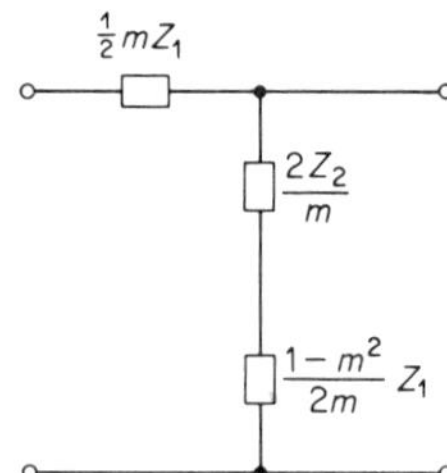

Fig. 4.51. For problem 36

140

5 Filter Networks

Section 1 deals with the principles of filter design based on the image method, whilst section 2 deals with the design obtained by network synthesis methods.

5.1. Theory Summary

Section 1: Application of the Image Parameters to Filter Design

1. INTRODUCTION

In the design of passive element filter networks, we are concerned with constructing 2-port networks composed of L–C elements, which pass certain bands of frequencies with ideally zero transmission loss, whilst attenuating other bands.

Some typical filter characteristics are shown in fig. 5.1.

The term *pass band(s)* for filters based on the image method of design refers to the frequency range(s) over which α the attenuation constant is zero, the filter being terminated in its image impedances. Attenuation or stop band is used to define those bands of frequencies for which $\alpha > 0$.

The *cut-off frequencies* refer to those frequencies defining the limits of the pass bands.

The *design impedance* (also known as the nominal impedance) of a filter section is the value of the characteristic impedance of the section at a frequency well within the pass-band. In filter design this is normally made equal to the generator or load impedance.

2. CONDITIONS FOR THE EXISTENCE OF PASS AND STOP BANDS

The condition for a pass band is that the image impedances of the filter section should be purely real and positive over the band of frequencies constituting the pass band. It is a sufficient condition that

$$Z_{01} = \sqrt{[Z_{OC_1} Z_{SC_1}]} = \sqrt{[-X_{OC_1} X_{SC_1}]}$$

is real and positive since if Z_{01} is real so is Z_{02}.

Note that the open-circuit input reactance X_{OC_1} and the short-circuit input reactance X_{SC_1} must have opposite signs for a pass band to exist.

141

Attenuation
$\alpha = \ln \left| \dfrac{V_1}{V_2} \right|$
ω
ω_C
$\frac{1}{2}L$
$\frac{1}{2}C$
Z_{01}
V_1
V_2
Z_{02}
(a) Low pass half-section
α
ω
ω_C
C
$2L$
$2L$
$Z_{C\pi}$
$Z_{C\pi}$
(b) High pass symmetrical π section (constant k)
α
ω
ω_{C_1}
ω_{C_2}
$\frac{mL}{2n}$
$\frac{2nC}{m}$
$\frac{mL}{2n}$
$\frac{nL}{m}$
$\frac{m}{n}C$
$\frac{1-m^2}{4mn}L$
$\frac{4mnC}{1-m^2}$
Z_C
Z_C
$\left[n = \frac{1}{2}\sqrt{LC}\,(\omega_{C_2} - \omega_{C_1}) \right]$
(c) m-derived symmetrical band-pass T section
α
ω
ω_{C_1}
ω_{C_2}
$\frac{1}{2}nL$
$\frac{1}{2}nL$
$\frac{2C}{n}$
$\frac{L}{n}$
$\frac{2C}{n}$
nC
Z_C
Z_C
$n = 2\sqrt{LC}\,(\omega_{C_2} - \omega_{C_1})$
(d) Constant k symmetrical band-stop T section

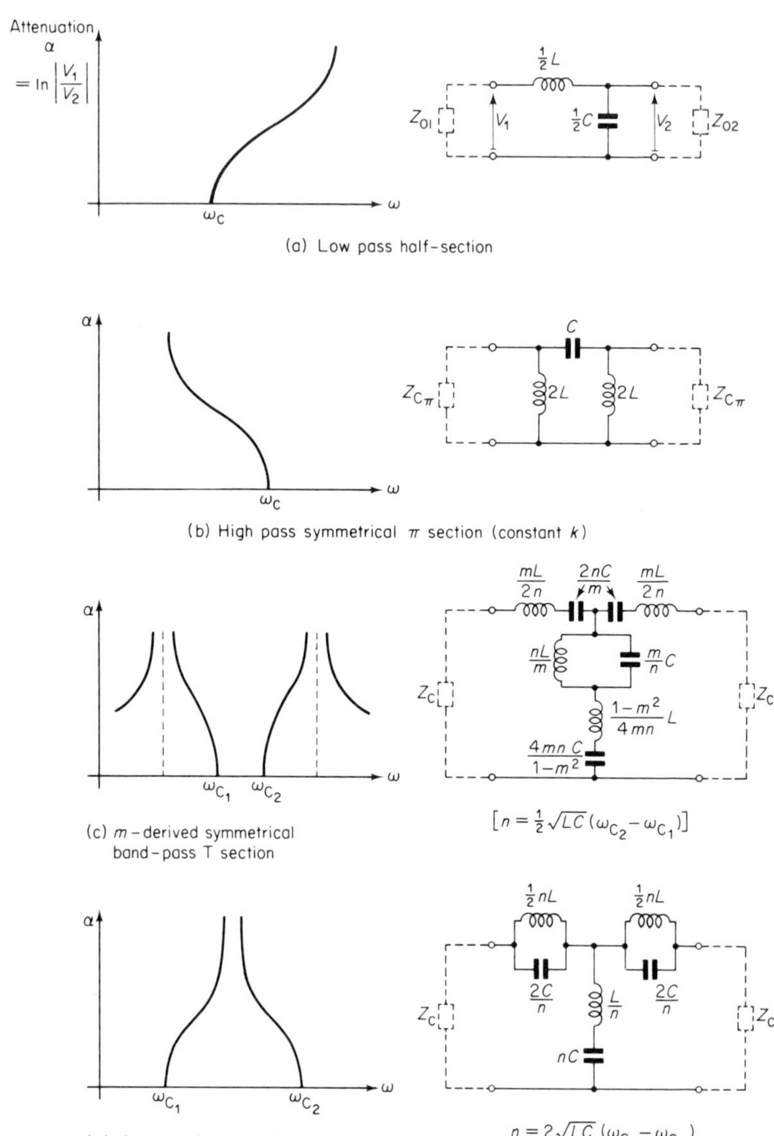

Fig. 5.1

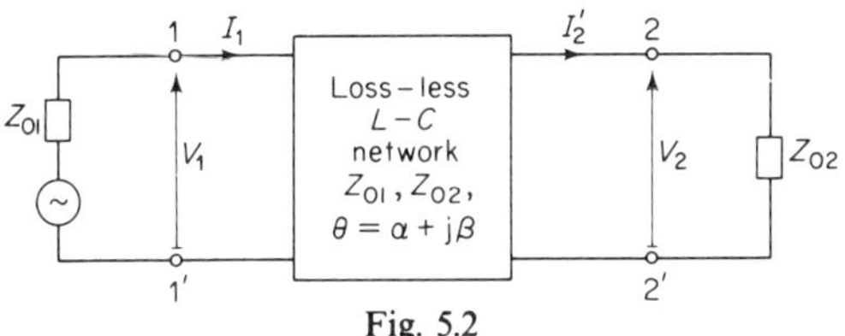

Fig. 5.2

Under these conditions, we have on using

$$\theta = \tanh^{-1}\sqrt{\left(\frac{X_{\mathrm{SC_1}}}{X_{\mathrm{OC_1}}}\right)} = \coth^{-1}\sqrt{\left(\frac{X_{\mathrm{OC_1}}}{X_{\mathrm{SC_1}}}\right)}$$

$$\mathrm{Re}\,(\theta) = \alpha = 0,\ \mathrm{Im}\,(\theta) = \beta = \tan^{-1}\sqrt{\left|\frac{X_{\mathrm{SC_1}}}{X_{\mathrm{OC_1}}}\right|}$$

$$= -\cot^{-1}\sqrt{\left|\frac{X_{\mathrm{OC_1}}}{X_{\mathrm{SC_1}}}\right|} = \left[\tan^{-1}\sqrt{\left|\frac{X_{\mathrm{OC_1}}}{X_{\mathrm{SC_1}}}\right|}\right] - \frac{\pi}{2}$$

The condition for a stop band is that the image impedances should be purely imaginary, i.e.

$$Z_{01} = \sqrt{[-X_{\mathrm{OC_1}}X_{\mathrm{SC_1}}]} = jX_{01}\ ,\ Z_{02} = \sqrt{[-X_{\mathrm{OC_2}}X_{\mathrm{SC_2}}]} = jX_{02}$$

and therefore X_{OC} and X_{SC} should be of the same sign. In the stop band

$$\alpha = \coth^{-1}\sqrt{\left(\frac{X_{\mathrm{OC_1}}}{X_{\mathrm{SC_1}}}\right)},\ \beta = n\pi\ \text{if}\ X_{\mathrm{OC_1}} > X_{\mathrm{SC_1}}$$

$$\alpha = \tanh^{-1}\sqrt{\left(\frac{X_{\mathrm{OC_1}}}{X_{\mathrm{SC_1}}}\right)},\ \beta = (2n-1)\pi\ \text{if}\ X_{\mathrm{OC_1}} < X_{\mathrm{SC_1}}\ ,\ n = 1, 2, 3 \ldots$$

3. CONSTANT k FILTER SECTIONS

(a) *Definition*: constant k filter sections are sections of a uniform ladder network for which the product of the series and shunt impedances $Z_1 Z_2 = k^2$ where k^2 is a real positive constant independent of frequency. For example the filter sections of fig. 5.1(a), (b) and (d) are constant k sections, since

for the low pass half section of fig. 5.1(a),

$$Z_1 = j\omega L,\ Z_2 = 1/j\omega C,\ Z_1 Z_2 = L/C$$

for the high pass section of fig. 5.1(b),

$$Z_1 = 1/j\omega C,\ Z_2 = j\omega L,\ Z_1 Z_2 = L/C$$

143

for the band stop section of fig. 5.1(d),

$$Z_1 = \frac{j\omega nL \times 1/(j\omega C/n)}{j\omega nL + 1/(j\omega C/n)}, \; Z_2 = j\omega L/n + 1/(j\omega Cn), \; Z_1 Z_2 = L/C$$

(b) *The low pass constant k T, Π and half filter sections*

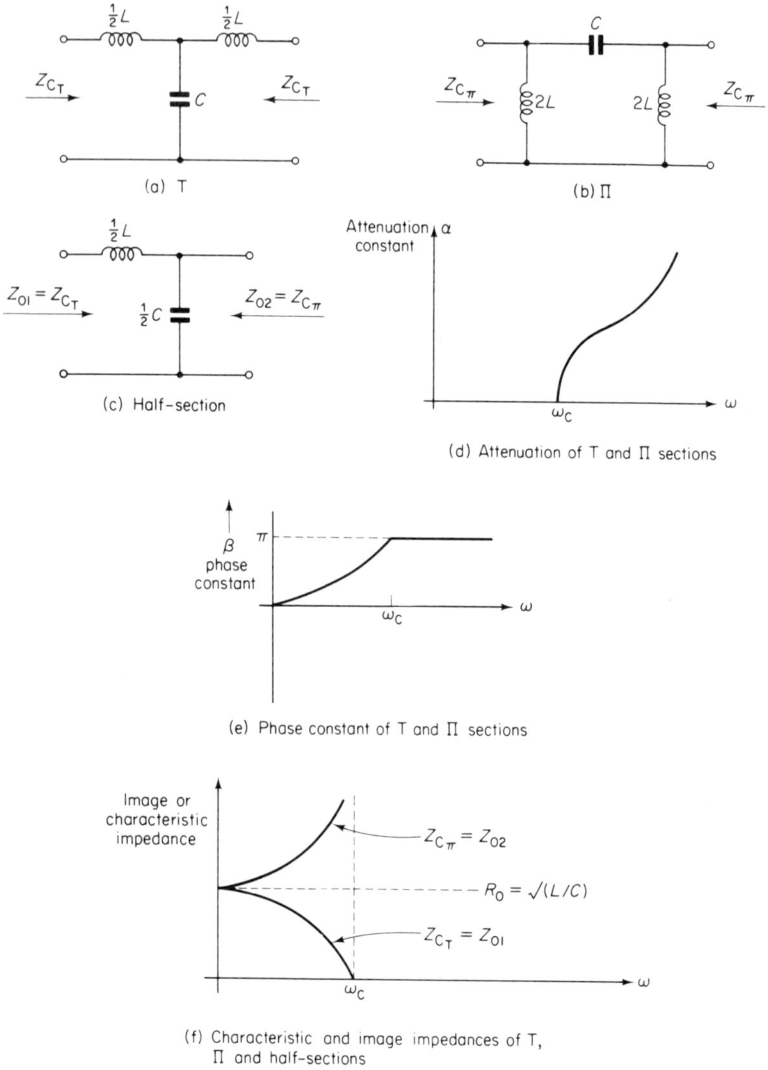

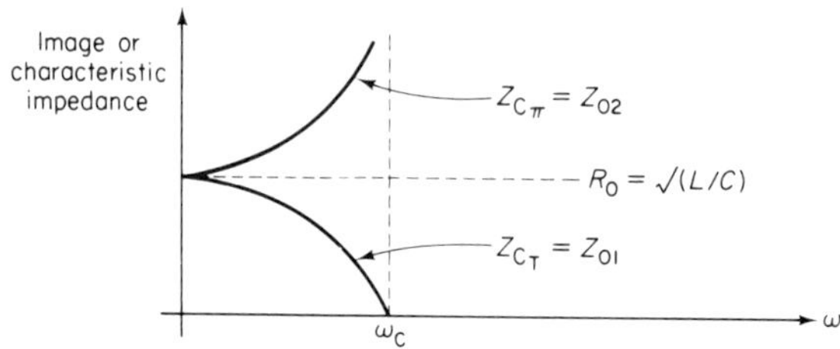

Fig. 5.3

144

The characteristics of the low pass constant k filter sections are summarized below and also sketched in fig. 5.3(d), (e) and (f).

$$Z_{C_T} = Z_{01} = \sqrt{\left(\frac{L}{C}\right)}\sqrt{[1 - \tfrac{1}{4}LC\omega^2]}$$

$$= \sqrt{\left(\frac{L}{C}\right)}\sqrt{\left[1 - \frac{\omega^2}{\omega_C^2}\right]} \ldots \text{ characteristic impedance of T}$$

$$Z_{C_\pi} = Z_{02} = \frac{\sqrt{\left(\frac{L}{C}\right)}}{\sqrt{\left[1 - \frac{\omega^2}{\omega_C^2}\right]}} \ldots \text{ characteristic impedance of } \Pi$$

$\omega_C = 2\pi f_C = 2/\sqrt{(LC)}, f_C = 1/[\pi\sqrt{(LC)}] \ldots$ cut-off frequency

$\beta = \cos^{-1}[1 - 2\omega^2/\omega_C^2] = 2\sin^{-1}(\omega/\omega_C) \ldots$ phase constant of T and Π sections in pass band

$\alpha = 2\cosh^{-1}(\omega/\omega_C), \beta = \pi \ldots$ stop band attenuation and phase constant of T and Π sections

For half section: $Z_{01} = Z_{C_T}, Z_{02} = Z_{C_\pi}$;

$$\beta = \sin^{-1}(\omega/\omega_C) \ldots \text{ in pass band}$$
$$\beta = \pi/2, \alpha = \cosh^{-1}(\omega/\omega_C) \ldots \text{ in stop band}$$

The cut-off frequency may be easily found using the condition that in the pass band the image or characteristic impedances must be real. The phase and attenuation constant may be derived using the results of fig. 4.13, e.g. $\gamma_T = \gamma_\pi = \cosh^{-1}(1 + Z_1/2Z_2), Z_1 = j\omega L, Z_2 = -j/\omega C$, or fig. 4.15(a).

The design formulae for the L and C elements of the above low pass filter sections are

$$L = \frac{R_0}{\pi f_C}\text{ henries} \quad C = \frac{1}{\pi R_0 f_C}\text{ farads}$$

where R_0 is the required design impedance ($R_0 = \sqrt{(L/C)}$ ohms) and f_C is the required cut-off frequency ($f_C = 1/[\pi\sqrt{(LC)}]$ hertz).

4. m-DERIVED FILTER SECTIONS

(a) There are two major disadvantages of constant k filter sections: the attenuation-frequency characteristic only rises relatively slowly close to cut-off; and the image impedances vary substantially with frequency in the pass band thereby causing mismatch losses when the filter is used, as

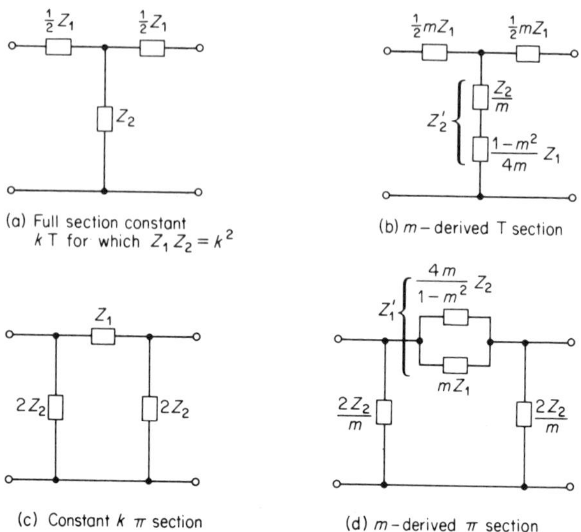

Fig. 5.4

in practice, with constant magnitude resistive terminations. To overcome these difficulties m-derived sections are used.

The m-derived T section may be considered to be derived from a constant k T section as follows:

(1) The series arm impedance Z_1 is changed to mZ_1, $0<m<1$.
(2) The shunt arm impedance Z_2 is changed to Z_2' where Z_2' is evaluated subject to the condition that the complete m-derived section has the same characteristic impedance as the constant k section.

Similarly in the case of the Π section Z_2 is changed to Z_2/m and Z_1 to Z_1' where Z_1' is evaluated so that the complete m-derived section has the same characteristic impedance as the constant k Π.

(b) *Low pass m-derived sections*
(i) *Full sections*

The most significant feature of the m-derived characteristics is that the attenuation rises to infinity when,

$$\omega_\infty = 1\Big/\sqrt{\left\{mC \times \frac{1-m^2}{4m}L\right\}} = 1\Big/\sqrt{\left\{\frac{1-m^2}{4m}C \times mL\right\}} = \frac{\omega_C}{\sqrt{(1-m^2)}}$$

where $\omega_C = 2/\sqrt{(LC)}$ . . . the cut-off frequency, identical to the corresponding constant k low pass sections.

146

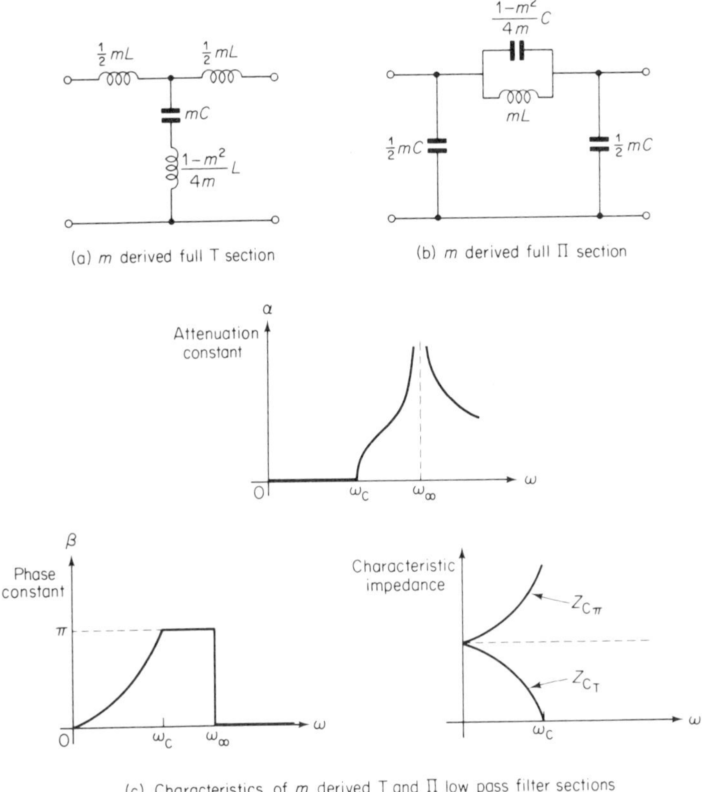

(c) Characteristics of m derived T and Π low pass filter sections

Fig. 5.5

Infinite attenuation occurs at $\omega = \omega_\infty$ since the shunt arm of the T is series resonant thereby presenting a short-circuit across the transmission path, whilst the Π section series arm is parallel resonant thereby causing an effective open circuit in the transmission path.

The propagation constant of the sections is

$$\gamma = j\beta = j\cos^{-1}\left[1 - \frac{2m^2}{\left(\dfrac{\omega_C}{\omega}\right)^2 - (1-m^2)}\right] \text{ in the pass band } 0 \leqslant \omega \leqslant \omega_C$$

$$\gamma = \alpha + j\beta = \cosh^{-1}\left[\frac{2m^2}{\left(\dfrac{\omega_C}{\omega}\right)^2 - (1-m^2)} - 1\right] + j\pi$$

$$\text{in the stop band, } \omega_C \leqslant \omega \leqslant \omega_\infty$$

$$\gamma = \alpha + j\beta = \cosh^{-1}\left[1 - \frac{2m^2}{\left(\dfrac{\omega_C}{\omega}\right)^2 - (1 - m^2)}\right]$$

$$+ j0 \text{ in the stop band, } \omega_\infty \leqslant \omega \leqslant \infty$$

The characteristic impedances are

$$Z_C = \sqrt{\frac{L}{C}}\sqrt{\left[1 - \frac{\omega^2}{\omega_C^2}\right]} \dots \text{ for the T}$$

$$Z_C = \sqrt{\frac{L}{C}}\Bigg/\sqrt{\left[1 - \frac{\omega^2}{\omega_C^2}\right]} \dots \text{ for the } \Pi$$

(ii) *Half sections*

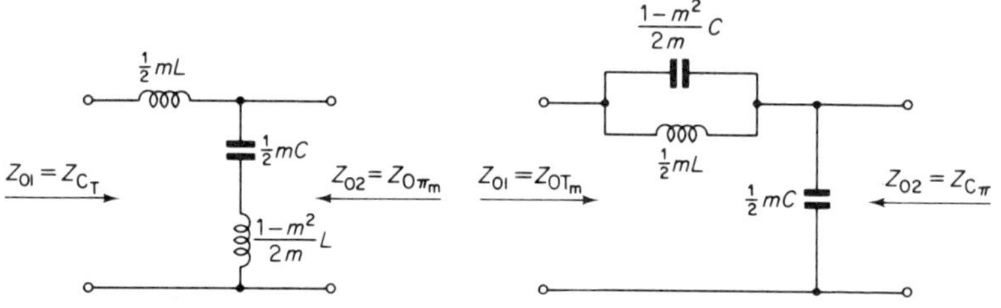

(a) Series *m*–derived half section   (b) Shunt *m*–derived half section

Fig. 5.6

The image impedances of the series *m*-derived half section are:

$$Z_{01} = Z_{C_T} = \sqrt{\left(\frac{L}{C}\right)}\sqrt{\left(1 - \frac{\omega^2}{\omega_C^2}\right)}$$

$$Z_{02} = Z_{0\pi_m} = \sqrt{\left(\frac{L}{C}\right)}\frac{1 - \dfrac{\omega^2}{\omega_\infty^2}}{\sqrt{\left(1 - \dfrac{\omega^2}{\omega_C^2}\right)}}$$

and for the shunt *m*-derived half section:

$$Z_{01} = Z_{0T_m} = \sqrt{\left(\frac{L}{C}\right)}\frac{\sqrt{\left(1 - \dfrac{\omega^2}{\omega_C^2}\right)}}{\left(1 - \dfrac{\omega^2}{\omega_\infty^2}\right)}$$

$$Z_{02} = Z_{C_\pi} = \frac{\sqrt{\left(\dfrac{L}{C}\right)}}{\sqrt{\left(1 - \dfrac{\omega^2}{\omega_C^2}\right)}}$$

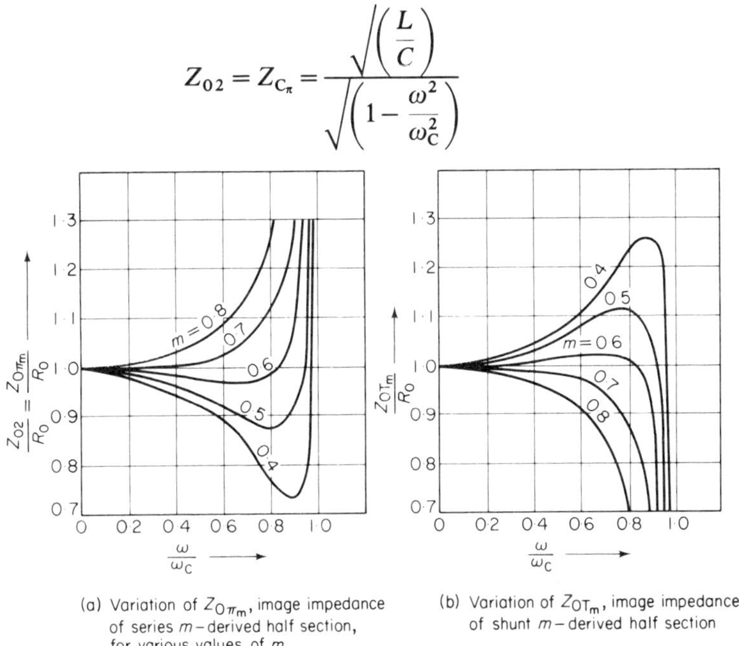

(a) Variation of $Z_{0\pi_m}$, image impedance
of series m–derived half section,
for various values of m

(b) Variation of Z_{0T_m}, image impedance
of shunt m–derived half section

Fig. 5.7

Note that the image impedances of the half section have the extremely important property of matching to full sections since respectively for the half section T and Π, $Z_{01} = Z_{C_T}$, $Z_{02} = Z_{C_\pi}$, whilst for $m = 0.6$ the second image impedance is approximately constant over the majority of the pass band. $Z_{0\pi_m}$ and Z_{0T_m} are within $\pm 4\%$ of $R_0 = \sqrt{(L/C)}$ ohms for over 90% of the pass band frequency range when $m = 0.6$. Thus these sections provide a very useful means of matching constant k and m-derived full sections to constant magnitude resistive terminations.

5. A COMPLETE LOW-PASS FILTER DESIGNED BY THE IMAGE METHOD

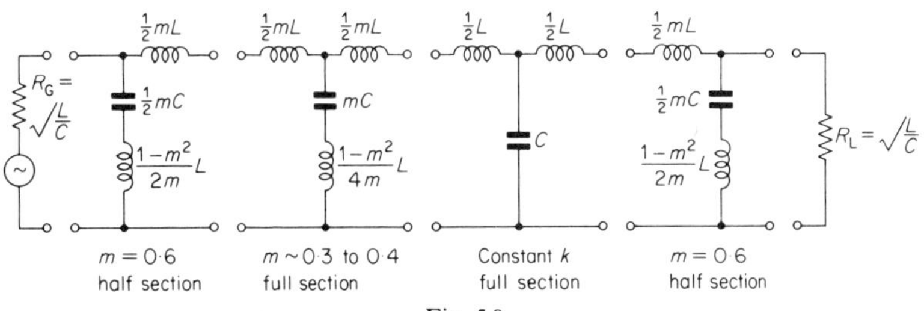

Fig. 5.8

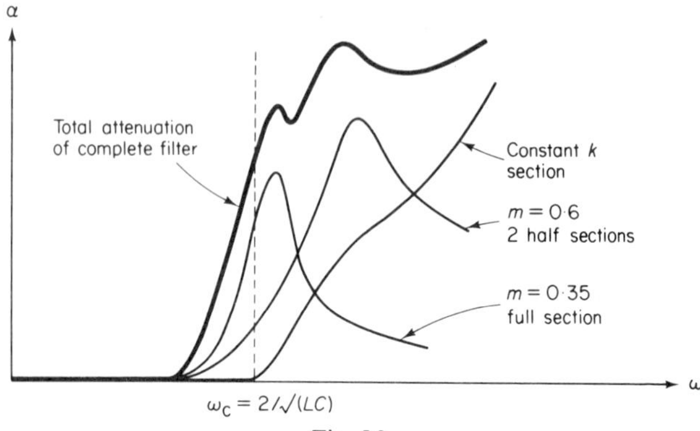

Fig. 5.9

An example of a composite low pass filter is shown in fig. 5.8 and its attenuation versus frequency characteristic in fig. 5.9. The latter includes the effect of resistive losses associated with practical L and C components. The action of the various sections is as follows:

to the generator and load impedances.

(2) The $m = 0.35$ full section ensures a steep rise in attenuation close to the cut-off frequency.

(3) The constant k full section ensures substantial attenuation at frequencies far above cut-off.

6. NORMALIZED FORMS FOR FILTER DESIGN AND THEIR DENORMALIZATION

To enable universal application of filter designs to any cut-off frequency and impedance level, filter data is often presented in normalized form. In conventional normalized form a filter is specified as working between 1 Ω impedances (i.e. at an impedance level of 1 Ω) and

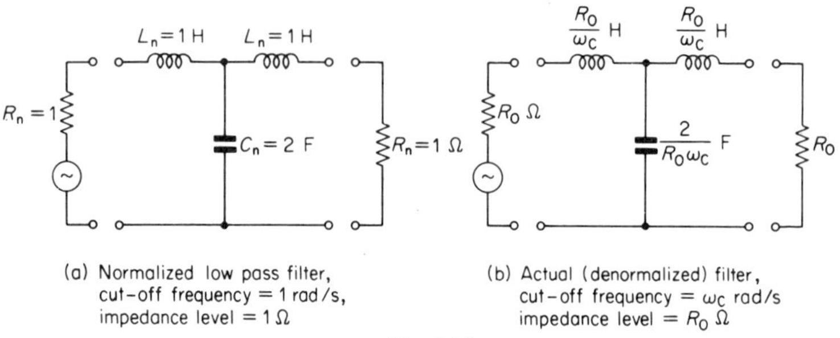

(a) Normalized low pass filter,
cut–off frequency = 1 rad/s,
impedance level = 1 Ω

(b) Actual (denormalized) filter,
cut–off frequency = ω_c rad/s
impedance level = $R_0\ \Omega$

Fig. 5.10

150

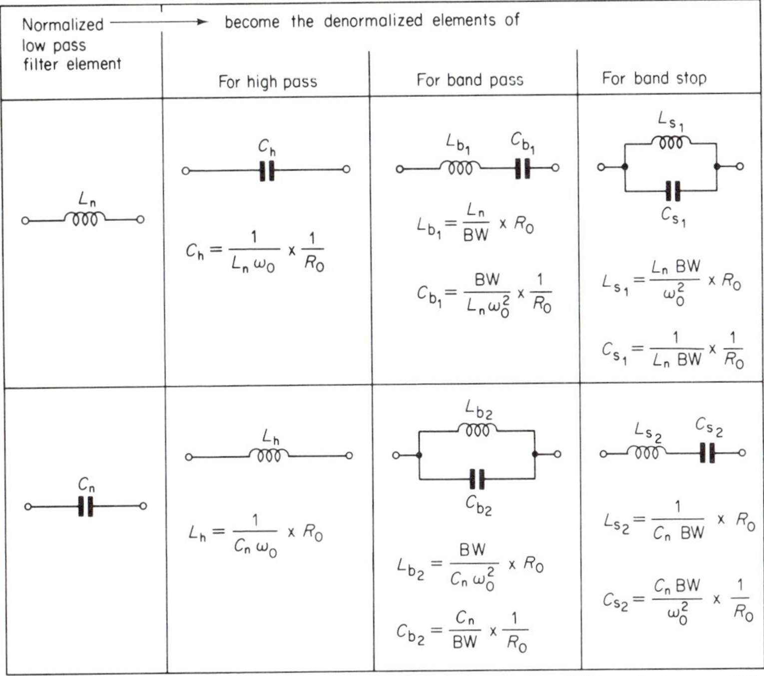

Fig. 5.11. Determination of elements for high, band, stop band filters in terms of the normalized elements of a low pass filter

$\omega_C, \omega_{C_1}, \omega_{C_2} \ldots$ cut-off frequencies (rad/s)
$BW = \omega_{C_2} - \omega_{C_1} \ldots$ bandwidth of band pass or band stop (rad/s)
$\omega_0 = \sqrt{(\omega_{C_1} \omega_{C_2})}$ rad/s

for low and high pass filters to have a cut-off frequency of 1 rad/s. A normalized constant k low pass T section is shown in fig. 5.10(a).

To facilitate denormalization, i.e. to convert the normalized parameters R_n, L_n, C_n given in a normalized filter design to the required values R, L, C so that the filter works at an impedance level of R_0 ohms with a cut-off frequency of ω_C rad/s, we use

$$R = R_0 R_n\,,\; C = \frac{C_n}{R_0 \omega_C}\,,\; L = \frac{R_0 L_n}{\omega_C}$$

7. FREQUENCY TRANSFORMATIONS: GENERATION OF HIGH-PASS, BAND PASS AND BAND STOP FILTERS FROM A NORMALIZED LOW-PASS FILTER

So far we have concentrated solely on the design of low pass filters. However, the normalized data obtained for these may be used with the

151

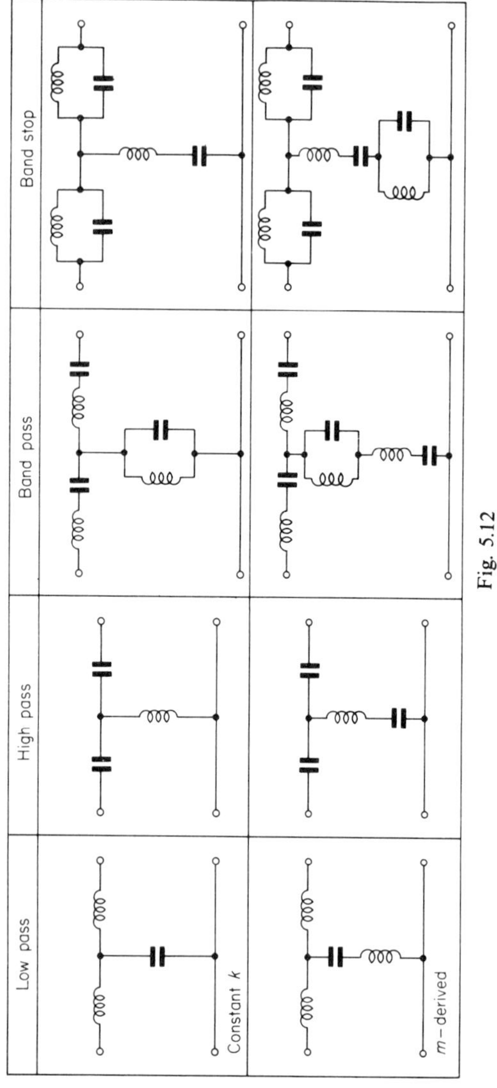

Fig. 5.12

frequency transformation table given in fig. 5.11 to determine the corresponding elements of high pass, band pass and band stop filters.

Examples of the generation of constant k and m derived high, band-pass, and stop-band filters are shown in fig. 5.12.

Section 2: Network Synthesis Methods of Filter Design

8. INTRODUCTION

(a) A major disadvantage of filters designed on the classical image theory basis is that the designs permit matching to the specified terminations only over a limited frequency range and thus the ensuing reflection effects may seriously degrade the overall performance of a filter. A good deal of intuition must be applied to produce a precision classical filter with low pass-band loss, accurately defined cut-off frequencies, and sharp stop-band edge attenuation.

(b) In network synthesis methods the desired transfer function of the filter including the effect of the terminations is first specified. From this function the input impedance (or admittance) is found and then by means of a continued or partial fraction expansion, the input impedance is expanded so that the individual values of the L and C elements may be evaluated. In this way a filter may be synthesized to give a specified transfer function and therefore eliminate much of the 'guess work' inherent in some image filter designs. In general, a filter designed by network synthesis methods will have somewhat similar characteristics to its image counterpart, although it is not possible to design the latter to meet a precise specification.

In considering network synthesis design methods it should be understood that the specified transfer function must be physically realizable. Below we examine briefly some ways in which the ideal filter response may be approximated to in practice. We consider the synthesis of low-pass filter section working between $1\,\Omega$ terminations. By the application of the denormalization results and frequency transformation tables given in section 1, parts 6 and 7, the normalized low-pass filter designs may be converted into filters with high-pass, band-pass or band-stop characteristics working at given impedance levels.

9. THE MAXIMALLY FLAT (BUTTERWORTH) APPROXIMATION FOR LOW-PASS FILTERS

The general form of a realizable transfer function from which the maximally flat filter characteristic is derived is

$$|H(j\omega)| = \left|\frac{V_2}{V_1}\right| = \frac{K}{\left[1+\varepsilon\left(\dfrac{\omega}{\omega_C}\right)^{2n}\right]^{1/2}}$$

where K is a constant and the other parameters are defined below. The characteristic (also known as Butterworth) derives its name from the

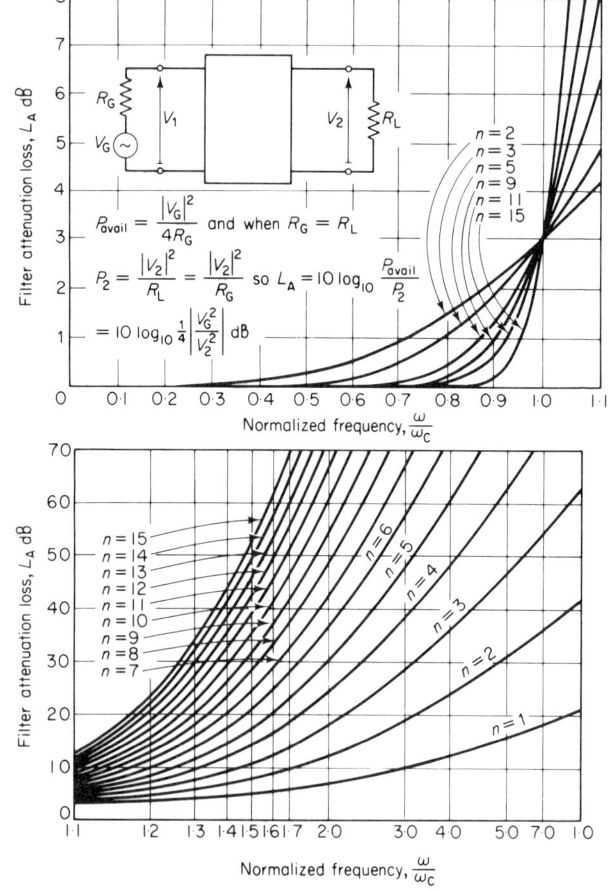

Fig. 5.13. Attenuation loss versus frequency characteristics of Butterworth type filters with a 3 dB band edge frequency attenuation

154

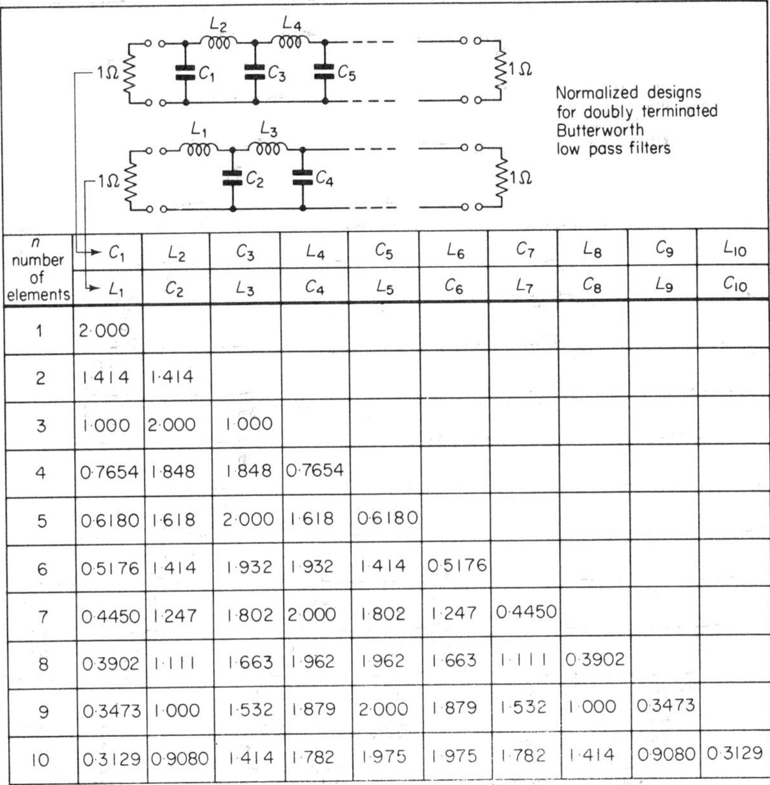

n number of elements	C_1 / L_1	L_2 / C_2	C_3 / L_3	L_4 / C_4	C_5 / L_5	L_6 / C_6	C_7 / L_7	L_8 / C_8	C_9 / L_9	L_{10} / C_{10}
1	2·000									
2	1·414	1·414								
3	1·000	2·000	1·000							
4	0·7654	1·848	1·848	0·7654						
5	0·6180	1·618	2·000	1·618	0·6180					
6	0·5176	1·414	1·932	1·932	1·414	0·5176				
7	0·4450	1·247	1·802	2·000	1·802	1·247	0·4450			
8	0·3902	1·111	1·663	1·962	1·962	1·663	1·111	0·3902		
9	0·3473	1·000	1·532	1·879	2·000	1·879	1·532	1·000	0·3473	
10	0·3129	0·9080	1·414	1·782	1·975	1·975	1·782	1·414	0·9080	0·3129

Fig. 5.14. Design data in normalized form for Butterworth filters working between 1Ω terminations, band edge frequency $\omega_C = 1$ rad/s, $L_{ref} = 3$ dB

fact that the first $(2n-1)$ derivatives of $|H(j\omega)|$ are zero at $\omega = 0$ thus providing maximal flatness in the pass-band.

Basing the design on the above transfer function, filters may be synthesized with a power loss ratio*

$$\frac{P_{avail}}{P_2} = \left[1 + \varepsilon\left(\frac{\omega}{\omega_C}\right)^{2n}\right]$$

or as what will now be defined as the filter attenuation loss*

* The power loss ratio is defined in Chapter 4, section 1, para. 3. For equal generator and load resistive terminations the power loss ratio equals the power insertion loss ratio of a given 2-port network and hence for the case $R_G = R_L$, L_A is identical to the general definition of insertion loss. Note, however, for image filter design the attenuation coefficient α is defined specifically for image impedance terminations and hence only represents a measure of the filter insertion loss under these conditions.

$$L_A(\omega) = 10\log_{10}\left[1+\varepsilon\left(\frac{\omega}{\omega_C}\right)^{2n}\right] \text{ decibels}$$

where $\varepsilon = \text{antilog } (L_{ref}/10) - 1$

$\omega = $ radian frequency

$\omega_C = $ pass band edge (cut-off) radian frequency

where $L_A(\omega_C) = L_{ref}$, in most cases

L_{ref} is taken as 3 dB so $\varepsilon = 1$

$n = $ number of reactive elements in the filter network.

Curves showing the form of the maximally flat characteristic are plotted in fig. 5.13 for the case $L_{ref} = 3$ dB, $\varepsilon = 1$, i.e. a 3 dB point at the band edge frequency.

Normalized designs derived using network synthesis methods are summarized in the tables of fig. 5.14 and 5.15. In fig. 5.14 the data refers to doubly-terminated filters with a band edge attenuation at

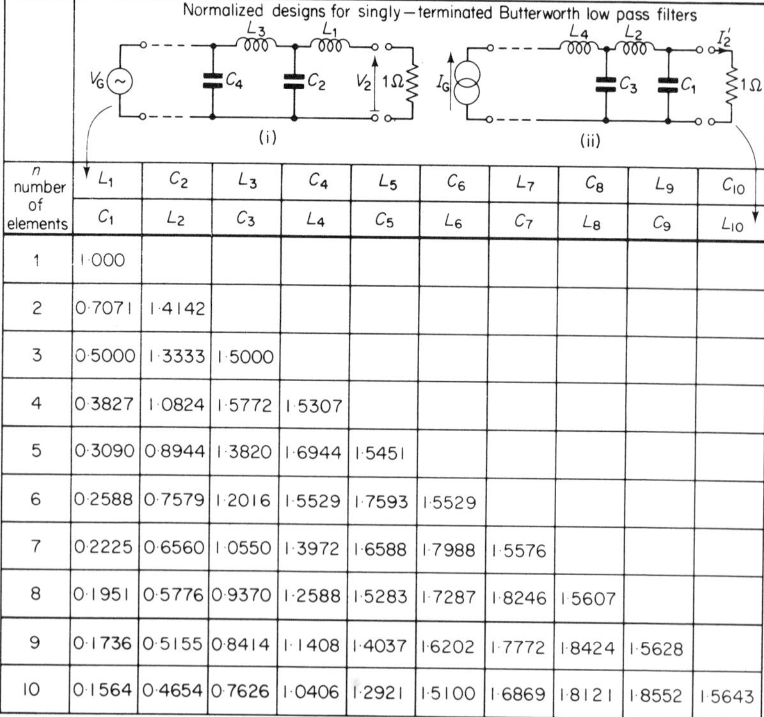

n number of elements	L_1 C_1	C_2 L_2	L_3 C_3	C_4 L_4	L_5 C_5	C_6 L_6	L_7 C_7	C_8 L_8	L_9 C_9	C_{10} L_{10}
1	1·000									
2	0·7071	1·4142								
3	0·5000	1·3333	1·5000							
4	0·3827	1·0824	1·5772	1·5307						
5	0·3090	0·8944	1·3820	1·6944	1·5451					
6	0·2588	0·7579	1·2016	1·5529	1·7593	1·5529				
7	0·2225	0·6560	1·0550	1·3972	1·6588	1·7988	1·5576			
8	0·1951	0·5776	0·9370	1·2588	1·5283	1·7287	1·8246	1·5607		
9	0·1736	0·5155	0·8414	1·1408	1·4037	1·6202	1·7772	1·8424	1·5628	
10	0·1564	0·4654	0·7626	1·0406	1·2921	1·5100	1·6869	1·8121	1·8552	1·5643

Fig. 5.15. Design data for singly-terminated Butterworth filters. Load = 1 Ω, (i) constant voltage source (zero impedance), (ii) constant current source (infinite impedance); band edge frequency $\omega_C = 1$ rad/s at which insertion loss,

$$20\log_{10}|V_G/V_2| = 20\log_{10}|I_G/I_2| = 3 \text{ dB}$$

156

$\omega_C = 1$ rad/s of 3 dB. In fig. 5.15 the designs refer to singly-terminated filters with a load of $1\,\Omega$ fed either from a constant voltage (zero impedance) or a constant current (infinite impedance) source.

10. THE TCHEBYSCHEFF (EQUAL RIPPLE) APPROXIMATION FOR LOW-PASS FILTERS

The maximally flat approximation to a low pass filter is best as ω approaches zero, but becomes successively poorer as ω approaches the band edge frequency. In the Tchebyscheff approximation a maximum

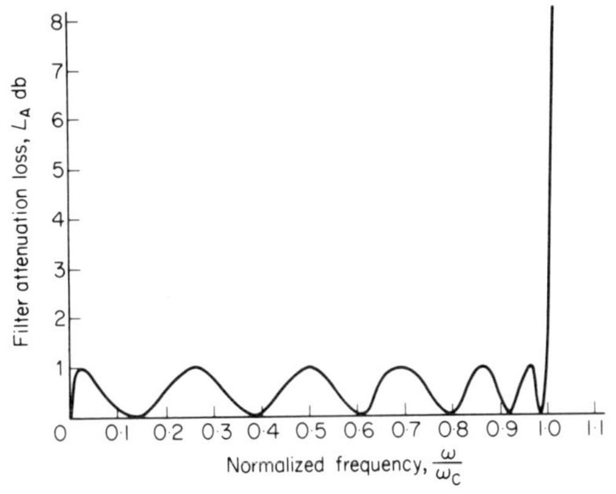

(a) Attenuation characteristic of a 12 element Tchebyscheff filter with a 1 db ripple in the pass band $- \left[R_G = R_0,\ R_L = 2{\cdot}66\,R_0 \right]$

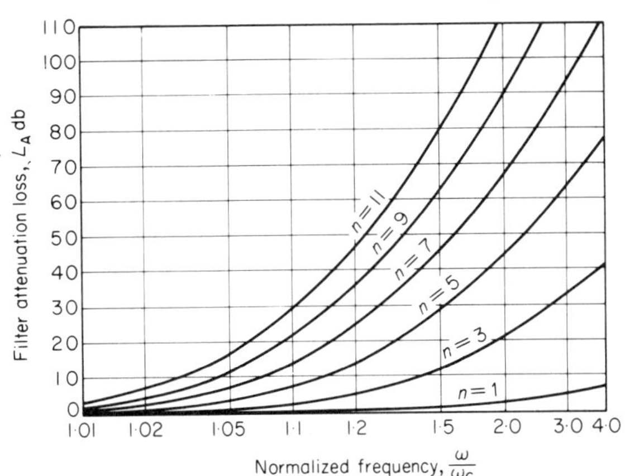

(b) Attenuation characteristics of $n = 1, 3, 5, 7, 9, 11$ doubly–terminated Tchebyscheff filters with a 1 db ripple and band edge frequency point

Fig. 5.16

permissible attenuation loss L_{ref} in the pass band is specified, and although the attenuation varies between 0 and L_{ref} it never exceeds the latter. The Tchebyscheff approximation ensures equally good performance as ω approaches zero and the band edge frequency, and the attenuation characteristic also climbs rapidly in the stop band producing a much sharper pass band edge than a Butterworth filter of the same number of elements.

The form of the transfer function from which Tchebyscheff filters are derived is

$$|H(j\omega')| = \frac{K}{[1 + \varepsilon C_n^2(\omega')]^{1/2}}$$

where $\omega' = \omega/\omega_C$ is the normalized frequency

ω_C = the band edge frequency where loss $= L_{ref}$ dB

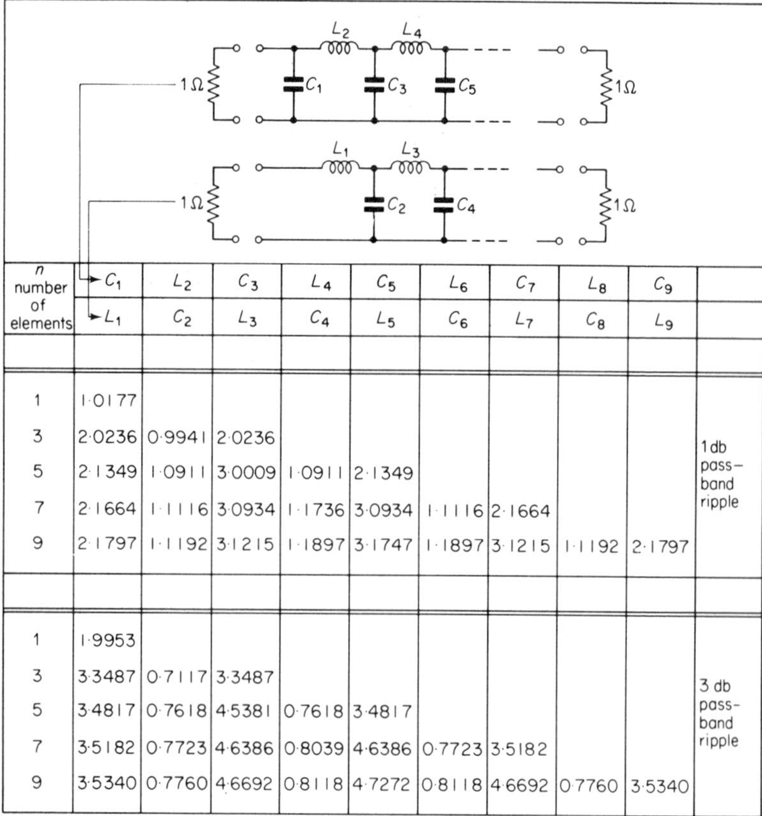

n number of elements	C_1 / L_1	L_2 / C_2	C_3 / L_3	L_4 / C_4	C_5 / L_5	L_6 / C_6	C_7 / L_7	L_8 / C_8	C_9 / L_9	
1	1·0177									
3	2·0236	0·9941	2·0236							1 db pass-band ripple
5	2·1349	1·0911	3·0009	1·0911	2·1349					
7	2·1664	1·1116	3·0934	1·1736	3·0934	1·1116	2·1664			
9	2·1797	1·1192	3·1215	1·1897	3·1747	1·1897	3·1215	1·1192	2·1797	
1	1·9953									
3	3·3487	0·7117	3·3487							3 db pass-band ripple
5	3·4817	0·7618	4·5381	0·7618	3·4817					
7	3·5182	0·7723	4·6386	0·8039	4·6386	0·7723	3·5182			
9	3·5340	0·7760	4·6692	0·8118	4·7272	0·8118	4·6692	0·7760	3·5340	

Fig. 5.17. Design data in normalized form for Tchebyscheff filters working between 1 Ω terminations, band edge frequency $\omega_C = 1$ rad/s, pass-band ripples 1 dB and 3 dB

158

| Normalized designs for singly-terminated Tchebyscheff filters | | | | | | | | | |
| n number of elements | | | | | | | | | |
L_1 / C_1	C_2 / L_2	L_3 / C_3	C_4 / L_4	L_5 / C_5	C_6 / L_6	L_7 / C_7	C_8 / L_8	L_9 / C_9	C_{10} / L_{10}
1 0·5088									
2 0·9110	0·9957								
3 1·0118	1·3332	1·5088							
4 1·0495	1·4126	1·9093	1·2817						
5 1·0674	1·4441	1·9938	1·5908	1·6652					
6 1·0773	1·4601	2·0270	1·6507	2·0491	1·3457				
7 1·0832	1·4694	2·0437	1·6736	2·1192	1·6489	1·7118			
8 1·0872	1·4751	2·0537	1·6850	2·1453	1·7021	2·0922	1·3691		
9 1·0899	1·4790	2·0601	1·6918	2·1583	1·7213	2·1574	1·6707	1·7317	
10 1·0918	1·4817	2·0645	1·6961	2·1658	1·7306	2·1803	1·7215	2·1111	1·3801

Fig. 5.18. Design data for singly-terminated Tchebyscheff filters. Load $= 1\,\Omega$, $\omega_\mathrm{C} = 1$ rad/s, pass-band ripple $= 1$ dB, (i) constant voltage source, (ii) constant current source

$$C_\mathrm{n} = \cos\left(n\cos^{-1}\omega'\right) \text{ is the Tchebyscheff}$$
$$\text{cosine polynomial of order } n$$
$$n = \text{number of elements}$$
$$\varepsilon = \text{antilog}_{10}\left(L_\mathrm{ref}/10\right) - 1$$

giving a filter attenuation loss characteristic:

$$L_\mathrm{A}(\omega') = 10\log_{10}\left[1 + \varepsilon\cos^2\left\{n\cos^{-1}\omega'\right\}\right] \text{dB}, \quad \omega' \leqslant 1$$

$$= 10\log_{10}\left[1 + \varepsilon\cosh^2\left\{n\cosh^{-1}\omega'\right\}\right] \text{dB}, \quad \omega' > 1$$

Curves showing the form of the attenuation characteristic are plotted in fig. 5.16, for the case of a 1 dB pass band ripple.

Normalized designs for doubly and singly terminated Tchebyscheff filters are summarized in the tables of fig. 5.17 and fig. 5.18, respectively. Note in the doubly terminated case filters containing an even number of elements are omitted since these are not realizable for equal value terminations.

5.2. Worked Problems

1. By reference to the infinite ladder network of fig. 5.19(a) show that the propagation constant γ of an individual Π section is given by $\gamma = \cosh^{-1}\{1 + Y_2/(2Y_1)\}$. Hence find the phase constant β in the pass-band and the attenuation constant α in the attenuation band of the filter section of fig. 5.19(b). Evaluate α at $\omega = \tfrac{1}{2}\omega_C$, where ω_C is the cut-off radian frequency.

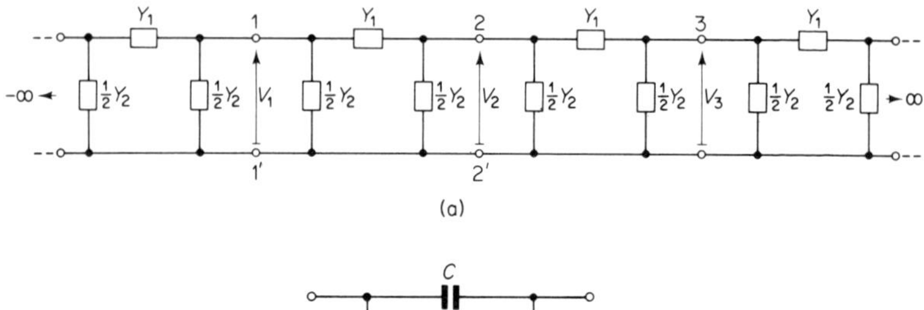

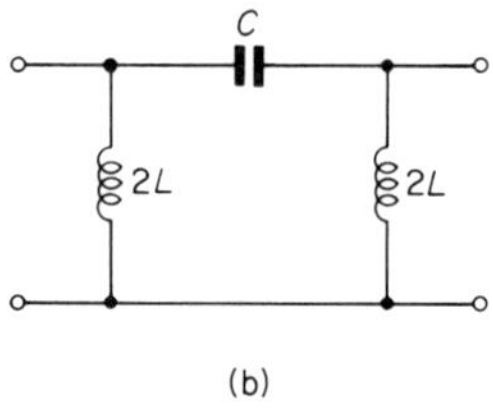

(b)

Fig. 5.19

Solution

On applying the current law at node 2 we obtain

$$(2Y_1 + Y_2)V_2 - Y_1 V_1 - Y_1 V_3 = 0$$

and dividing throughout by V_2,

$$(2Y_1 + Y_2) - Y_1(V_1/V_2 + V_3/V_2) = 0 \qquad \dots (1)$$

but by the definition of the propagation constant,

$$V_1/V_2 = e^{\gamma}, \quad V_3/V_2 = e^{-\gamma}$$

and so (1) provides the relation:

$$Y_1(e^{\gamma} + e^{-\gamma}) = 2Y_1 + Y_2$$

i.e. $Y_1 2\cosh\gamma = 2Y_1 + Y_2$

$$\cosh\gamma = 1 + Y_2/(2Y_1), \quad \gamma = \cosh^{-1}\{1 + Y_2/(2Y_1)\} \qquad \dots (2)$$

On substituting $Y_1 = j\omega C$, $Y_2 = -j/(\omega L)$ in (2) we have

$$\cosh\gamma = \cosh(\alpha + j\beta) = 1 - 1/(2\omega^2 LC) \qquad \dots (3)$$

160

but $\cosh(\alpha+j\beta) = \cosh\alpha\cos\beta + j\sinh\alpha\sin\beta$

and so on equating respectively real and imaginary parts of (3):

$$\cosh\alpha\cos\beta = 1 - 1/(2\omega^2 LC) \qquad \ldots 4(a)$$

$$\sinh\alpha\sin\beta = 0 \qquad \ldots 4(b)$$

In the pass-band $\alpha = 0$, $\cosh\alpha = \cosh 0 = 1$ so 4(a) reduces to

$$\cos\beta = 1 - 1/(2\omega^2 LC), \ \beta = \cos^{-1}\{1 - 1/(2\omega^2 LC)\} \text{ radians}$$
$$\ldots (5)$$

and, since $-1 \leqslant \cos\beta \leqslant +1$, (5) shows that the pass-band is defined by

$$-1 \leqslant 1 - 1/(2\omega^2 LC), \text{ i.e. } \omega \geqslant \omega_C = 1/\{2\sqrt{(LC)}\}$$

At cut-off $\beta = -\pi$ and this value also satisfies 4(b) in the attenuation band. Thus substituting $\cos\beta = \cos -\pi = -1$ in 4(a) we have

$$-\cosh\alpha = 1 - 1/(2\omega^2 LC) = 1 - 2\omega_C^2/\omega^2$$

i.e. $\cosh\alpha = 2\omega_C^2/\omega^2 - 1$, $\alpha = \cosh^{-1}\{2\omega_C^2/\omega^2 - 1\}$ nepers $\quad \ldots (6)$

On substituting $\omega = \tfrac{1}{2}\omega_C$ in (6) we have

$$\cosh\alpha = \tfrac{1}{2}(e^\alpha + e^{-\alpha}) = 7$$

i.e. $e^{2\alpha} - 14 e^\alpha + 1 = 0$

so $e^\alpha = \tfrac{1}{2}\{14 \pm \sqrt{(14^2 - 4)}\}$ yielding $\alpha = 2{\cdot}634$ nepers

2. Fig. 5.20 shows the low-pass constant k T section in normalized form, i.e. the filter has a design impedance of $1\,\Omega$ and a cut-off frequency of $1\,\text{rad/s}$. Using the results given in the frequency translation table of fig. 5.11 design the following filters:

   (a) a band-pass filter with cut-off frequencies $20\,\text{kHz}$ and $24\,\text{kHz}$ and design impedance $600\,\Omega$,

   (b) a band-stop filter to reject frequencies in the range $50\,\text{kHz}$ to $60\,\text{kHz}$, with a design impedance of $50\,\Omega$.

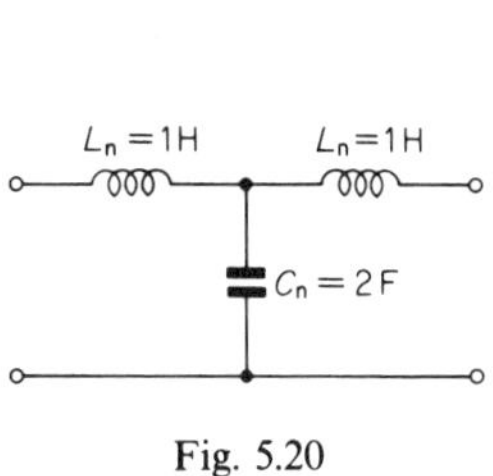

Fig. 5.20

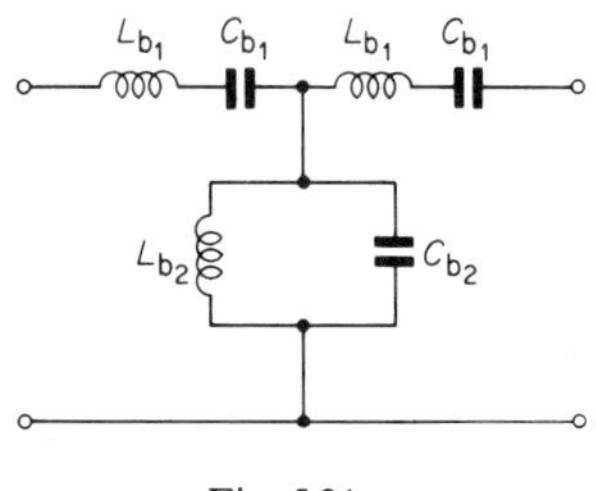

Fig. 5.21

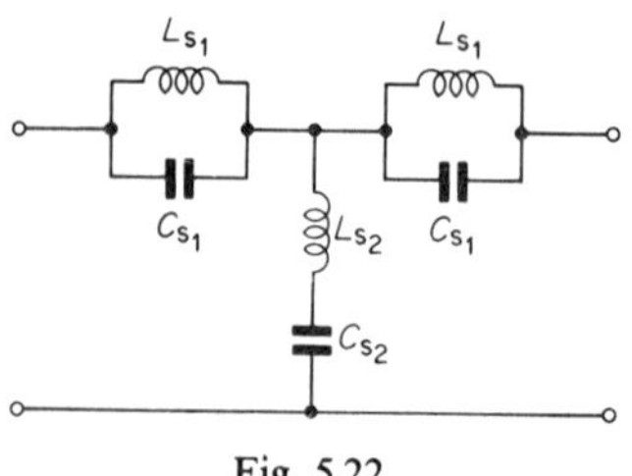

Fig. 5.22

Solution

(a) In the band-pass case the series elements are translated into a series *L–C* network where

$$L_{b_1} = \frac{L_n}{BW}\,R_0 = \frac{1}{2\pi(24-20)10^3} \times 600 = 0{\cdot}02387\,\text{H} = 23{\cdot}87\,\text{mH}$$

$$C_{b_1} = \frac{BW}{L_n\omega_0^2 R_0} = \frac{BW}{L_n\omega_{C_1}\omega_{C_2}R_0} = \frac{4000}{2\pi \times 20 \times 24 \times 10^6 \times 600}\,\text{F} = 2210\,\text{pF}$$

whilst the shunt C_n element is translated into a parallel *L–C* network where

$$L_{b_2} = \frac{BW}{C_n\omega_0^2}\,R_0 = \frac{BW\,R_0}{C_n\omega_{C_1}\omega_{C_2}} = 2{\cdot}211\,\text{mH}$$

$$C_{b_2} = \frac{C_n}{BW\,R_0} = 0{\cdot}1326\,\mu\text{F}$$

The full band-pass filter is shown in fig. 5.21.

(b) The band-stop filter elements are shown in fig. 5.22. Their values are

$$L_{S_1} = \frac{L_n\,BW\,R_0}{\omega_0^2} = \frac{L_n\,BW\,R_0}{\omega_{C_1}\omega_{C_2}} = \frac{(60-50)10^3 \times 50}{2\pi \times 50 \times 60 \times 10^6} = 26{\cdot}53\,\mu\text{H}$$

$$C_{S_1} = \frac{1}{L_n\,BW\,R_0} = 0{\cdot}3183\,\mu\text{F}$$

$$L_{S_2} = \frac{R_0}{C_n\,BW} = 0{\cdot}3979\,\text{mH}$$

$$C_{S_2} = \frac{C_n\,BW}{\omega_{C_1}\omega_{C_2}R_0} = 0{\cdot}0212\,\mu\text{F}$$

3. The diagram of figure 5.23 shows a multiple section low-pass filter designed to work between 75 Ω terminations with a cut-off frequency

of 200 kHz. Describe the function of each section and calculate the values of L and C for the constant k-section. Draw also network diagrams of the m-derived full section and of the input and output $m = 0.6$ half sections.

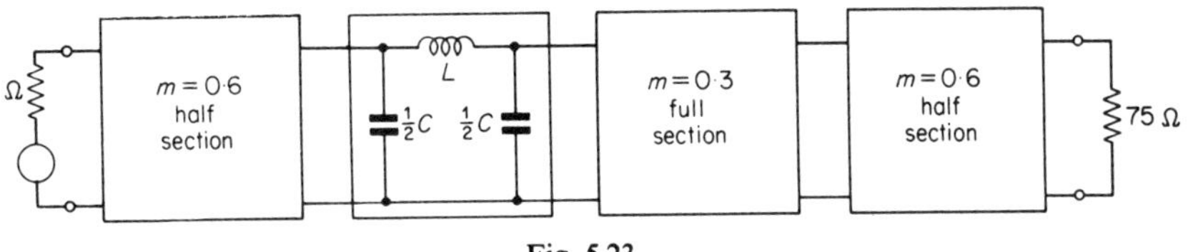

Fig. 5.23

Solution

The functions of the sections are described briefly in section 1, para. 5 of the theory summary.

The values of L and C for the constant k-section may be derived as follows. The characteristic impedance,

$$Z_C = Z_1 Z_2 (\tfrac{1}{4} Z_1^2 + Z_1 Z_2)^{-1/2}$$
$$= (L/C)(-\tfrac{1}{4}\omega^2 L^2 + L/C)^{-1/2}$$
$$= \sqrt{(L/C)}(1 - \omega^2 LC/4)^{-1/2} \text{ as } Z_1 = j\omega L, \ Z_2 = 1/(j\omega C)$$

The design impedance $R_0 =$ value of Z_C well into pass-band

$$= \lim_{\omega \to 0} Z_C = \sqrt{(L/C)} \qquad \ldots (1)$$

and the cut-off frequency (occurring when Z_C changes from a real to an imaginary quantity) is found from

$$1 - \omega^2 LC/4 = 0 \text{ i.e. } \omega_C = 2/\sqrt{(LC)} \qquad \ldots (2)$$

Thus solving for L and C using (1) and (2) we obtain:

$$L = \frac{2R_0}{\omega_C} = \frac{R_0}{\pi f_C} = 119.4\,\mu\text{H}$$

$$C = \frac{2}{\omega_C R_0} = \frac{1}{\pi f_C R_0} = 21221\,\text{pF}$$

as $f_C = 200\,\text{kHz}, \ R_0 = 75\,\Omega$.

Diagrams of the m-derived full section and $m = 0.6$ matching sections are shown in fig. 5.24.

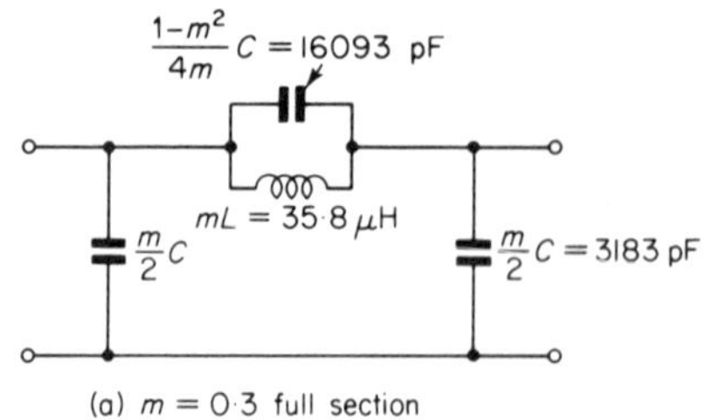

(a) $m = 0\cdot3$ full section

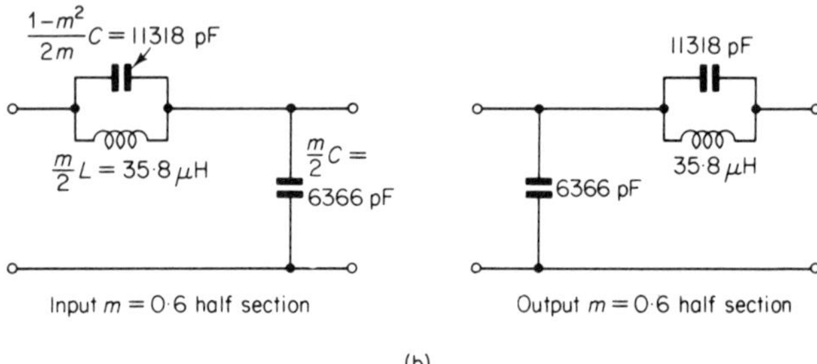

Input $m = 0\cdot6$ half section   Output $m = 0\cdot6$ half section

(b)

Fig. 5.24

4. Design a 5-element maximally flat low-pass filter working between $600\,\Omega$ terminations with a 3 dB band edge frequency of 20 kHz. Calculate the filter attenuation loss at 10 kHz, 30 kHz and 40 kHz.

Solution

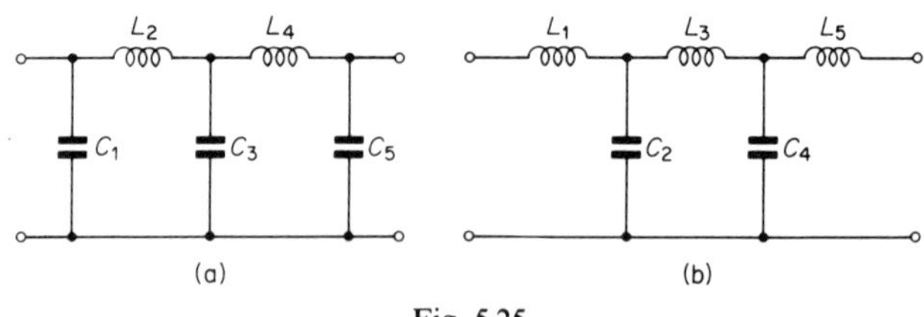

Fig. 5.25

The normalized values for the elements of the maximally flat filter, obtained from the table of fig. 5.14 are for filter fig. 5.25(a):

$$C_1 = 0\cdot6180, \; L_2 = 1\cdot6180, \; C_3 = 2,$$

$$L_4 = 1\cdot6180, \; C_5 = 0\cdot6180$$

and on denormalizing by multiplying C elements by $1/(R_0\,\omega_C)$ and L

164

elements by R_0/ω_C where $R_0 = 600\,\Omega$, $\omega_C = 2\pi \times 20\,000$ rad/s we obtain the values:

$$C_1 = 8196 \cdot 5\,\text{pF}, \quad L_2 = 7 \cdot 725\,\text{mH}, \quad C_3 = 0 \cdot 026526\,\mu\text{F},$$

$$L_4 = 7 \cdot 725\,\text{mH}, \quad C_5 = 8196 \cdot 5\,\text{pF};$$

whilst for the alternative form of filter shown in fig. 5.25(b): the normalized values are,

$$L_1 = 0 \cdot 6180, \quad C_2 = 1 \cdot 6180, \quad L_3 = 2,$$

$$C_4 = 1 \cdot 6180, \quad L_5 = 0 \cdot 6180$$

and the actual values on denormalization:

$$L_1 = 2 \cdot 951\,\text{mH}, \quad C_2 = 0 \cdot 02146\,\mu\text{F}, \quad L_3 = 9 \cdot 549\,\text{mH},$$

$$C_4 = 0 \cdot 02146\,\mu\text{F}, \quad L_5 = 2 \cdot 951\,\text{mH}$$

The filter attenuation loss,

$$L_A(\omega) = 10\log_{10}\left[1 + \left(\frac{\omega}{\omega_C}\right)^{10}\right]\text{dB}$$

has the values $0 \cdot 004$ dB, $17 \cdot 68$ dB, $30 \cdot 11$ dB at $10\,\text{kHz}$, $30\,\text{kHz}$, $40\,\text{kHz}$ respectively.

5. Show that the single-terminated filter network of fig. 5.26 has a Butterworth characteristic when fed from a source of infinite impedance and terminated in a $1\,\Omega$ load.

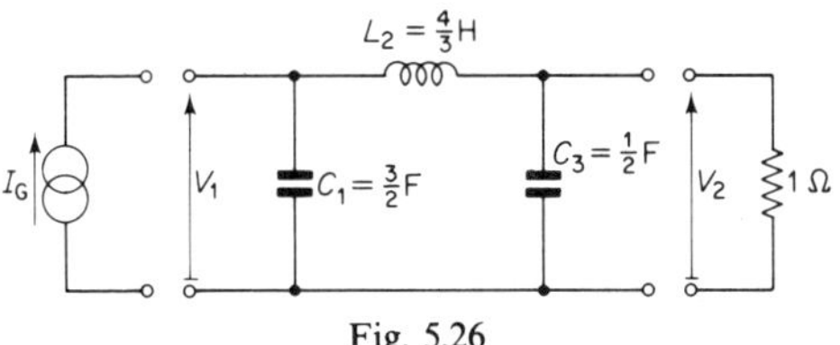

Fig. 5.26

Solution
This may be accomplished by showing that the transfer function $H(s) = V_2/I_G$ is of the form $|H(\mathrm{j}\omega)|^2 = [1 + (\omega^2)^3]^{-1}$. The transfer function may be found from the nodal equations of the circuit which are:

$$[sC_1 + 1/(sL_2)]V_1 - 1/(sL_2)V_2 = I_G$$

$$-1/(sL_2)V_1 + [sC_3 + 1/(sL_2) + 1]V_2 = 0$$

and solving for V_2, i.e.

$$V_2 = \frac{\{1/(sL_2)\}I_G}{[sC_1+1/(sL_2)][sC_3+1/(sL_2)+1]-[1/(sL_2)]^2}$$

$$= \frac{I_G}{s^3(L_2C_1C_3)+s^2(C_1L_2)+(C_1+C_3)s+1}$$

so $H(s) = V_2/I_G = [s^3+2s^2+2s+1]^{-1}$ on substituting for C_1, L_2, C_3.

Further $H(j\omega) = [(1-2\omega^2)+j(2\omega-\omega^3)]^{-1}$ on substituting $s = j\omega$,

and $|H(j\omega)|^2 = [(1-2\omega^2)^2+(2\omega-\omega^3)^2]^{-1}$
$$= (1+\omega^6)^{-1}$$

6. Fig. 5.27(a) shows a 3-element Tchebyscheff low-pass filter which has a 1 dB ripple in the pass-band and a band edge frequency of 1 rad/s when working between 1 Ω terminations. Using this data design,
   (a) a low-pass filter with a 1 dB pass-band ripple and a 10 kHz band edge frequency when working between 600 Ω terminations,
   (b) a 10 kHz to 14 kHz band-pass filter with a 1 dB pass-band ripple when working between 600 Ω terminations.

   Determine also the filter attenuation loss in terms of the respective L–C elements and sketch the attenuation versus frequency characteristics.

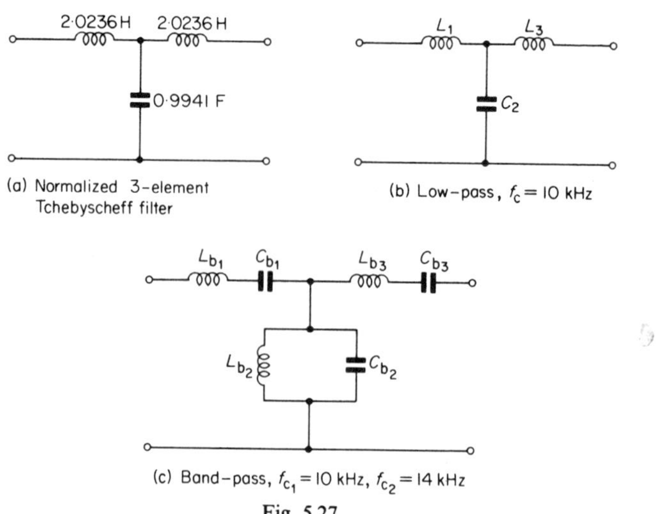

(a) Normalized 3-element Tchebyscheff filter

(b) Low-pass, $f_c = 10$ kHz

(c) Band-pass, $f_{c_1} = 10$ kHz, $f_{c_2} = 14$ kHz

Fig. 5.27

Solution

(a) The element values of the low-pass filter shown in fig. 5.27(b) are

$$L_1 = L_3 = 2{\cdot}0236 \times \frac{R_0}{\omega_C} = \frac{2{\cdot}0236 \times 600}{2\pi \times 10\,000} = 19{\cdot}324\,\text{mH}$$

$$C_2 = 0{\cdot}9941 \times \frac{1}{R_0\omega_C} = \frac{0{\cdot}9941}{600 \times 2\pi \times 10\,000} = 0{\cdot}02637\,\mu\text{F}$$

(b) Using the results of the frequency translation table of fig. 5.11, the elements of the band-pass filter shown in fig. 5.27(c) are:

$$L_{b_1} = L_{b_3} = \frac{2{\cdot}0236}{\text{BW}} \times R_0 = \frac{2{\cdot}0236 \times 600}{2\pi(14-10)1000} = 48{\cdot}31\,\text{mH}$$

$$C_{b_1} = C_{b_2} = \frac{\text{BW}}{2{\cdot}0236\omega_{C_1}\omega_{C_2}R_0}$$

$$= \frac{2\pi \times 4000}{2{\cdot}0236 \times 4\pi^2 \times 10 \times 14 \times 10^6 \times 600} = 3745\,\text{pF}$$

$$L_{b_2} = \frac{\text{BW} \times R_0}{0{\cdot}9941\omega_{C_1}\omega_{C_2}} = 2{\cdot}745\,\text{mH}, \quad C_{b_2} = \frac{0{\cdot}9941}{\text{BW} \times R_0} = 65923{\cdot}3\,\text{pF}$$

To determine the filter attenuation loss, consider the general symmetrical T terminated in R_0 ohms, shown in fig. 5.28.

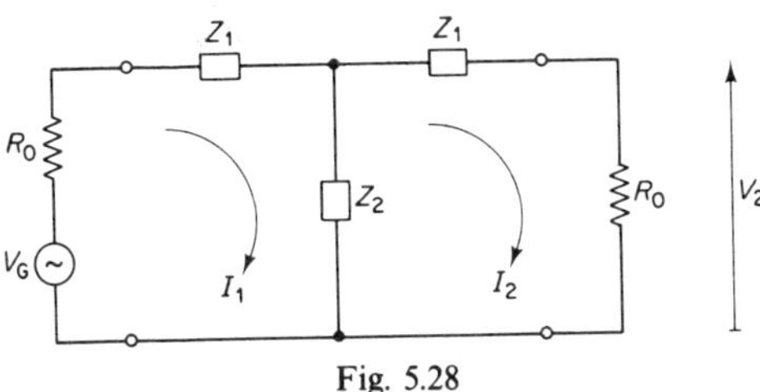

Fig. 5.28

The mesh equations of this circuit are:

$$(R_0 + Z_1 + Z_2)I_1 \qquad\qquad - Z_2 I_2 = V_G$$

$$- Z_2 I_1 + (Z_2 + Z_1 + R_0)I_2 = 0$$

$$\text{from which } I_2 = \frac{Z_2 V_G}{(R_0 + Z_1 + Z_2)^2 - Z_2^2}$$

$$= \frac{Z_2 V_G}{R_0^2 + 2R_0(Z_1 + Z_2) + Z_1^2 + 2Z_1 Z_2}$$

By definition the filter power loss is

$$\frac{P_{\text{avail}}}{P_2} = \left|\frac{V_{\text{G}}^2}{4R_0}\bigg/\frac{V_2^2}{R_0}\right| = \frac{1}{4}\left|\frac{V_{\text{G}}}{V_2}\right|^2 = \frac{1}{4}\left|\frac{V_{\text{G}}}{R_0 I_2}\right|^2$$

$$= \frac{1}{4}\left|\frac{R_0}{Z_2} + 2\frac{Z_1}{Z_2} + 2 + \frac{Z_1^2}{Z_2 R_0} + \frac{2Z_1}{R_0}\right|^2$$

$$= \tfrac{1}{4}|(2 - 2\omega^2 L_1 C_1) + j(\omega C_2 R_0 - \omega^3 L_1^2 C_2 / R_0 + 2\omega L_1 / R_0)|$$

for the low-pass filter of fig. 5.27(b) where $Z_1 = j\omega L_1$, $Z_2 = 1/(j\omega C_2)$; whilst for the band-pass filter of fig. 5.27(c), $Z_1 = j\{\omega L_{b_1} - 1/\omega C_{b_1}\}$, $Z_2 = j\omega L_{b_2}/(1 - \omega^2 L_{b_2} C_{b_2})$ and so

$$\frac{P_{\text{avail}}}{P_2} = \tfrac{1}{4}|\{2 + 2(\omega L_{b_1} - 1/\omega C_{b_1})(1 - \omega^2 L_{b_2} C_{b_2})/(\omega L_{b_2})\}$$

$$+ j\{2(\omega L_{b_1} - 1/\omega C_{b_1})^2(1 - \omega^2 L_{b_2} C_{b_2})/(R_0 \omega L_{b_2})$$

$$+ 2(\omega L_{b_1} - 1/\omega C_{b_1})/R_0 - R_0(1 - \omega^2 L_{b_2} C_{b_2})\}|^2$$

The filter attenuation loss, $L_{\text{A}} = 10 \log_{10}(P_{\text{avail}}/P_2)\,\text{dB}$, for the low- and band-pass filters is plotted in fig. 5.29(a), (b) respectively.

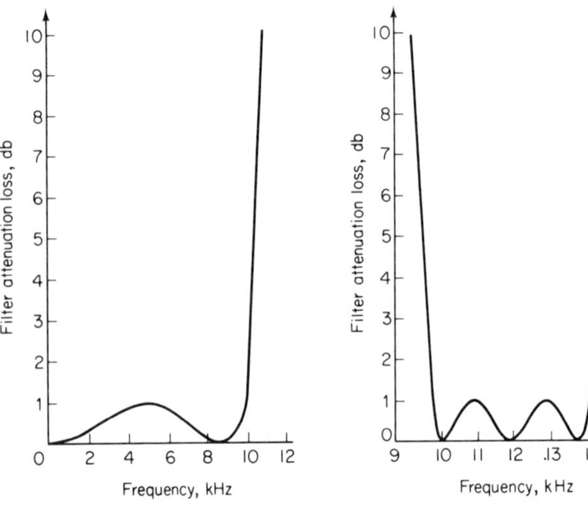

Fig. 5.29

168

5.3. Exercise Problems

Note. Questions 1 to 13 refer to constant k and m derived filters (section 1), whilst 14 to 23 refer to modern filter design (section 2).

1. Determine the pass-band and cut-off frequencies of the filter sections shown in fig. 5.30.

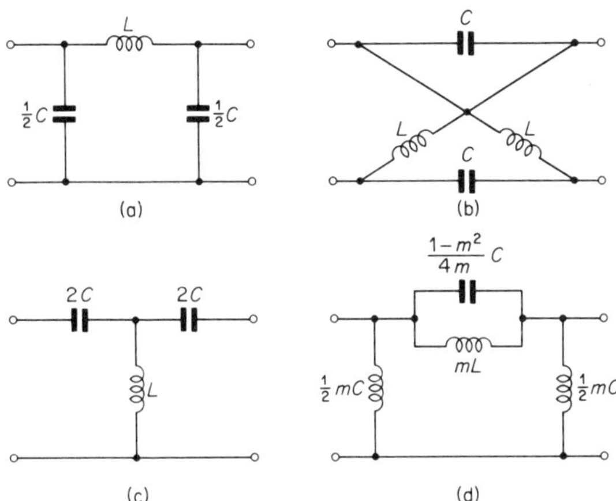

Fig. 5.30. For problem 1

2. Determine the cut-off frequencies of the pass-band and stop-band filters shown in fig. 5.31.

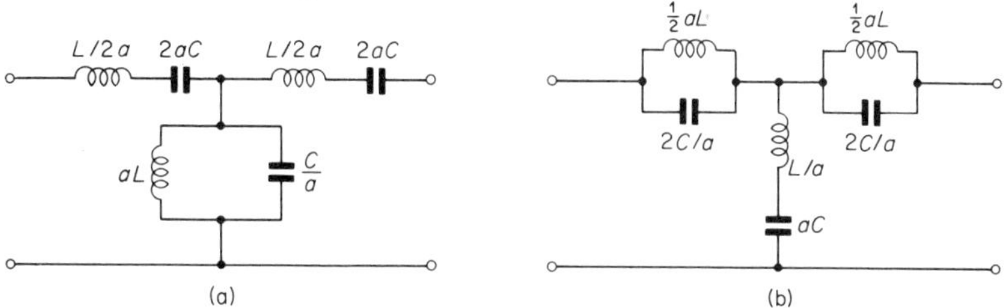

Fig. 5.31. For problem 2

3. If the symmetrical Π section of fig. 5.32 is a constant k-section, i.e. $X_1 X_2 = -k^2$ where k^2 is a real positive constant, show that the section possesses a pass-band provided $-2k < X_1 < 2k$.

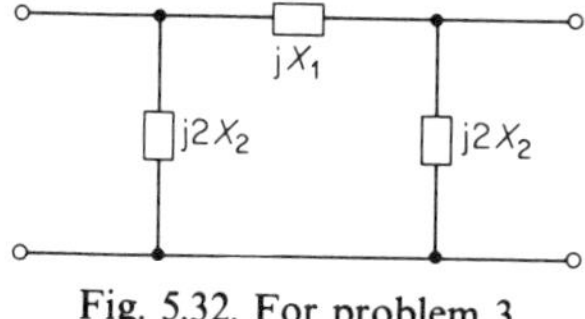

Fig. 5.32. For problem 3

4. Design a high pass constant k Π section with a cut-off frequency of 25 kHz and a design impedance of 600 Ω. Design also matching half sections so that the filter is well matched to 600 Ω terminations over the majority of the pass-band.

5. Design a low-pass m-derived T section with a cut-off frequency of 2·5 kHz, a frequency of infinite attenuation at 2·6 kHz, and a design impedance of 600 Ω.

6. In a certain measurement system working with coaxial line of 70 Ω characteristic impedance a 30 dB attenuator pad and a constant k low-pass filter with a 200 kHz cut-off frequency are required. Design these components deriving any formulae used.

7. Design an m-derived low-pass T filter section with a design impedance of 600 Ω, a cut-off frequency of 15 kHz, and a frequency of infinite attenuation at 15·75 kHz. At what frequency will the section match 300 Ω terminations.

8. Each arm of a symmetrical T low-pass filter consists of a 6 mH inductor, whilst the shunt arm is a 0·03 μF capacitor. Calculate the design impedance and cut-off frequency of the section and sketch the insertion loss versus frequency characteristic for the case of characteristic impedance and design impedance terminations. Suppose it is now required that the filter should produce infinite attenuation at a frequency 5% above cut-off. Redesign the filter so that this may be theoretically achieved.

9. The constant k filter section of fig. 5.33 is terminated in (a) its characteristic impedance (b) its design impedance, $R_0 = \sqrt{(L/C)}$. Determine the insertion phase shift and the input impedance at the cut-off frequency $\omega_C = 2/\sqrt{(LC)}$ for these two cases.

10. If the section of fig. 5.33 is modified by the introduction of mutual magnetic coupling between the two inductors, show that the resulting network is equivalent to a series m-derived T section with $m = \sqrt{[(1-k)/(1+k)]}$, where k is the coefficient of coupling between the two inductors.

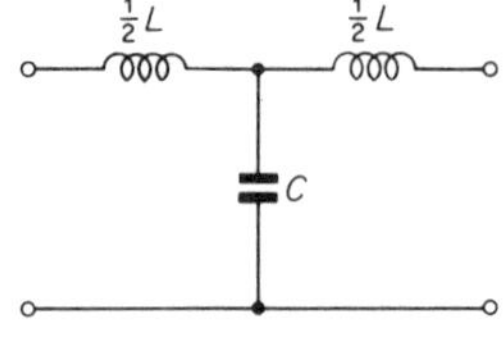

Fig. 5.33. For problems 9 and 10

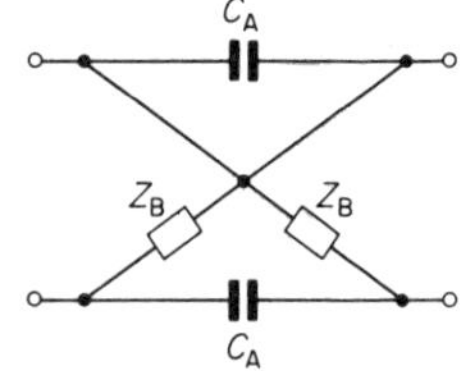

Fig. 5.34. For problem 11

11. Calculate the component values of the symmetrical lattice shown in
fig. 5.34 such that the section acts as a high-pass filter with a cut-off
frequency of 3 MHz and which is to be used in a 600 Ω balanced
transmission line system.

12. The diagram of fig. 5.35 shows a composite low-pass filter designed
to work between 600 Ω terminations with a cut-off frequency of
10 kHz. Describe in full the function of each section and calculate
the values of L and C for the constant k-section. Draw also network
diagrams of the m-derived full section and of the input and output
$m = 0.6$ half sections.

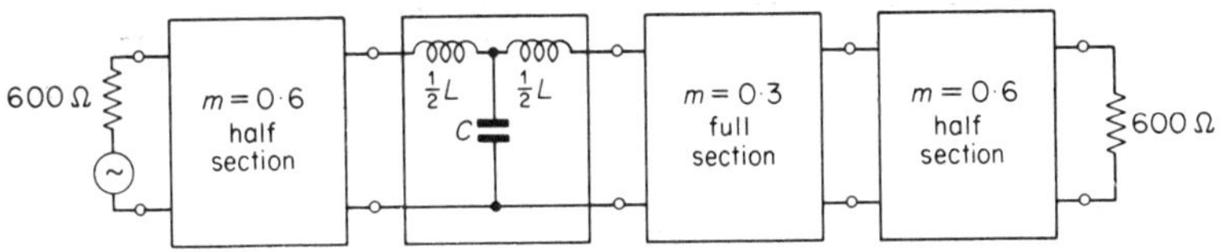

Fig. 5.35. For problem 12

13. Determine the insertion loss ratio of the symmetrical T, shown in
fig. 5.36(a), when it is connected between a generator of internal
resistance R_0 ohms and a load of the same value. Hence show that if
$R_0 = \sqrt{(L/C)}$ ohms the insertion loss of the high-pass filter section
of fig. 5.36(b) working between R_0 terminations is
$10 \log_{10}[1 + (\omega_C/\omega)^6]$ dB where $\omega_C = 1/[2\sqrt{(LC)}] \leqslant \omega$.

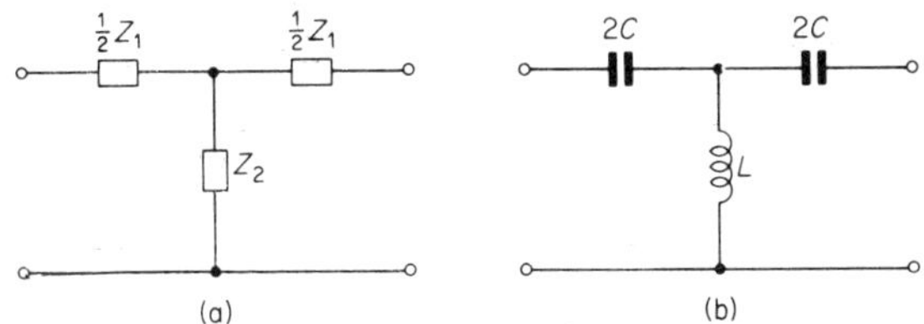

Fig. 5.36. For problem 13

14. Design a 3-element low-pass filter with a maximally flat characteristic and a 3 dB band edge frequency at 50 kHz, when working
between 70 Ω resistive terminations. Calculate also the attenuation
loss produced by the filter at 25 kHz and 75 kHz.

15. Design a 7-element low-pass Butterworth filter with a 3 dB band
edge frequency at 100 kHz when working between 50 Ω terminations. What attenuation loss would this filter produce at 230 kHz.

16. Design a 3-element Tchebyscheff low-pass filter with a maximum attenuation loss of 1 dB in the pass-band and a band edge frequency of 20 kHz when working between 600 Ω terminations. Calculate the attenuation loss of the filter at 40 kHz.

17. Design the following Tchebyscheff filters:
    (a) a 5-element high-pass with a band edge frequency of 30 kHz, a 3 dB pass-band ripple when working between 75 Ω terminations,
    (b) a 10-element band-pass with pass band 200 kHz to 250 kHz, a 1 dB pass-band ripple when working between 50 Ω terminations.

18. It is desired to design a low-pass Butterworth filter which produces an attenuation loss of at least 30 dB at a frequency 50% above a 3 dB band edge frequency. Evaluate the minimum number of elements required and calculate the maximum loss of such a filter in the pass-band between the range 0 to 90% of the band edge frequency.

19. Design a 6-element Tchebyscheff band-stop filter to work between 600 Ω terminations, to have a maximum pass-band ripple of 3 dB with band edge frequencies of 8 kHz and 12 kHz. Calculate at what frequency within the rejection band the filter produces infinite attenuation.

20. Show that the transfer impedance $Z_{21}(s) = V_2/I_G$ of the single-terminated $n = 3$ optimum filter shown in fig. 5.37 is given by

$$Z_{21} = \frac{0 \cdot 577}{s^3 + 1 \cdot 31 s^2 + 1 \cdot 359 s + 0 \cdot 577}$$

and calculate its insertion loss $(20 \log_{10} I_G/V_2)$ at $\omega = 1$ and $\omega = 2$ rad/s. How much greater is its insertion loss at $\omega = 2$ than a single-terminated $n = 3$ Butterworth filter.

21. Design a low-pass 5-element Butterworth filter with a 3 dB band edge frequency at 1 MHz to be used between a generator of zero internal impedance and a load of 50 Ω.

22. Design a low-pass Tchebyscheff filter to feed a 600 Ω load from a high impedance source and to have a 1 dB ripple in the pass-band, a

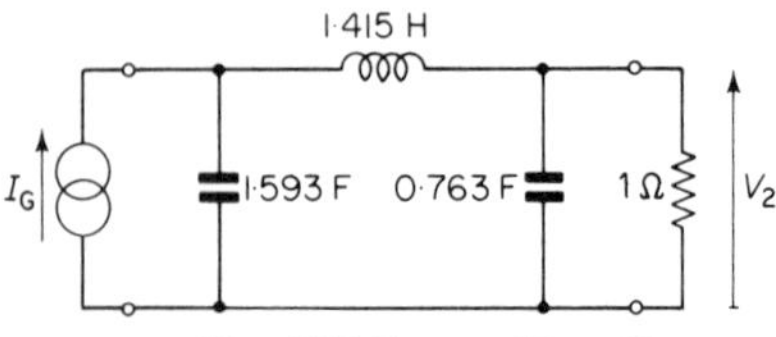

Fig. 5.37. For problem 20

172

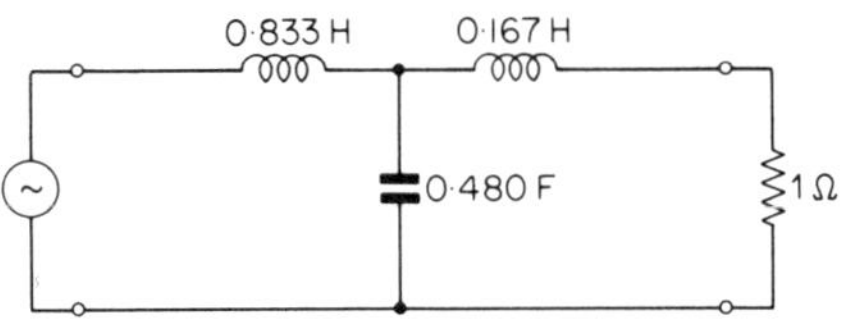

Fig. 5.38. For problem 23

band edge frequency of 500 kHz with the requirement that at 1000 kHz the filter attenuation is greater than 45 dB and that the filter contains the minimum number of L–C elements.

23. Fig. 5.38 shows in normalized form an $n = 3$ single terminated Bessel filter. Determine its insertion phase shift as a function of normalized frequency $\omega' = \omega/\omega_C$. What special property does this characteristic exhibit.

6 Introduction to Some Further Topics

A. INTRODUCTION TO PASSIVE NETWORK SYNTHESIS

B. THE SCATTERING MATRIX

C. INTRODUCTION TO STATE VARIABLE ANALYSIS

6A INTRODUCTION TO PASSIVE NETWORK SYNTHESIS

6.1A. Theory Summary

1. PHYSICAL REALIZABILITY: POSITIVE REAL FUNCTIONS

A driving point impedance or admittance of a passive 1-port (2-terminal) network may be specified in the general form (see also Chapter 1, para. 5) of

$$H(s) = Z(s) \text{ or } Y(s) = \frac{b_m s^m + b_{m-1} s^{m-1} + \ldots + b_1 s + b_0}{a_n s^n + a_{n-1} s^{n-1} + \ldots + a_1 s + a_0}$$

$$= \frac{b_m}{a_n} \frac{(s - z_1)(s - z_2) \ldots (s - z_m)}{(s - p_1)(s - p_2) \ldots (s - p_n)}$$

If this function is to be physically realizable in terms of R, L, C elements, $H(s)$ must be a positive real function. $H(s)$ is termed a positive real function when

(1) $H(s)$ is real when s is real

(2) $\text{Re}[H(s)] \geqslant 0$ when $\text{Re}(s) \geqslant 0$

A more practical and useful set of necessary and sufficient conditions for $H(s)$ to be positive real is

(1) $H(s)$ has no poles in the right-hand s-plane, i.e. its poles can only be of the form $-\alpha, \pm j\omega, -\alpha \pm j\omega$ where α and ω are real and positive.

(2) $H(s)$ may have only simple poles on the $j\omega$ axis with real positive residues,

e.g. (i) If $H(s) = H_0/(s^2 + \omega_1^2)^2$, then $H(s)$ has a double pole at $s = j\omega_1$ and $-j\omega_1$ and is therefore not a positive real function

174

(ii) If $H(s) = H_0 \dfrac{s^m + k_{m-1} s^{m-1} + \ldots + k_1 s + k_0}{(s-p_1)(s-p_2) \ldots (s-p_n)}$

$$= \frac{K_1}{(s-p_1)} + \frac{K_2}{(s-p_2)} + \ldots + \frac{K_n}{(s-p_n)}$$

and $p_1 = j\omega_1$, $p_2 = -j\omega_1$ then the residues K_1 and K_2 must be real and positive.

(3) $\mathrm{Re}\,[H(j\omega)] \geqslant 0$ for all ω.

Some further important properties of positive real functions are listed below:

(i) The sum of a number of positive real functions is positive real, e.g. if $Z_1(s)$, $Z_2(s)$ are positive real so is $Z(s) = Z_1(s) + Z_2(s)$.

(ii) If $H(s)$ is positive real so is $1/H(s)$.

(iii) If $H_1(s)$, $H_2(s)$ are positive real so is $H_1(s)H_2(s)/(H_1(s) + H_2(s))$.

(iv) The highest power of the numerator and denominator polynomials of $H(s)$ can at the most differ by unity. Similarly the lowest power terms can at the most only differ by one.

2. SYNTHESIS OF L–C 1-PORT NETWORKS

(a) *Foster's theorem*

The driving point immittance of a network composed solely of L–C elements may be expressed in the form

$$H(s) = H \frac{s(s^2 + \omega_1^2)(s^2 + \omega_3^2) \ldots}{(s^2 + \omega_0^2)(s^2 + \omega_2^2) \ldots} \qquad \ldots (1)$$

where H (the scale factor), ω_1, $\omega_3 \ldots$, $\omega_0 \geqslant 0$, $\omega_2 \ldots$ are positive real numbers. When $s = j\omega$

$$H(j\omega) = H \frac{j\omega(\omega_1^2 - \omega^2)(\omega_3^2 - \omega^2) \ldots}{(\omega_0^2 - \omega^2)(\omega_2^2 - \omega^2) \ldots} = jX(\omega)$$

The following important properties may be deduced:

(i) *Interlacing property.* The poles $(\pm j\omega_0, \pm j\omega_2, \ldots)$ and zeros $(\pm j\omega_1, \pm j\omega_3, \ldots)$ of $H(s)$ occur alternately along the $j\omega$ axis. In reactance (susceptance) versus frequency curve sketching this property is extremely useful since the poles $(\omega_0, \omega_2, \ldots)$ and zeros $(\omega_1, \omega_3, \ldots)$ of $X(\omega)$ must be interlaced along the ω frequency axis. The slope $\mathrm{d}X/\mathrm{d}\omega$ is always positive and $X(\omega)$ must change from $-$ to $+$ as the frequency is increased through a zero and jump discontinuously from $+\infty$ to $-\infty$ as the frequency is increased through a pole.

(ii) *Behaviour of $X(\omega)$ as ω approaches 0 and ∞. $X(\omega)$ must either rise from 0 or $-\infty$ as the frequency is increased from zero. As ω tends to infinity, $X(\omega)$ must either rise to 0 or $+\infty$.*

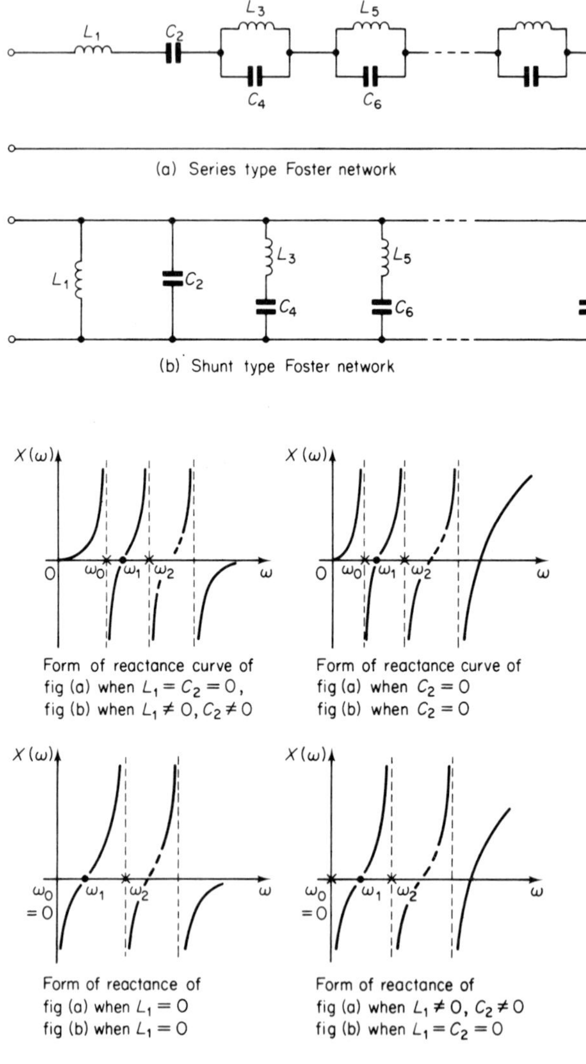

(a) Series type Foster network

(b) Shunt type Foster network

Form of reactance curve of
fig (a) when $L_1 = C_2 = 0$,
fig (b) when $L_1 \neq 0$, $C_2 \neq 0$

Form of reactance curve of
fig (a) when $C_2 = 0$
fig (b) when $C_2 = 0$

Form of reactance of
fig (a) when $L_1 = 0$
fig (b) when $L_1 = 0$

Form of reactance of
fig (a) when $L_1 \neq 0$, $C_2 \neq 0$
fig (b) when $L_1 = C_2 = 0$

(c) The 4-basic forms of reactance-frequency curves of $L-C$ networks

Fig. 6.1A

(iii) *The 4 basic forms of reactance (susceptance) curves.* As a consequence of (ii) and (iii) the reactance and susceptances curves of an *L–C* network must conform to one of the 4 basic form shown in fig. 6.1A(c).

(iv) The total number of internal poles and zeros of $X(\omega)$, i.e. excluding those occurring at $\omega = 0$ and as $\omega \to \infty$, is always at least one less than the number of *L–C* elements in the network. Thus in order to synthesize a given $X(\omega)$ of m internal poles and zeros, the minimum number of *L–C* elements required is $m+1$. Such networks containing no redundant elements are referred to as canonic networks. The series and shunt Foster, and Cauer ladder networks considered below are examples of canonic networks.

(b) *Series and shunt Foster L–C networks*

The component values of an *L–C* Foster network required to synthesize a given immittance function are obtained by expanding the general expression (1) in terms of a series of partial fractions, i.e.

$$H(s) = \frac{K_1}{s} + K_2 s + \frac{K_3 s}{s^2 + \omega_0^2} + \frac{K_4 s}{s^2 + \omega_2^2} + \cdots \qquad \cdots (2)$$

(noting that if $\omega_0 > 0$, $K_1 = 0$, whilst if $\omega_0 = 0$, $K_3 = 0$).

If $H(s) = Z(s)$ is regarded as a driving point impedance the method leads to the synthesis of series type Foster networks, shown in fig. 6.1A(a). The first two terms in (2) may be recognized respectively as a C element and an L element ($C = 1/K_1$ farads, $L = K_2$ henries), whilst each of the remaining terms may be identified as the impedance of a parallel *L–C* combination, i.e.

$$\frac{Ls \times 1/(Cs)}{Ls + 1/(Cs)} = \frac{(1/C)s}{s^2 + 1/(LC)} = \frac{Ks}{s^2 + \omega_r^2}$$

so $C = 1/K$, $\omega_r^2 = 1/(LC)$ and $L = 1/(\omega_r^2 C) = K/\omega_r^2$.

If $H(s) = Y(s)$ is taken as an admittance then the synthesis yields the shunt type networks shown in fig. 6.1A(b). In this case the first two terms respectively represent an L and a C element, whilst each of the succeeding terms may be identified as the admittance of a series *L–C* network, i.e.

$$\frac{1}{Ls + 1/(Cs)} = \frac{(1/L)s}{s^2 + 1/(LC)} = \frac{Ks}{s^2 + \omega_r^2}$$

so $L = 1/K$, $C = K/\omega_r^2$.

(c) *Cauer ladder networks (continued fraction expansions)*

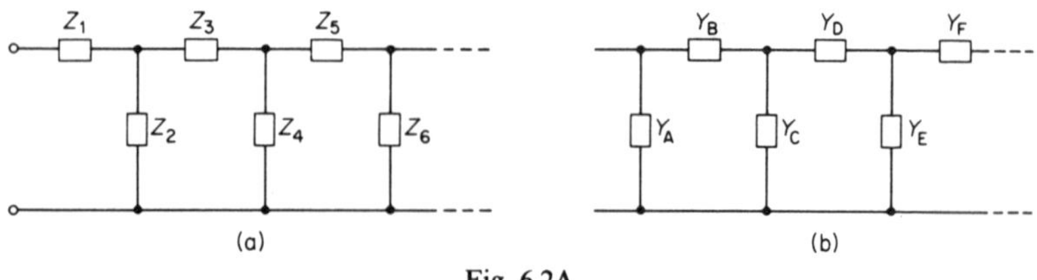

Fig. 6.2A

In the Cauer method of synthesis the immittance function is expanded in the form of a continued fraction. In this way the immittance may be synthesized by a ladder network.

For example the driving point impedance of the ladder network of fig. 6.2A(a) may be written in the continued fraction form of

$$Z(s) = Z_1 + \cfrac{1}{\cfrac{1}{Z_2} + \cfrac{1}{Z_3 + \cfrac{1}{\cfrac{1}{Z_4} + \cfrac{1}{Z_5 + \cfrac{1}{\cfrac{1}{Z_6} + \cdots}}}}}$$

and the driving point admittance of fig. 6.2A(b) as:

$$Y(s) = Y_A + \cfrac{1}{\cfrac{1}{Y_B} + \cfrac{1}{Y_C + \cfrac{1}{\cfrac{1}{Y_D} + \cfrac{1}{Y_E + \cfrac{1}{\cfrac{1}{Y_F} + \cdots}}}}}$$

Thus if the required immittance function is expressed in a similar form and compared term by term with the above expressions, the elements of a ladder network may be evaluated. A continued fraction method of L–C impedance synthesis is illustrated in worked problem 2.

3. SYNTHESIS OF R–C AND R–L 1-PORT NETWORKS

(a) *Series and shunt type Foster networks*

In a similar manner to Foster L–C network synthesis the driving point immittance is expanded in the form of a series of partial fractions in the following ways:

(i) *For the synthesis of series type R–C impedances and shunt type R–L admittances*

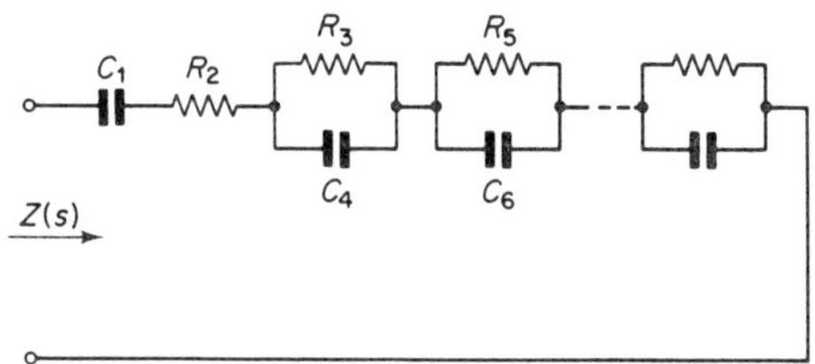
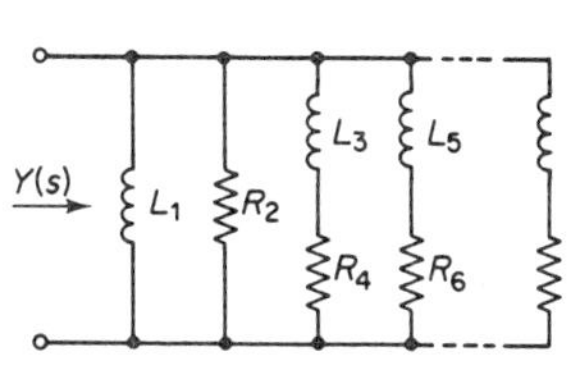

(a) Series type network for the synthesis of an R–C impedance

(b) Shunt type network for the synthesis of an R–L admittance

Fig. 6.3A

$$H(s) = \frac{K_1}{s} + K_2 + \frac{K_3}{s+\sigma_1} + \frac{K_4}{s+\sigma_2} + \ldots \qquad \ldots (3)$$

where $H(s) = Z(s)$, the impedance of a required R–C network
$\qquad\quad = Y(s)$, the admittance of a required R–L network

In the R–C case the partial fraction expansion (3) of the given impedance leads to the series type network shown in fig. 6.3A(a), where $C_1 = 1/K_1$, $R_2 = K_2$ and the other elements are found by equating the impedance of a parallel R–C network to the respective partial fraction terms, i.e.

$$\frac{1}{1/R + Cs} = \frac{1/C}{s + 1/(RC)} = \frac{K}{s + \sigma_r}$$

so $C = 1/K$, $1/(RC) = \sigma_r$, $R = 1/(C\sigma_r) = K/\sigma_r$.

In the R–L case the expansion of $Y(s)$ in form (3) leads to the shunt type network of fig. 6.3A(b), where $L_1 = 1/K_1$, $R_2 = 1/K_2$, and the other elements are found by equating the admittance of a series R–L network to each of the other terms in (3), i.e.

$$\frac{1}{R + Ls} = \frac{K}{s + \sigma_r} = \frac{1}{\sigma_r/K + (1/K)s}$$

so $R = \sigma_r/K$ and $L = (1/K)$.

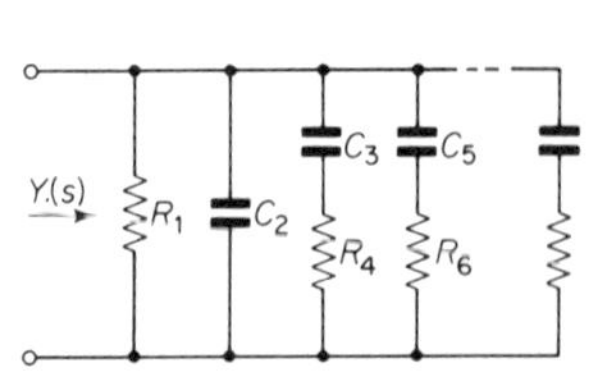
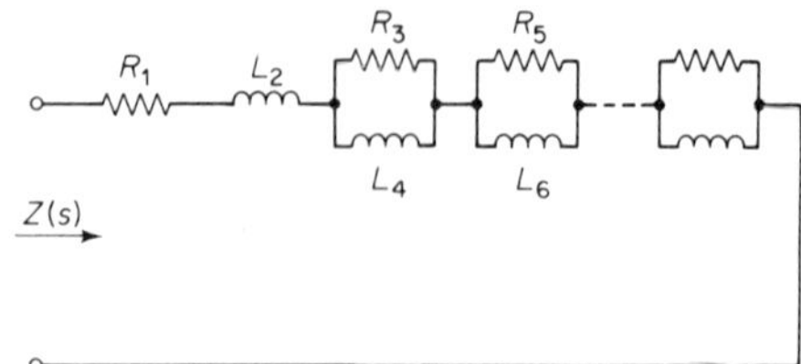

(a) Shunt type network for the synthesis of an R–C admittance

(b) Series type network for the synthesis of an R–L impedance

Fig. 6.4A

(ii) *For the synthesis of shunt type R–C admittances and series type R–L impedances*

In these cases we expand $H(s)/s$ (not $H(s)$) in a series of partial fractions:

$$\frac{H(s)}{s} = \frac{K_1}{s} + K_2 + \frac{K_3}{s+\sigma_1} + \frac{K_4}{s+\sigma_2} + \dots$$

and then multiply by s, i.e.

$$\left[\frac{H(s)}{s}\right] \times s = K_1 + K_2 s + \frac{K_3 s}{s+\sigma_1} + \frac{K_4 s}{s+\sigma_2} + \dots \qquad \dots (4)$$

This procedure must be adopted, since a partial fraction expansion of $H(s)$ alone will yield partial fraction terms with negative coefficients and not in a form which is readily indentifiable with simple series R–C and parallel R–L combinations.

Form (4), however, leads to the shunt type network of fig. 6.4A(a) for an R–C admittance function, where

$$R_1 = 1/K_1 , \; C_2 = K_2$$

and the remaining elements are found from,

$$\frac{1}{R+1/(Cs)} = \frac{(1/R)s}{s+1/(RC)} = \frac{Ks}{s+\sigma_r}$$

so $R = 1/K$ and $C = 1/(R\sigma_r) = K/\sigma_r$; whilst it leads to the series type network of fig. 6.4A(b) for an R–L impedance function, where

$$R_1 = K_1 , \; L_2 = K_2$$

and the other elements are found from,

$$\frac{1}{1/R+1/(Ls)} = \frac{Rs}{s+(R/L)} = \frac{Ks}{s+\sigma_r}$$

so $R = K$ and $L = R/\sigma_r = K/\sigma_r$

180

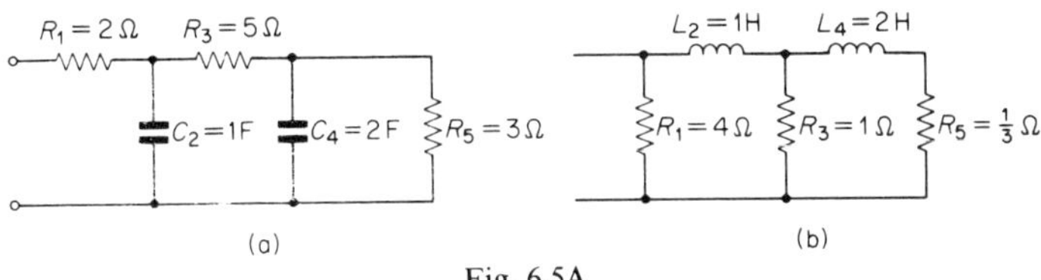

Fig. 6.5A

(b) *R–C and R–L ladder networks*

R–C and *R–L* immittances may also be synthesized by ladder networks using a continued fraction expansion. Two examples are shown in fig. 6.5A. Fig. 6.5A(a) is the ladder realization of the impedance,

$$Z(s) = \frac{60s^2 + 58s + 10}{30s^2 + 14s + 1}$$

which in continued fraction form becomes,

$$Z(s) = 2 + \frac{30s + 8}{30s^2 + 14s + 1}$$

$$= 2 + \cfrac{1}{\cfrac{s}{30s+8\,\overline{)30s^2 + 14s + 1}}}$$

$$\begin{array}{r}
30s + 8\,\overline{)30s^2 + 14s + 1} \\
\underline{30s^2 + 8s} \\
6s + 1\,\overline{)30s + 8} \\
\underline{30s + 5} \\
3\,\overline{)6s + 1} \\
\underline{6s} \\
1\,\overline{)3} \\
\underline{3} \\
0
\end{array}$$

$$= 2 + \cfrac{1}{s + \cfrac{1}{5 + \cfrac{1}{2s + \cfrac{1}{3}}}}$$

$$\equiv R_1 + \cfrac{1}{C_2 s + \cfrac{1}{R_3 + \cfrac{1}{C_4 s + \cfrac{1}{R_5}}}}$$

Fig. 6.5A(b) is a ladder realization of the admittance,

$$Y(s) = \frac{6s^2 + 34s + 17}{24s^2 + 40s + 4}$$

$$= \frac{1}{4} + \cfrac{1}{s + \cfrac{1}{1 + \cfrac{1}{2s + \cfrac{1}{3}}}} \equiv \frac{1}{R_1} + \cfrac{1}{L_2 s + \cfrac{1}{\left(\dfrac{1}{R_3}\right) + \cfrac{1}{L_4 s + \left(\dfrac{1}{R_5}\right)}}}$$

(c) *Some properties of R–C and R–L networks*

The impedance of an *R–C* and the admittance of an *R–L* network have the following characteristics:

 (i) Their poles and zeros are on the negative real s plane $(-\sigma)$ axis and they alternate.

 (ii) The residues of the poles are real and positive.

 (iii) The singularity at or nearest to the origin must be a pole, and the singularity at or nearest to $\sigma \to -\infty$ must be a zero.

The admittance of an *R–C* and the impedance of an *R–L* network possess the following characteristics:

 (i) Their poles and zeros are on the $-\sigma$ axis and alternate.

 (ii) The residues of the poles are real and negative.

 (iii) The singularity at or nearest to $\sigma = 0$ is a zero, and the singularity at or nearest to $\sigma \to -\infty$ is a pole.

4. SYNTHESIS OF *L–C* LADDER NETWORKS WITH A SPECIFIED TRANSFER IMMITTANCE

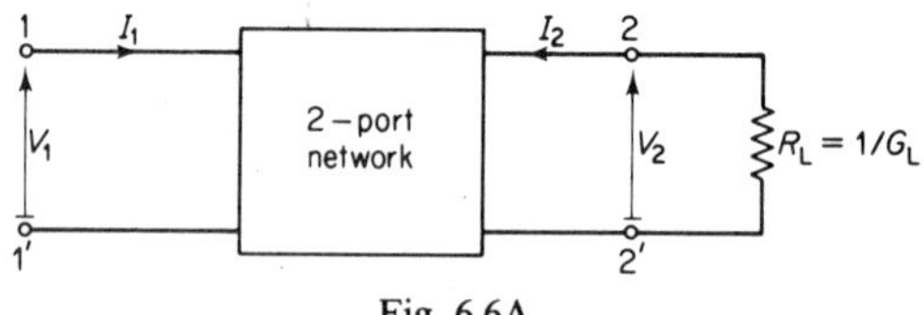

Fig. 6.6A

The transfer immittance of a 2-port terminated in a load resistance R_L or conductance G_L is,

in terms of its z-parameters: $\dfrac{V_2}{I_1} = Z_{21} = \dfrac{z_{21} R_L}{z_{22} + R_L}$

in terms of the y parameters: $\dfrac{I_2}{V_1} = Y_{21} = \dfrac{y_{21} G_L}{y_{22} + G_L}$

182

The synthesis of an L–C network with a specified transfer immittance and a $1\,\Omega$ termination (i.e. $R_{\mathrm{L}} = G_{\mathrm{L}} = 1$) is considered below.

Suppose the given transfer immittance is

$$H_{21}(s) = \frac{N(s)}{D(s)}$$

$$= \frac{N(s)}{E(s)+O(s)} \equiv \frac{h_{21}(s)}{h_{22}(s)+1} \qquad \dots (1)$$

where $N(s)$, $D(s)$ are the numerator, denominator polynomials and $E(s)$ is the even part (i.e. containing even powers of s) and $O(s)$ is the odd part of $D(s)$.

The open-circuit and short-circuit transfer immittances $h_{21} = z_{21}$ or y_{21} of L–C networks are always odd functions and this important fact is used to determine h_{21} and h_{22} from the identity inferred by (1).

If $N(s)$ is an even function then

$$\frac{h_{21}}{h_{22}+1} = \frac{N/O}{E/O+1}, \text{ i.e. } h_{21} = N/O, \; h_{22} = E/O$$

If $N(s)$ is an odd function,

$$\frac{h_{21}}{h_{22}+1} = \frac{N/E}{O/E+1}, \text{ i.e. } h_{21} = N/E, \; h_{22} = O/E$$

Thus from a specified transfer immittance we may find h_{21} and h_{22}, but the problem still remains as to how we may synthesize the required network. One method of attack is to synthesize h_{22} in terms of a ladder and then check that the ladder gives the correct h_{21}. For example, synthesize an L–C ladder network with the transfer impedance,

$$Z_{21}(s) = \frac{3}{3s^3 + 2s^2 + 6s + 3},$$

when terminated in $1\,\Omega$.

$$Z_{21} = \frac{[3/(3s^3+6s)]}{[(2s^2+3)/(3s^3+6s)]+1} = \frac{z_{21}}{z_{22}+1}$$

so $\quad z_{22} = (2s^2+3)/(3s^3+6s)$

$$= \cfrac{1}{\frac{3}{2}s + \cfrac{1}{\frac{4}{3}s + \cfrac{1}{\frac{1}{2}s}}} \equiv \cfrac{1}{C_1 s + \cfrac{1}{L_2 s + \cfrac{1}{C_3 s}}}$$

on expanding z_{22} as a continued fraction.

The actual network, which also is constistent with $z_{21} = 3/(3s^3 + 6s)$, is shown in fig. 6.7A

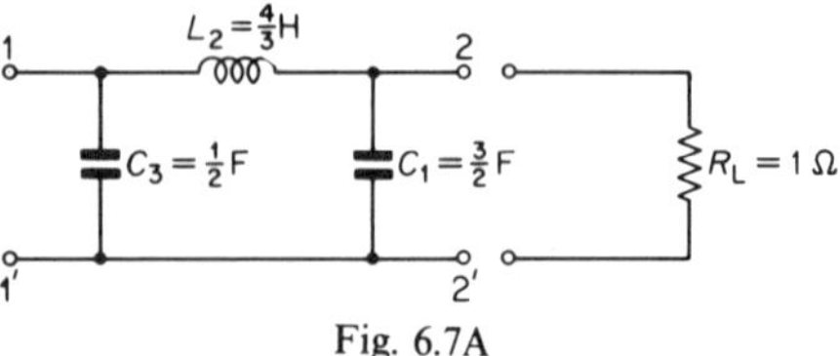

Fig. 6.7A

5. DARLINGTON'S METHOD FOR THE SYNTHESIS OF L–C LOW-PASS FILTER NETWORKS

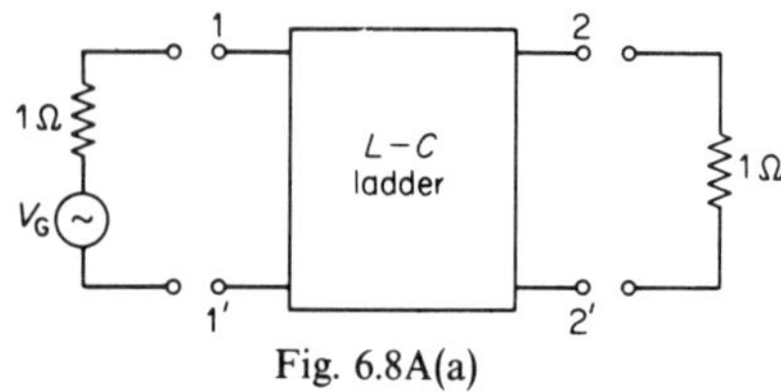

Fig. 6.8A(a)

This method may be used to synthesize doubly terminated L–C low-pass filters, some of which were discussed in Chapter 5.

The method uses the following results obtained from the scattering-matrix analysis of a loss-less 2-port (see next section 6B).

(i) The power insertion loss ratio of a 2-port terminated in $1\,\Omega$ is

$$\frac{P_{20}}{P} = \frac{1}{|S_{21}(j\omega)|^2}$$

(ii) The driving-point impedance at port $11'$ is

$$Z_{in}(s) = \frac{1 + S_{11}(s)}{1 - S_{11}(s)}$$

(iii) The relation between the scattering parameters S_{11} and S_{21} for a loss-less 2-port is

$$|S_{11}(j\omega)|^2 = 1 - |S_{21}(j\omega)|^2$$

Thus if the required insertion loss, i.e. $|S_{21}(j\omega)|^2$, is specified, we first find $|S_{11}(j\omega)|^2$ and subsequently by letting $j\omega = s$ we may determine $S_{11}(s)$ and hence $Z_{in}(s)$. $Z_{in}(s)$ is then synthesized in the form of ladder network using a continued fraction expansion.

For example, synthesize a low-pass filter with the power insertion loss ratio,

$$\frac{P_{20}}{P_2} = \frac{1}{|S_{21}(j\omega)|^2} = 1 + \omega^4 \quad (n = 2, \text{ Butterworth filter}).$$

184

$$|S_{11}(j\omega)|^2 = 1 - |S_{21}(j\omega)|^2 = \frac{\omega^4}{\omega^4 + 1}$$

and substituting $j\omega = s$, we obtain

$$S_{11}(s)S_{11}(-s) = \frac{s^4}{s^4 + 1}, \quad \text{which on factorization yields}$$

$$= \left(\frac{s^2}{s^2 + \sqrt{2}s + 1}\right)\left(\frac{(-s)^2}{s^2 - \sqrt{2}s + 1}\right)$$

so
$$S_{11}(s) = \frac{s^2}{s^2 + \sqrt{2}s + 1}$$

and
$$Z_{in}(s) = \frac{1 + S_{11}}{1 - S_{11}} = \frac{2s^2 + \sqrt{2}s + 1}{\sqrt{2}s + 1}$$

$$= \sqrt{2}s + \frac{1}{\sqrt{2}s + 1} = L_1 s + \frac{1}{C_2 s + 1}$$

The actual network is shown in fig. 6.8A(b).

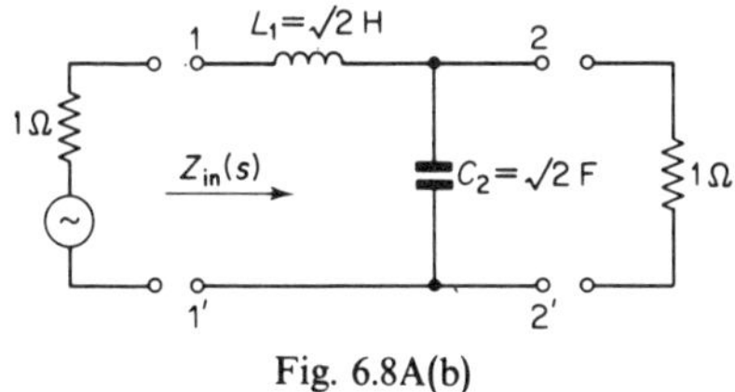

Fig. 6.8A(b)

6.2A. Worked Problems

1. Using the conditions given in para. 1 of the Theory summary show which of the following immittances are physically realizable. In the realizable cases synthesize an appropriate network.

(a) $Z(s) = \dfrac{12s^2 + 5}{2s^3 + s}$, (b) $Y(s) = \dfrac{s^2 + 16}{s(s^2 + 4)}$

(c) $Z(s) = \dfrac{s^2 + 7}{(s+1)^3}$, (d) $Y(s) = \dfrac{20s^2 + 32s + 11}{(4s + 3)(5s + 2)}$

Solution

(a) $Z(s) = \dfrac{12s^2 + 5}{2s^3 + s} = \dfrac{12s^2 + 5}{s(2s^2 + 1)}$

$$= \frac{K_1}{s} + \frac{K_2 s + K_3}{2s^2 + 1}$$

and on evaluating the partial fraction constants, we have

$$Z(s) = \frac{5}{s} + \frac{2s}{2s^2 + 1}$$

On making the three tests to see whether $Z(s)$ is a positive real function and therefore realizable, we obtain:
(1) The poles of $Z(s)$ are 0, $\pm j\tfrac{1}{2}$ and therefore not in the R.H. s plane.
(2) The poles are simple and the residues $K_1 = 5$, $K_2 = 2$, $K_3 = 0$ are real and positive.

$$(3) \ \mathrm{Re}\,[Z(j\omega)] = \mathrm{Re}\left[\frac{5 - 12\omega^2}{j(\omega - 2\omega^3)}\right] = 0 \ \text{for all } \omega.$$

Thus $Z(s)$ is a positive real function.
Using the partial fraction expansion,

$$Z(s) = \frac{1}{\frac{1}{5}s} + \frac{1}{s + \dfrac{1}{2s}}$$

$$= \frac{1}{C_1 s} + \frac{1}{C_2 s + \dfrac{1}{Ls}}$$

showing that $Z(s)$ may be synthesized by a capacitor $C_1 = \tfrac{1}{5}\,\mathrm{F}$ in series with a parallel combination of a capacitor $C_2 = 1\,\mathrm{F}$ and inductor $L = 2\,\mathrm{H}$, as shown in fig. 6.9A(a).

$$(b) \ Y(s) = \frac{s^2 + 16}{s(s^2 + 4)} = \frac{4}{s} + \frac{-3s}{(s^2 + 4)}$$

on expanding as a sum of partial fractions. Although the poles, $\pm j2$, of the second term are simple, the residue is -3 (i.e. not positive) and so $Y(s)$ is not a positive real function and therefore not realizable.

$$(c) \ Z(s) = \frac{s^2 + 7}{(s+1)^3} \ \text{has poles at } s = -1$$

in the L.H. plane. However, on making the test

$$\mathrm{Re}\,[Z(j\omega)] = \mathrm{Re}\left[\frac{7 - \omega^2}{(1 + j\omega)^3}\right] = \mathrm{Re}\left[\frac{7 - \omega^2}{(1 - 3\omega^2) + j(3\omega - \omega^3)}\right]$$

$$= \frac{(7 - \omega^2)(1 - 3\omega^2)}{(1 - 3\omega^2)^2 + (3\omega - \omega^3)^2} = \frac{3\omega^4 - 22\omega^2 + 21}{(1 - 3\omega^2)^2 + (3\omega - \omega^3)^2}$$

is not greater than zero for all ω, i.e. for $1/\sqrt{3}<\omega<\sqrt{7}$, $\mathrm{Re}\,[Z(\mathrm{j}\omega)]$ is negative. Thus $Z(s)$ is not a positive real function and so is not physically realizable.

$$\text{(d)} \quad Y(s) = \frac{20s^2+32s+11}{(4s+3)(5s+2)}$$

$$= 1 + \frac{9s+5}{(4s+3)(5s+2)} \quad \text{on first dividing out,}$$

$$= 1 + \frac{1}{4s+3} + \frac{1}{5s+2} \quad \text{on splitting into partial fractions.}$$

On checking the three conditions we find each individual term is a positive real function and therefore so must be $Y(s)$. From the above form it may be deduced that $Y(s)$ can be synthesized by $R_1 = 1\,\Omega$, in parallel with $R_2 = 3\,\Omega$ and $L_2 = 4\,\mathrm{H}$, in parallel with $R_3 = 2\,\Omega$ and $L_3 = 5\,\mathrm{H}$, as shown in fig. 6.9A(b).

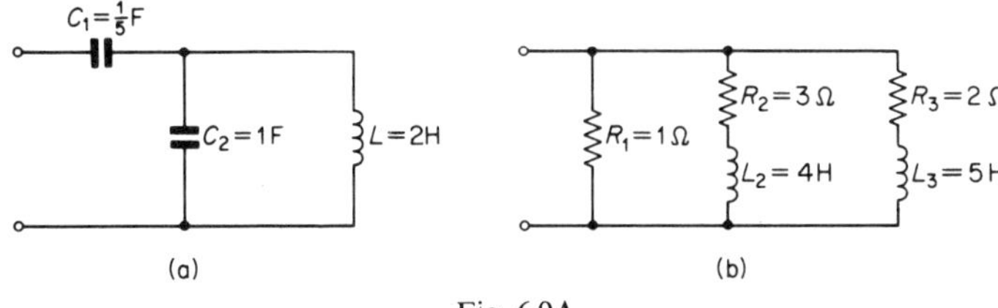

Fig. 6.9A

2. Write down an expression for the input impedance of the network of fig. 6.10A in the form of a continued fraction and hence synthesize an L–C ladder network to meet the following specification: a zero at $2\,\mathrm{Mrad/s}$, poles at $\sqrt{3}\,\mathrm{Mrad/s}$ and $\sqrt{5}\,\mathrm{Mrad/s}$, and an input impedance of $\mathrm{j}1000\,\Omega$ at $1\,\mathrm{Mrad/s}$.

Solution

The input impedance of the network is

$$Z = \cfrac{1}{Y_2 + \cfrac{1}{Z_3 + \cfrac{1}{Y_4 + \cfrac{1}{Z_5}}}} \qquad \cdots(1)$$

in continued fraction form.

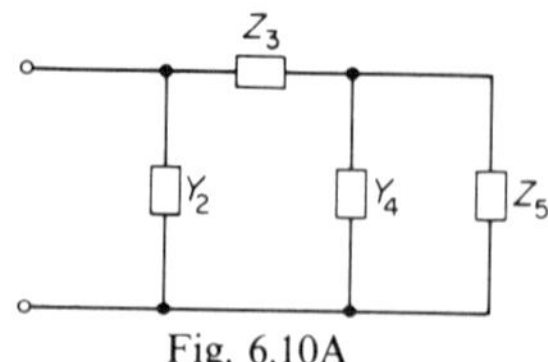

Fig. 6.10A

Using Foster's theorem the input impedance of the network to be synthesized may be written as

$$Z(s) = H \frac{s(s^2 + \omega_1^2)}{(s^2 + \omega_0^2)(s^2 + \omega_2^2)}$$

where $\omega_1 = 2 \times 10^6$, $\omega_0 = \sqrt{3} \times 10^6$, $\omega_2 = \sqrt{5} \times 10^6$, and the scale factor H may be evaluated using $Z(j\omega) = j1000\,\Omega$ when $\omega = 10^6$ rad/s, i.e.

$$j10^3 = jH \frac{10^6(4-1) \times 10^{12}}{(3-1)10^{12} \times (5-1)10^{12}}, \text{ so } H = \tfrac{8}{3} \times 10^9$$

Thus
$$Z(s) = \frac{H(s^3 + 4 \times 10^{12}s)}{s^4 + 8 \times 10^{12}s^2 + 15 \times 10^{24}}$$
$$= \frac{hs'^3 + 4hs'}{s'^4 + 8s' + 15} \qquad \cdots (2)$$

on substituting $s' = s \times 10^{-6}$, $h = H \times 10^{-6}$.

We now expand (2) in the form of a continued fraction:

$$
\begin{array}{r}
\tfrac{1}{h}s' \\
hs'^3 + 4hs')\overline{s'^4 + 8s'^2 + 15} \\
\underline{s'^4 + 4s'^2} \qquad\qquad \tfrac{1}{4}hs' \\
4s'^2 + 15)\overline{hs'^3 + 4hs'} \\
hs'^3 + 3\tfrac{3}{4}hs' \qquad \dfrac{16}{h}s' \\
\underline{\qquad\qquad} \\
\tfrac{1}{4}hs')\overline{4s'^2 + 15} \\
4s'^2 \qquad \tfrac{1}{60}hs' \\
\underline{\qquad} \\
15)\tfrac{1}{4}hs' \\
\tfrac{1}{4}hs' \\
\underline{\qquad} \\
0
\end{array}
$$

and hence

$$Z = \cfrac{1}{\cfrac{1}{h}s' + \cfrac{1}{\cfrac{1}{4}hs' + \cfrac{1}{\cfrac{16}{h}s' + \cfrac{1}{\cfrac{1}{60}hs'}}}}$$

On comparing with expression (1) we obtain the L and C elements of the required ladder network:

$$Y_2 = \frac{1}{h} s' = \frac{1}{H} s \equiv C_2 s \qquad \text{so } C_2 = \frac{1}{H} = 375 \text{ pF}$$

$$Z_3 = \tfrac{1}{4} h s' = \tfrac{1}{4} H \times 10^{-12} s \equiv L_3 s \qquad \text{so } L_3 = \tfrac{1}{4} H \times 10^{-12} = 0.667 \text{ mH}$$

$$Y_4 = \frac{16}{h} s' = \frac{16}{H} s \equiv C_4 s \qquad \text{so } C_4 = 6000 \text{ pF}$$

$$Z_5 = \tfrac{1}{60} h s' = \tfrac{1}{60} H \times 10^{-12} s \equiv L_5 s \quad \text{so } L_5 = 44.4 \ \mu\text{H}$$

6.3A. Exercise Problems

1. The impedance of a network to be synthesized from L–C elements is given by

$$Z(s) = s \frac{(s^2 + a^2)}{s^2 + b^2}$$

   What conditions must be imposed on a and b for the network to be physically realizable? If these conditions are satisfied sketch the $Z(j\omega)$ versus ω graph and synthesize the network with the provisor that it contains a series inductor.

2. By means of reactance-frequency sketches determine which of the following functions represent L–C driving point impedances:

   (a) $\dfrac{s(s^2+4)}{(s^2+1)(s^2+9)}$ (b) $\dfrac{s(s^2+1)(s^2+4)}{(s^2+2)(s^2+9)}$ (c) $\dfrac{(s^2+1)(s^2+9)}{s(s^2+4)(s^2+16)}$

3. Using a continued fraction expansion synthesize the following impedances:

   $$\text{(a) } Z(s) = \frac{s^3 + 6s}{s^2 + 2} \quad \text{(b) } Z(s) = \frac{10s + 4}{s + \frac{1}{5}}$$

4. By means of a partial fraction expansion synthesize the following:

   $$\text{(a) } Z(s) = \frac{5s + 2}{2s(s + 3)} \quad \text{(b) } Y(s) = \frac{s(3s^2 + 7)}{(s^2 + 4)(2s^2 + 3)}$$

5. Synthesize a Cauer ladder network for the impedance

   $$Z(s) = \frac{24s^4 + 44s^2 + 3}{32s^3 + 12s}$$

6. Synthesize in the form of a ladder network the admittance

$$Y(s) = \frac{6s^3 + 17s + 14s + 2}{24s^3 + 36s^2 + 8s + 8}$$

7. Synthesize the impedance

$$Z(s) = \frac{5s^3 + 24s}{s^4 + 18s^2 + 48}$$

in the form shown in fig. 6.11A.

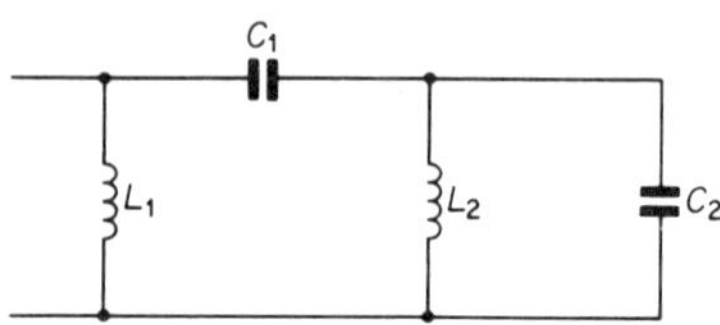

Fig. 6.11A. For problem 7

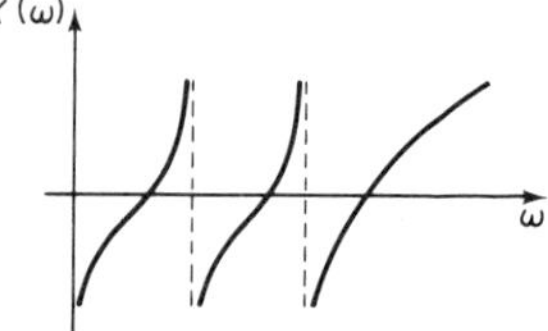

Fig. 6.12A. For problem 8

8. Construct an *L–C* ladder network which would produce the same form of reactance versus frequency characteristic as shown in fig. 6.12A.
9. Synthesize a series type Foster network with the following specification: parallel resonant at 2 Mrad/s and 5 Mrad/s, series resonant at 4 Mrad/s, input impedance j1000 Ω at 1 Mrad/s.
10. The driving point impedance of an *L–C* network is

$$Z(s) = 2 \cdot 67 \times 10^9 \, \frac{s(s^2 + 4 \times 10^{12})}{(s^2 + 3 \times 10^{12})(s^2 + 5 \times 10^{12})}$$

Sketch its reactance versus frequency curve and suggest 4 alternative ways by which such an impedance could be synthesized.
11. Synthesize a series type Foster network for the impedance

$$Z(s) = \frac{120s^2 + 43s + 1}{5s(8s + 1)}$$

12. Synthesize a shunt type Foster network for the *R–L* admittance

$$Y(s) = \frac{5s^2 + 29s + 8}{2(s + 1)(5s + 2)}$$

13. Synthesize the *R–L* impedance

$$Z(s) = \frac{s^2 + 12s + 9}{s + 3}$$

in the form of (a) a series Foster, (b) a ladder network.

14. Identify which of the following functions are R–L, R–C or LC immittances:

$$\text{(a)}\quad Y(s) = \frac{3s+2}{2s^2+3s+1}$$

$$\text{(b)}\quad Z(s) = \frac{2s^4+4s^2+1}{s(2s^2+1)}$$

$$\text{(c)}\quad Z(s) = \frac{s^2+2s+2}{s^2+2}$$

$$\text{(d)}\quad Z(s) = \frac{3s+2}{s(s+2)}$$

and sketch an appropriate network to realize the function in each case.

15. Synthesize the admittance,

$$Y(s) = \frac{12s^3+18s^2+13s+2}{s(12s^2+13s+2)}$$

in the form shown in fig. 6.13A.

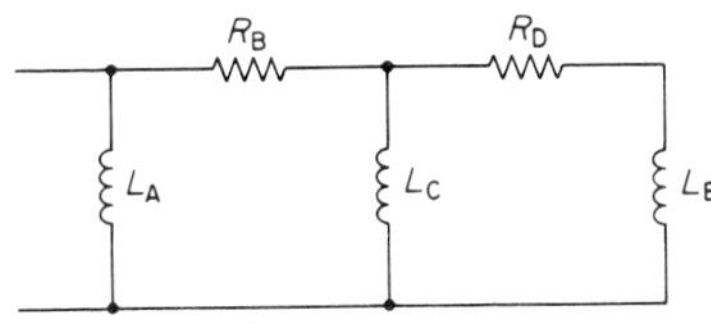

Fig. 6.13A. For problem 15

16. The transfer admittance of the 2-port network shown in fig. 6.14A is required to be

$$Y_{21}(s) = \frac{I_2}{V_1} = \frac{-1}{60s^3+20s^2+7s+1}$$

when terminated in a $1\,\Omega$ load.

Determine the y_{22} and y_{21} parameters of the network and hence synthesize an L–C T network to realize $Y_{21}(s)$.

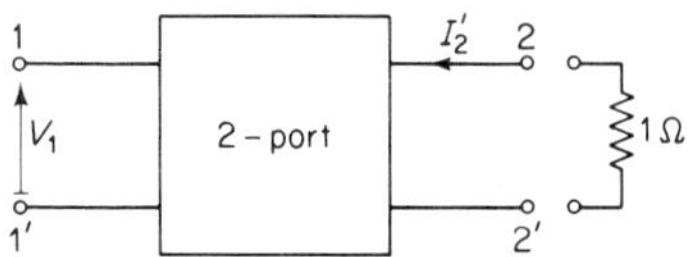

Fig. 6.14A. For problem 16

191

17. The power insertion loss ratio of an $n = 3$ Butterworth filter working between $1\,\Omega$ terminations is $(1 + \omega^6)$. Determine the scattering parameter $S_{11}(s)$ and the driving point impedance at the input port. Synthesize the filter in terms of both a T and a Π network.

6B THE SCATTERING MATRIX

6.1B. Theory Summary

1. THE IMPEDANCE, ADMITTANCE, AND SCATTERING MATRIX DESCRIPTIONS OF A MULTI-PORT NETWORK

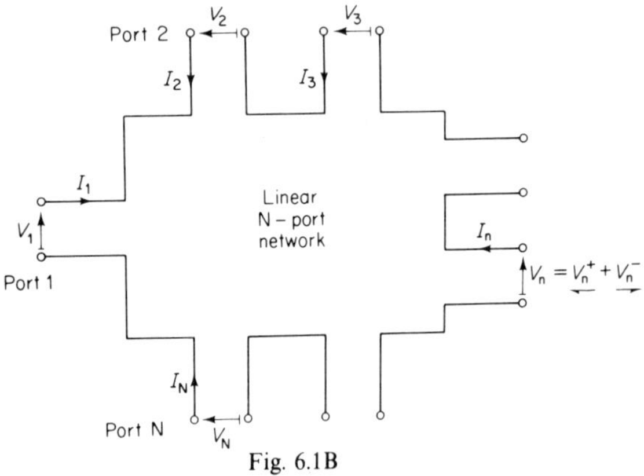

Fig. 6.1B

In the majority of circuit theory and in particular in the loop and nodal methods of analysis* we define the network equations in terms of voltage and current variables and impedance and admittance elements. For example the loop analysis of an N-loop, N-port network leads to the equations:

$$\begin{bmatrix} Z_{11} & Z_{12} & Z_{13} & \cdots & Z_{1n} & \cdots & Z_{1N} \\ Z_{21} & Z_{22} & Z_{23} & \cdots & Z_{2n} & \cdots & Z_{2N} \\ \vdots & & & & & & \\ Z_{n1} & Z_{n2} & Z_{n3} & \cdots & Z_{nn} & \cdots & Z_{nN} \\ \vdots & & & & & & \\ Z_{N1} & Z_{N2} & Z_{N3} & \cdots & Z_{Nn} & \cdots & Z_{NN} \end{bmatrix} \begin{bmatrix} I_1 \\ I_2 \\ \vdots \\ I_n \\ \vdots \\ I_N \end{bmatrix} = \begin{bmatrix} V_1 \\ V_2 \\ \vdots \\ V_n \\ \vdots \\ V_N \end{bmatrix}$$

* See Volume 1, Chapter 5.

192

whilst the nodal analysis leads to the equations:

$$\begin{bmatrix} Y_{11} & Y_{12} & Y_{13} & \cdots & Y_{1n} & \cdots & Y_{1N} \\ Y_{21} & Y_{22} & Y_{23} & \cdots & Y_{2n} & \cdots & Y_{2N} \\ \vdots & & & & & & \\ Y_{n1} & Y_{n2} & Y_{n3} & \cdots & Y_{nn} & \cdots & Y_{nN} \\ \vdots & & & & & & \\ Y_{N1} & Y_{N2} & Y_{N3} & \cdots & Y_{Nn} & \cdots & Y_{NN} \end{bmatrix} \begin{bmatrix} V_1 \\ V_2 \\ \vdots \\ V_n \\ \vdots \\ V_N \end{bmatrix} = \begin{bmatrix} I_1 \\ I_2 \\ \vdots \\ I_n \\ \vdots \\ I_N \end{bmatrix}$$

where $Z_{11}, Z_{12} \ldots Z_N$ are impedances and $Y_{11}, Y_{12} \ldots Y_{NN}$ are admittances (consisting of linear combinations of the actual elements making up the network) and $V_1, V_2 \ldots V_N$ and $I_1, I_2 \ldots I_N$ are the terminal voltages and currents at the N-ports.

One major practical disadvantage of the impedance and admittance descriptions of a network particularly in the higher r.f. and microwave frequency ranges is that there are not necessarily unique definitions of voltage and current and consequently the Z and Y parameters of a network cannot be directly measured. For example, this is so in waveguide transmission line systems. However such parameters as reflection coefficient, transmission or power transfer coefficient can be measured using slotted line, reflectometry and power measurement techniques. It is for this reason that the scattering matrix description of a multi-port network is particularly important.

In the scattering matrix description we define the circuit behaviour of a network in terms of incident waves carrying power into the network and reflected waves transporting power out of the network. Thus if we select the incident wave voltage components V_n^+ as the independent variables and the reflected wave voltages V_n^- as the dependent variables, we may write down the circuit equations for a linear N-port network in the following manner:

$$\begin{bmatrix} V_1^- \\ V_2^- \\ V_3^- \\ \vdots \\ V_N^- \end{bmatrix} = \begin{bmatrix} S_{11} & S_{12} & S_{13} & \cdots & S_{1n} & \cdots & S_{1N} \\ S_{21} & S_{22} & S_{23} & \cdots & S_{2n} & \cdots & S_{2N} \\ S_{31} & S_{32} & S_{33} & \cdots & S_{3n} & \cdots & S_{3N} \\ \vdots & & & & & & \\ S_{N1} & S_{N2} & S_{N3} & \cdots & S_{Nn} & \cdots & S_{NN} \end{bmatrix} \begin{bmatrix} V_1^+ \\ V_2^+ \\ V_3^+ \\ \vdots \\ V_N^+ \end{bmatrix}$$

where $S_{11}, S_{12}, S_{13} \ldots S_{NN}$ are known as the scattering-matrix coefficients, and the incident and reflected voltage components are related to the total terminal port voltage V_n by

$$V_n = V_n^+ + V_n^-$$

Note that the incident and reflected components of current I_n^+, I_n^- (not used in the above description) are related to the total terminal current I_n by

$$I_n = I_n^+ + I_n^- = (V_n^+ - V_n^-)/R_{0n}$$

where $R_{0n} = \dfrac{V_n^+}{I_n^+} = \dfrac{V_n^-}{-I_n^-}$ is the reference impedance level of port n.

In transmission line applications where the multi-port may consist of the junction of a number of lines, the respective ports of the network are

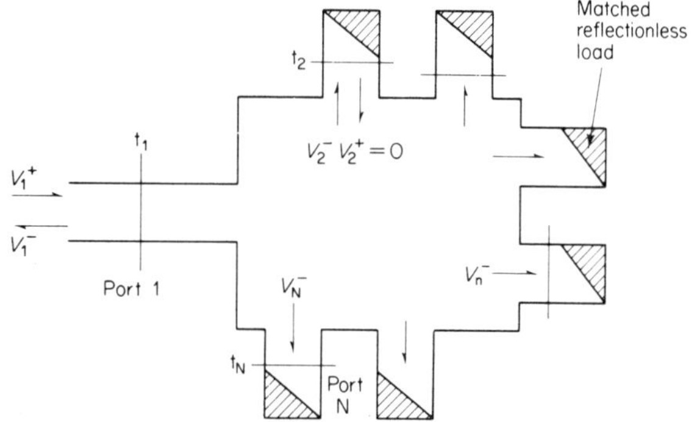

$S_{11} = \dfrac{V_1^-}{V_1^+}$ when all other ports are terminated in matched loads so

that $V_2^+ = V_3^+ = \cdots V_n^+ \cdots = V_N^+ = 0$

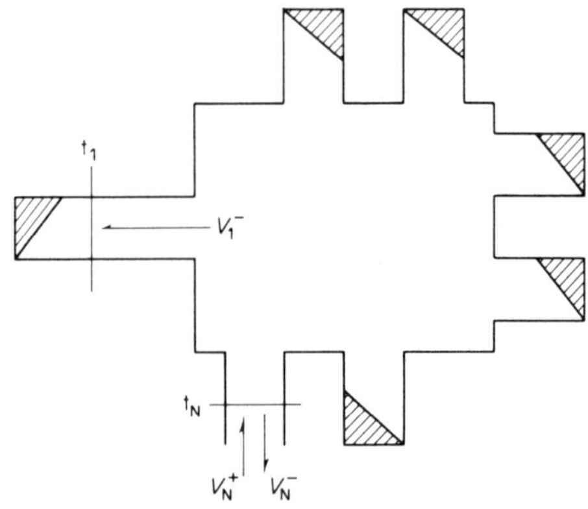

$S_{1N} = \dfrac{V_1^-}{V_N^+}$ when all ports, other than port N, are terminated
in matched loads

Fig. 6.2B. Diagram illustrating meaning of scattering coefficients S_{11} and S_{1N}

taken as suitably chosen reference planes in these lines and in this case R_{0n} would be identified as the characteristic impedance of line n.

We may consider the physical significance of the scattering coefficients by investigating the total reflected wave at a given port. For example at port 1,

$$V_1^- = S_{11}V_1^+ + S_{12}V_2^+ + S_{13}V_3^+ + \ldots + S_{in}V_n^+ + \ldots + S_{1N}V_N^+$$

i.e. the reflected wave V_1^- is made up of contributions due to incident waves at all N-ports, and the individual coefficients:

$$S_{11} = \left(\frac{V_1^-}{V_1^+}\right)_{V_n^+ = 0 \text{ for all } n \neq 1} \quad \ldots \text{ is the reflection coefficient at port 1}$$

$$S_{12} = \left(\frac{V_1^-}{V_2^+}\right)_{V_n^+ = 0 \text{ for all } n \neq 2} \quad \ldots \text{ is the voltage transmission}$$

coefficient from port 2 (defined by the terminal plane t_2) to port 1 (t_1)

$$\vdots$$

$$S_{1N} = \left(\frac{V_1^-}{V_N^+}\right)_{V_n^+ = 0 \text{ for all } n \neq N} \quad \ldots \text{ is the voltage transmission coefficient}$$

from port N (t_N) to port 1 (t_1).

If all the ports of the network have equal terminating impedances and if the network is reciprocal $S_{ij} = S_{ji}$ for all i, j. However, if this is not the case then, because of different impedance levels at the various ports, the scattering-matrix is no longer symmetrical. It is therefore common practice to use the normalized scatter variables and matrix discussed below.

2. NORMALIZED INCIDENT AND REFLECTED WAVES AND SCATTERING-MATRIX

In this case we define incident and reflected wave parameters a and b as respectively directly proportional to the incident and reflected voltages at each port, i.e.

$$a_n = k_n V_n^+, \quad b_n = k_n V_n^- \quad \text{for all } n = 1, 2, 3 \ldots N \qquad \ldots (1)$$

with the power qualification that

$$\tfrac{1}{2}a_n a_n^* = \tfrac{1}{2}|a_n|^2 = \text{incident power at port n flowing into the network}$$
$$\tfrac{1}{2}b_n b_n^* = \tfrac{1}{2}|b_n|^2 = \text{reflected power at port n flowing out of the network}$$

The constant of proportionality k_n in (1) may be evaluated using the power condition and knowing the terminating impedance (or reference level impedance) R_{0n} of port n:

power in incident wave at port n

$$= \tfrac{1}{2}V_n^+ I_n^{+*} = \tfrac{1}{2}V_n^+ (V_n^+/R_{0n})^*$$

$$= \tfrac{1}{2}|V_n^+|^2/R_{0n} = \tfrac{1}{2}|a_n|^2 = \tfrac{1}{2}k_n^2|V_n^+|$$

so $\quad k_n = 1/\sqrt{R_{0n}}$ $\qquad\qquad \cdots (2)$

We may now express the scatter variables a_n and b_n explicitly in terms of the terminal port voltage V_n and current I_n:

$$V_n = V_n^+ + V_n^- = \frac{1}{k_n}(a_n + b_n) = \sqrt{R_{0n}}(a_n + b_n) \qquad \cdots (3)$$

$$I_n = (V_n^+ - V_n^-)/R_{0n} = (a_n - b_n)/(R_{0n}k_n) = (a_n - b_n)/\sqrt{R_{0n}} \qquad \cdots (4)$$

and solving (3) and (4) we obtain

$$a_n = \tfrac{1}{2}k_n(V_n + R_{0n}I_n) = \tfrac{1}{2}(V_n/\sqrt{R_{0n}} + \sqrt{R_{0n}}I_n) \qquad \cdots (5)$$

$$b_n = \tfrac{1}{2}k_n(V_n - R_{0n}I_n) = \tfrac{1}{2}(V_n/\sqrt{R_{0n}} - \sqrt{R_{0n}}I_n) \qquad \cdots (6)$$

The scattering matrix description is of the same form:

$$\begin{bmatrix} b_1 \\ b_2 \\ \vdots \\ b_N \end{bmatrix} = \begin{bmatrix} S_{11} & S_{12} & S_{13} & \cdots & S_{1N} \\ S_{21} & S_{22} & S_{23} & & S_{2N} \\ \vdots & \vdots & \vdots & & \vdots \\ S_{N1} & S_{N2} & S_{N3} & & S_{NN} \end{bmatrix} \begin{bmatrix} a_1 \\ a_2 \\ \vdots \\ a_N \end{bmatrix} \qquad \cdots (7)$$

but has the important additional properties of

$$S_{ij} = S_{ji} \text{ for all i, j if the network is reciprocal}$$

$[S][S^*] = [1]$ if the network is loss-less, i.e. the product of $[S]$ and its complex conjugate matrix $[S^*]$ equals the unit matrix $[1]$.

For example, for a loss-less reciprocal ($S_{12} = S_{21}$) 2-port network:

$$\begin{bmatrix} S_{11} & S_{12} \\ S_{21} & S_{22} \end{bmatrix} \begin{bmatrix} S_{11}^* & S_{12}^* \\ S_{21}^* & S_{22}^* \end{bmatrix} = \begin{bmatrix} 1 & 0 \\ 0 & 1 \end{bmatrix}$$

which yields the following relations:

$$S_{11}S_{11}^* + S_{12}S_{21}^* = 1 \text{ or } |S_{11}|^2 + |S_{21}|^2 = 1$$
$$S_{11}S_{12}^* + S_{12}S_{22}^* = 0$$
$$S_{21}S_{11}^* + S_{22}S_{21}^* = 0$$
$$S_{21}S_{12}^* + S_{22}S_{22}^* = 1 \text{ or } |S_{21}|^2 + |S_{11}|^2 = 1$$

In particular the power supplied to port 2 via a loss-less 2-port from a source at port 1, provided port 2 is 'correctly' terminated in its reference level (reflectionless) impedance is

$$\tfrac{1}{2}|b_2|^2 = \tfrac{1}{2}|S_{21}a_1|^2 = |S_{21}|^2 P_1^+ = (1-|S_{11}|^2)P_1^+$$

where $P_1^+ = \tfrac{1}{2}|a_1|^2$ is the incident power at port 1.

3. RELATIONSHIPS BETWEEN $[S]$, $[Z]$ AND $[Y]$ MATRICES

(a) $[S]$ matrix in terms of $[Z]$ and $[Y]$ matrices

$$[S] = \{[1]+[k][Z][k]\}^{-1}\{[k][Z][k]-[1]\}$$
$$= \{[1]+[1/k][Y][1/k]\}^{-1}\{[1]-[1/k][Y][1/k]\}$$

where

$$[k] = \begin{bmatrix} 1/\sqrt{R_{01}} & 0 & \cdots & 0 \\ 0 & 1/\sqrt{R_{02}} & & 0 \\ & & & \\ 0 & 0 & \cdots & 1/\sqrt{R_{ON}} \end{bmatrix}, \quad [1/k] = \begin{bmatrix} \sqrt{R_{01}} & 0 & \cdots & 0 \\ 0 & \sqrt{R_{02}} & & 0 \\ & & & \\ 0 & 0 & \cdots & \sqrt{R_{ON}} \end{bmatrix}$$

whilst if all the impedance levels of the ports are normalized to unity, i.e. $[k] = [1/k] = [1]$, the relations simplify to

$$[S] = \{[1]+[Z]\}^{-1}\{[Z]-[1]\}$$
$$= \{[1]+[Y]\}^{-1}\{[1]-[Y]\}$$

(b) $[Z]$ and $[Y]$ matrices in terms of the $[S]$ matrix

$$[Z] = [[1/k]\{[1]+[S]\}][k]\{[1]-[S]\}]^{-1}$$
$$[Y] = [[k]\{[1]-[S]\}][[1/k]\{[1]+[S]\}]^{-1}$$

which simplify to

$$[Z] = \{[1]+[S]\}\{[1]-[S]\}^{-1}$$
$$[Y] = \{[1]-[S]\}\{[1]+[S]\}^{-1}$$

if all $R_{0n} = 1$.

6.2B. Worked Problems

1. Determine the scattering coefficient S_{11} of the 1-port network shown in fig. 6.3B where R_0 is reference level impedance and Z_{in} is the input

impedance at port 11′. Show also that the power P supplied to the load resistor R_L is given by:

$$P = \frac{V_G^2}{8R_0}\left[1 - \left|\frac{Z_{in} - R_0}{Z_{in} + R_0}\right|^2\right]$$

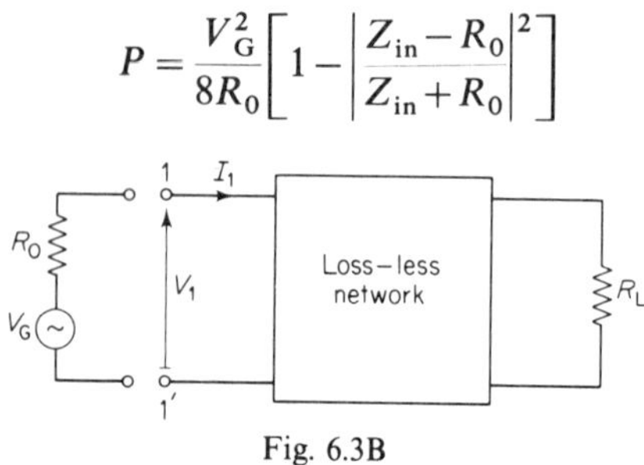

Fig. 6.3B

Solution

On expressing the voltage and current at port 11′ in terms of incident and reflected components we have

$$V_1 = V_1^+ + V_1^-$$

$$R_0 I_1 = V_1^+ - V_1^-$$

$$= R_0\frac{V_1}{Z_{in}} = \frac{R_0}{Z_{in}}(V_1^+ + V_1^-)$$

i.e.
$$V_1^+ - V_1^- = \frac{R_0}{Z_{in}}(V_1^+ + V_1^-)$$

from which
$$\frac{V_1^-}{V_1^+} = S_{11} = \frac{Z_{in} - R_0}{Z_{in} + R_0}.$$

If port 11′ is 'correctly' terminated so $Z_{in} = R_0$, $V_1^- = 0$, $V_1 = V_1^+ = (V_G/2R_0)\times R_0 = \frac{1}{2}V_G$, and the incident power (in this case, all transferred to the R_0 termination) is

$$P_1^+ = \tfrac{1}{2}|V_1^+|^2/R_0 = |V_G|^2/8R_0$$

If, however, $Z_{in} \neq R_0$, $V^- \neq 0$ and the power reflected at 11′ is

$$P_1^- = \tfrac{1}{2}|V_1^-|^2/R_0 = \tfrac{1}{2}|S_{11}V_1^+|/R_0 = |S_{11}|^2 P_1^+$$

and the net power flowing into the network at 11′ is

$$P_1 = P_1^+ - P_1^- = P^+(1 - |S_{11}|^2)$$

$$= \frac{|V_G|^2}{8R_0}\left[1 - \left|\frac{Z_{in} - R_0}{Z_{in} + R_0}\right|^2\right]$$

However since the part of the network preceeding the load R_L is loss-less and can therefore not dissipate power, the above expression for P_1 must represent the power transferred and dissipated in R_L.

2. Find the scattering-matrix parameters of a 2-port network in terms of the z-parameters z_{11}, z_{12}, z_{21}, z_{22} when both V^+, V^- and a, b variables are used, and when the reference level impedances at the 2 ports are R_1 and R_2. Show also, for a reciprocal network, that $S_{12} = S_{21}$ for the a, b variables but that $S_{12} R_2 = S_{21} R_1$ for the V^+, V^- case.

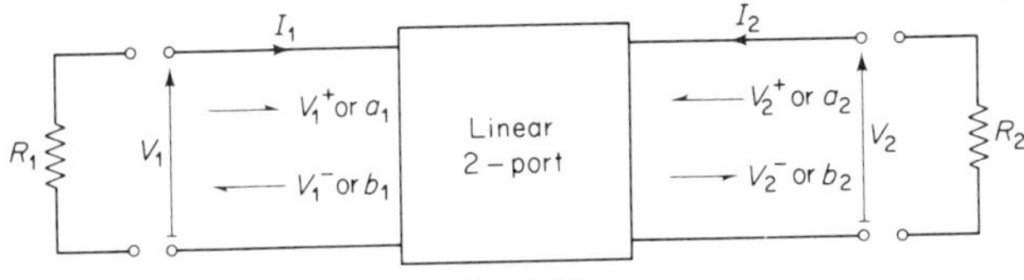

Fig. 6.4B

Solution
On substituting,

$$V_1 = V_1^+ + V_1^- = \sqrt{R_1}(a_1 + b_1)$$
$$V_2 = V_2^+ + V_2^- = \sqrt{R_2}(a_2 + b_2)$$
$$I_1 = (V_1^+ - V_1^-)/R_1 = (a_1 - b_1)/\sqrt{R_1}$$
$$I_2 = (V_2^+ - V_2^-)/R_2 = (a_2 - b_2)/\sqrt{R_2}$$

into the z parameter equations of the 2-port:

$$z_{11} I_1 + z_{12} I_2 = V_1$$
$$z_{21} I_1 + z_{22} I_2 = V_2$$

we obtain, in terms of V^+ and V^-:

$$(1 + z_{11}/R_1)V_1^- + (z_{12}/R_2)V_2^- = (z_{11}/R_1 - 1)V_1^+ + (z_{12}/R_2)V_2^+ \qquad \ldots (1)$$

$$(z_{21}/R_1)V_1^- + (1 + z_{22}/R_2)V_2^- = (z_{21}/R_1)V_1^+ + (z_{22}/R_2 - 1)V_2^+ \qquad \ldots (2)$$

or in terms of a and b

$$\sqrt{R_1}(1 + z_{11}/R_1)b_1 + (z_{12}/\sqrt{R_2})b_2$$
$$= \sqrt{R_1}(z_{11}/R_1 - 1)a_1 + (z_{12}/\sqrt{R_2})a_2 \qquad \ldots (3)$$

$$(z_{21}/\sqrt{R_1})b_1 + \sqrt{R_2}(1 + z_{22}/R_2)b_2$$
$$= (z_{21}/\sqrt{R_1})a_1 + \sqrt{R_2}(z_{22}/R_2 - 1)a_2 \qquad \dots (4)$$

On solving (1) and (2) for V_1^- and V_2^- we obtain

$$\begin{bmatrix} V_1^- \\ V_2^- \end{bmatrix} = \cfrac{1}{\left(1 + \cfrac{z_{11}}{R_1}\right)\left(1 + \cfrac{z_{22}}{R_2}\right) - \cfrac{z_{12}z_{21}}{R_1 R_2}}$$

$$\begin{bmatrix} \left(\dfrac{z_{11}}{R_1} - 1\right)\left(1 + \dfrac{z_{22}}{R_2}\right) - \dfrac{z_{12}z_{21}}{R_1 R_2} & \dfrac{2z_{12}}{R_2} \\[3mm] \dfrac{2z_{21}}{R_1} & \left(\dfrac{z_{22}}{R_2} - 1\right)\left(1 + \dfrac{z_{11}}{R_1}\right) - \dfrac{z_{12}z_{21}}{R_1 R_2} \end{bmatrix} \begin{bmatrix} V_1^+ \\ V_2^+ \end{bmatrix}$$

$$\equiv [S] \begin{bmatrix} V_1^+ \\ V_2^+ \end{bmatrix}$$

Alternatively on solving (3) and (4) for b_1 and b_2 we obtain

$$\begin{bmatrix} a_1 \\ a_2 \end{bmatrix} = \cfrac{1}{\sqrt{(R_1 R_2)}\left(1 + \cfrac{z_{11}}{R_1}\right)\left(1 + \cfrac{z_{22}}{R_2}\right) - \cfrac{z_{12}z_{21}}{\sqrt{(R_1 R_2)}}}$$

$$\begin{bmatrix} \sqrt{(R_1 R_2)}\left(\dfrac{z_{11}}{R_1} - 1\right)\left(1 + \dfrac{z_{22}}{R_2}\right) - \dfrac{z_{12}z_{21}}{\sqrt{(R_1 R_2)}} & 2z_{12} \\[3mm] 2z_{21} & \sqrt{(R_1 R_2)}\left(\dfrac{z_{22}}{R_2} - 1\right)\left(1 + \dfrac{z_{11}}{R_1}\right) - \dfrac{z_{12}z_{21}}{\sqrt{(R_1 R_2)}} \end{bmatrix} \begin{bmatrix} b_1 \\ b_2 \end{bmatrix}$$

$$\equiv [S] \begin{bmatrix} b_1 \\ b_2 \end{bmatrix}$$

For a reciprocal network $z_{12} = z_{21}$, hence in the $[V^-] = [S][V^+]$ case where $S_{12} \propto 2z_{12}/R_2$ and $S_{21} \propto 2z_{21}/R_1$, we have $R_2 S_{12} = R_1 S_{21}$ and in the $[b] = [S][a]$ case, $S_{12} \propto 2z_{12}$, $S_{21} \propto 2z_{21}$ so $S_{12} = S_{21}$.

3. Determine the power reflection (at port 11′) and transmission (port 11′ to port 22′) coefficients for the circuit shown in fig. 6.5B. The line connected to port 22′ is of length l, characteristic impedance Z_C, phase constant β, and is terminated in a load $R_L = 4Z_C$. The scattering matrix of 2-port is

$$\begin{bmatrix} S_{11} & S_{12} \\ S_{21} & S_{22} \end{bmatrix} = \begin{bmatrix} \frac{1}{10} & \frac{1}{2} \\ \frac{1}{2} & \frac{1}{10} \end{bmatrix}$$

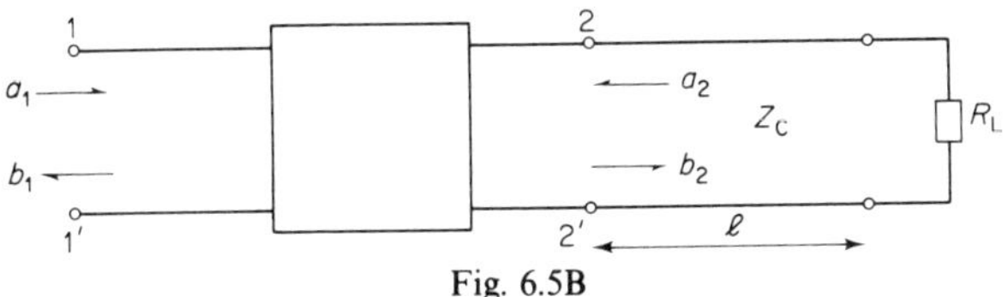

Fig. 6.5B

Solution

The scattering-matrix equations of the 2-port are

$$b_1 = S_{11}a_1 + S_{12}a_2$$

$$b_2 = S_{21}a_1 + S_{22}a_2 \qquad \ldots (1)$$

and the relation between a_2 and b_2 is

$$a_2 = \rho b_2 \, e^{-j2\beta l} \qquad \ldots (2)$$

where $\rho = \dfrac{R_L - Z_C}{R_L + Z_C} = \dfrac{4Z_C - Z_C}{4Z_C + Z_C} = \dfrac{3}{5}$

is the reflection coefficient at the load.

The amplitude reflection and transmission coefficients at port 11', denoted by R and T, are

$$R = \frac{b_1}{a_1} \text{ and } T = \frac{b_2}{a_1}$$

and from (1)

$$R = \frac{b_1}{a_1} = S_{11} + S_{12}\frac{a_2}{a_1} \qquad \ldots (3)$$

$$T = \frac{b_2}{a_1} = S_{21} + S_{22}\frac{a_2}{a_1} \qquad \ldots (4)$$

On substituting for b_2 in (4) using (2) we obtain

$$\frac{a_2}{a_1} = \frac{S_{21}}{\dfrac{1}{\rho}e^{-j2\beta l} - S_{22}}$$

and thus

$$R = S_{11} + \frac{S_{12}S_{21}}{e^{j2\beta l}/\rho - S_{22}} = \frac{10\,e^{j2\beta l} + 14\cdot4}{100\,e^{j2\beta l} - 6}$$

$$T = S_{21} + \frac{S_{22}S_{21}}{e^{j2\beta l}/\rho - S_{22}} = \frac{50\,e^{j2\beta l}}{100\,e^{j2\beta l} - 6}$$

The power reflection and transmission coefficients are, respectively,

$$|R|^2 = RR^* = \frac{10\,e^{j2\beta l} + 14{\cdot}4}{100\,e^{j2\beta l} - 6} \times \frac{10\,e^{-j2\beta l} + 14{\cdot}4}{100\,e^{-j2\beta l} - 6}$$

$$= \frac{100 + 144(e^{j2\beta l} + e^{-j2\beta l}) + 14{\cdot}4^2}{10\,000 - 600(e^{j2\beta l} + e^{-j2\beta l}) + 36}$$

$$= \frac{307{\cdot}36 + 288\cos 2\beta l}{10\,036 - 1200\cos 2\beta l}$$

$$|T|^2 = TT^* = \frac{50^2}{10\,036 - 1200\cos 2\beta l}$$

6.3B. Exercise Problems

1. Determine the incident wave variable a and the scattering coefficient S_{11} at port 11′ for the networks shown in fig. 6.6B.

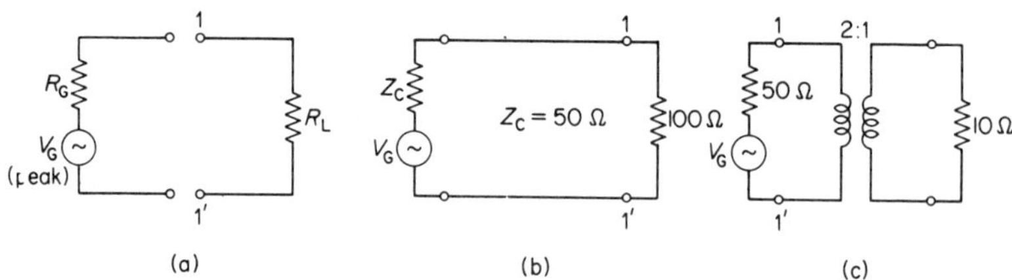

Fig. 6.6B. For problem 1

2. Determine the incident wave variable a and the scattering coefficient S_{11} for the 1-port network of fig. 6.7B. If
   (a) $R_0 = 50\,\Omega$, $R_1 = 30\,\Omega$, determine the value of R_2 for which $S_{11} = 0$.
   (b) $S_{11} = \tfrac{1}{2}$, determine the power supplied to the network.

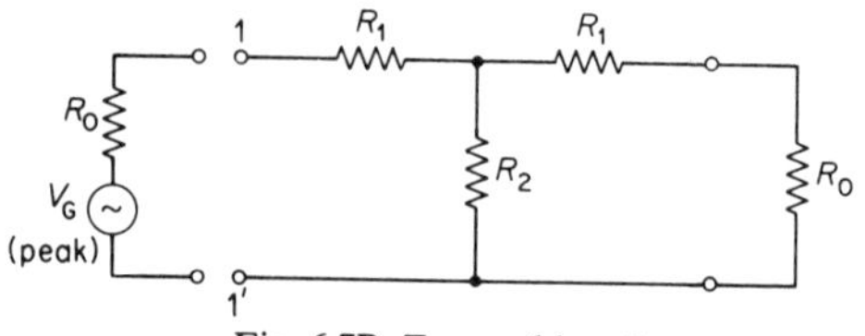

Fig. 6.7B. For problem 2

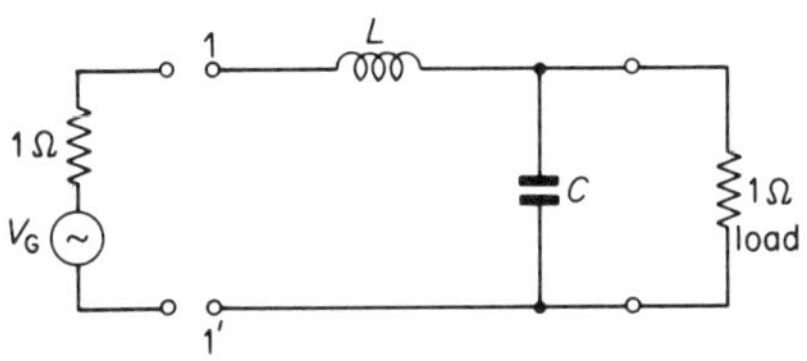

Fig. 6.8B. For problem 3

3. Determine $S_{11}(j\omega)$ for the 1-port shown in fig. 6.8B and if $L = C = \sqrt{2}$ calculate as a fraction of the available power of the generator the power supplied to the $1\,\Omega$ load.

4. Determine $|S_{11}(j\omega)|$ for the network of fig. 6.9B at $\omega = 0, 1, 2\,\text{rad/s}$. Calculate also as a fraction of the available generator power, the power supplied to the $1\,\Omega$ load.

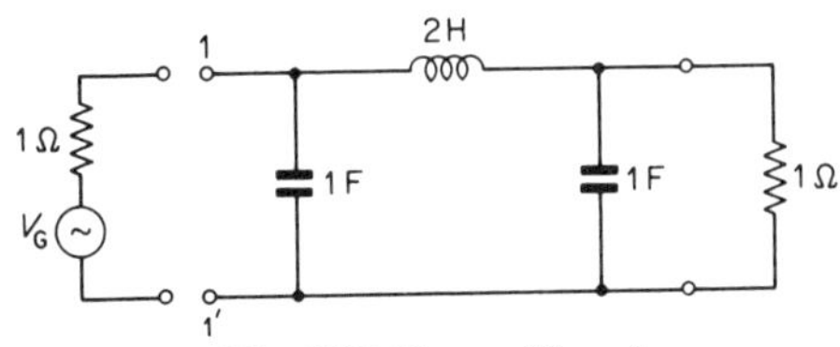

Fig. 6.9B. For problem 4

5. Find the scattering matrix, when normal incident and reflected voltages (V^+, V^-) and also when the scatter variables a and b are used, of
   (a) A length l of uniform transmission line of propagation constant γ.
   (b) An ideal transformer (defined by $V_1 = nV_2$, $I_1 = -I_2/n$).

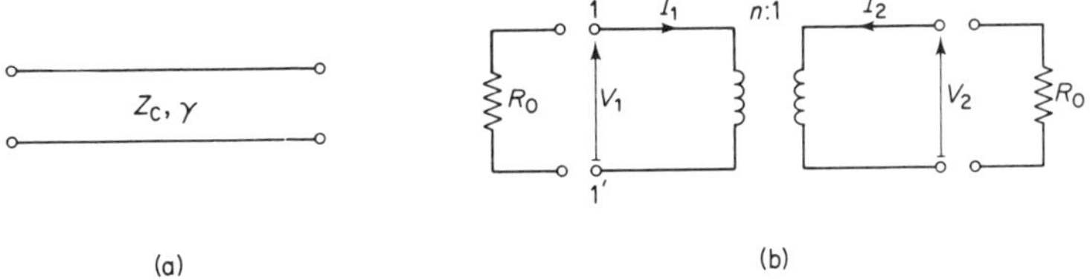

(a)

(b)

Fig. 6.10B. For problem 5

6. Find the scattering-matrix referred to terminal ports immediately to the left and immediately to the right of the junction of the two transmission lines shown in fig. 6.11B when:
   (a) Normal transmission line voltages (V^+, V^-) are used.
   (b) Normalized scatter variables (a, b) are used.

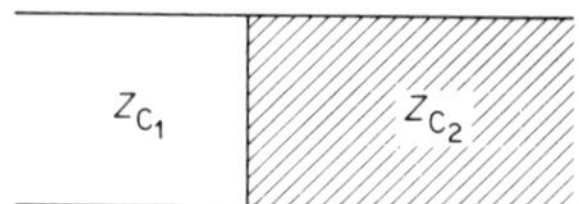

Fig. 6.11B. For problem 6

7. In an experiment to measure the scattering parameters of a reciprocal 2-port network the following results were obtained:

V.S.W.R. in line 1, $S_1 = 2\cdot33$

with line 2 terminated in a matched load,

V.S.W.R. in line 2, $S_2 = 1\cdot22$

with line 1 terminated in a matched load.

Also a bolometer placed at port 2 and matched to line 2 read 5 mW when a power of 20 mW was incident at port 1. From these results determine the magnitudes of the scattering matrix coefficients of the network.

8. Find the scattering matrix of (a) the series impedance Z, (b) the shunt admittance Y shown in fig. 6.12B when incident and reflected voltages (V^+, V^-) and the scatter variables (a, b) are used.

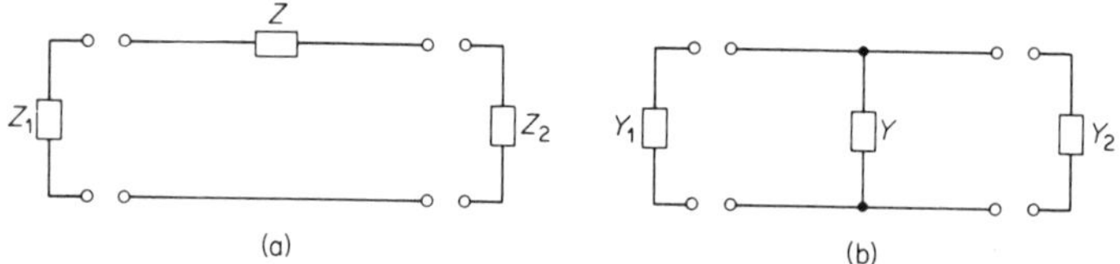

Fig. 6.12B. For problem 8

9. Find the $ABCD$ transmission parameters of a 2-port network in terms of the normalized scattering matrix parameters $S_{11}, S_{12}, S_{21}, S_{22}$. Find also the insertion loss of the network when inserted between a generator of normalized impedance z_G and a load of normalized impedance z_L, as shown in fig. 6.13B. Show, if $z_G = z_L = 1$, that the insertion loss ratio is $1/S_{21}$.

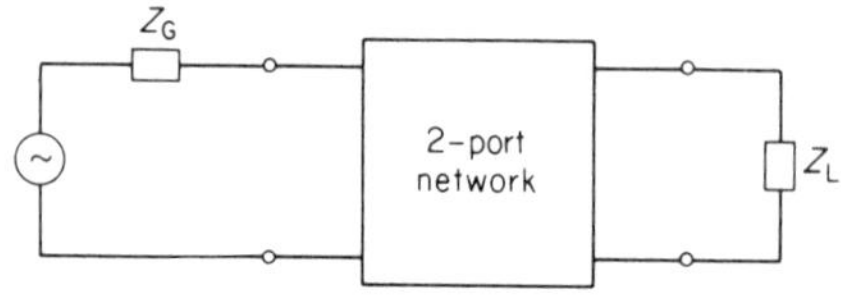

Fig. 6.13B. For problem 9

10. An ideal directional coupler is a 4-port reciprocal network, see fig. 6.14B, and has the following properties:

When any 3 ports are terminated in matched loads there is no reflection at the 4th. If power P enters port 1, C^2P and $(1-C^2)P$ emerges at ports 3 and 2, respectively and zero power emerges at port 4.

If P enters at port 2, C^2P, $(1-C^2)P$, 0 emerges at ports 4, 1, 3.
If P enters at port 3, C^2P, $(1-C^2)P$, 0 emerges at ports 1, 4, 2.
If P enters at port 4, C^2P, $(1-C^2)P$, 0 emerges at ports 2, 3, 1.
Using these properties write down the scattering matrix of the directional coupler. Assume the terminal ports are chosen so that the S_{12} and S_{14} parameters are real, whilst the S_{13} and S_{24} parameters are purely imaginary.

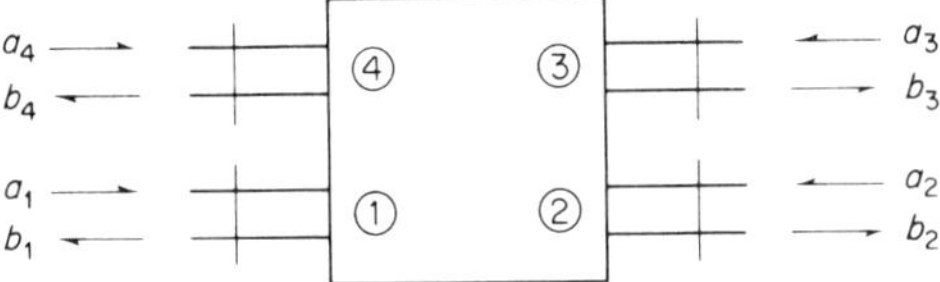

Fig. 6.14B. For problem 10

11. A loss-less, reciprocal 4-port network is defined by

$$
\begin{bmatrix} b_1 \\ b_2 \\ b_3 \\ b_4 \end{bmatrix} = \begin{bmatrix} 0 & S_{12} & 0 & S_{14} \\ S_{12} & 0 & S_{14} & 0 \\ 0 & S_{14} & 0 & S_{34} \\ S_{14} & 0 & S_{34} & 0 \end{bmatrix} \begin{bmatrix} a_1 \\ a_2 \\ a_3 \\ a_4 \end{bmatrix}
$$

If $|S_{14}| = \alpha$, determine $|S_{12}|$ and $|S_{34}|$.

12. Determine the power reflection coefficient $|R|^2 = |b_1/a_1|^2$ at port 11′ for the circuit of fig. 6.15B. The line length $l = \frac{3}{4}\lambda$ and is short-circuited at its far end. The scattering matrix of the 2-port is

$$
\begin{bmatrix} S_{11} & S_{12} \\ S_{21} & S_{22} \end{bmatrix} = \begin{bmatrix} 0.2 & 0.6 \\ 0.6 & 0.2 \end{bmatrix}
$$

Calculate also the value of $|R|^2$ when the line is open-circuited, and the value of a load at AB so that there is zero reflection at 11′.

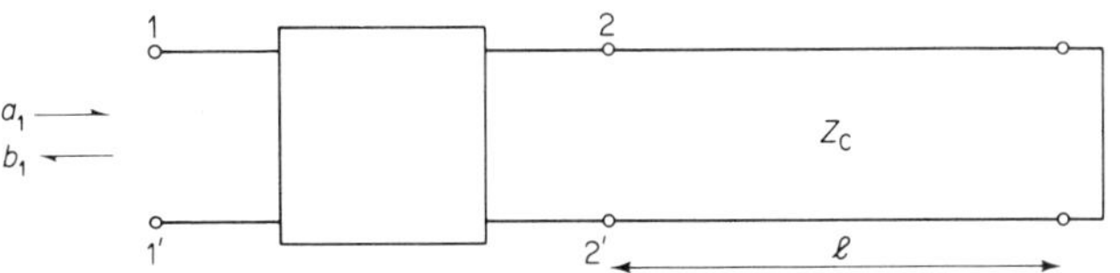

Fig. 6.15B. For problem 12

6C INTRODUCTION TO STATE VARIABLE ANALYSIS

6.1C. Theory Summary

1. INTRODUCTION

In state variable analysis circuit equations can be formulated to solve for both transient and steady state cases. The method is particularly useful in that it is amenable to computer solution, although for less complex circuits state variable analysis may be considerably more involved than transform methods.

2. CONCEPT AND EQUATIONS OF STATE

In previous transient analysis work the state of a passive element network was defined in terms of the currents in the inductors and the p.d.s. across the capacitors or alternatively by the energy stored in the L and C elements. This information together with the network transfer function provided sufficient data to determine the response of a network.

In state variable analysis the state of a network is defined in terms of a state vector $\mathbf{x}$ (or column matrix):

$$\mathbf{x} = \begin{bmatrix} i_{L_1} \\ i_{L_2} \\ \vdots \\ i_{L_n} \\ v_{C_1} \\ \vdots \\ v_{C_m} \end{bmatrix}$$

where $i_{L_1}, i_{L_2} \ldots i_{L_n}$ are the inductor currents,

$v_{C_1}, v_{C_2} \ldots v_{C_m}$ are the capacitor voltages,

Note that the elements of $\mathbf{x}$ may also be linearly independent combinations of the i_Ls and the v_Cs so that in general the state vector $\mathbf{x}$ is that vector whose linearly independent components uniquely describe the state of a network or system at a given or any future time.

The equations of state of a linear system are expressed in terms of a first order matrix differential equation of the form:

$$\frac{\mathrm{d}}{\mathrm{d}t}(\mathbf{x}) = \mathbf{A}\mathbf{x} + \mathbf{B}\mathbf{u} \qquad \ldots (1)$$

206

where $\quad$ $\mathbf{x}$ is the state vector,

$\quad\quad\quad$ $\mathbf{u}$ is the driving function vector (consisting of the applied voltage and current sources),

$\quad\quad\quad$ $\mathbf{A}$ is a square matrix often known as the system matrix (for the given state vector $\mathbf{x}$),

$\quad\quad\quad$ $\mathbf{B}$ is a square matrix often known as the associated matrix of the driving function vector $\mathbf{u}$.

The means of forming the equations of state for R–L–C networks is illustrated in the Worked problem section.

3. ANALYTICAL SOLUTIONS OF STATE EQUATIONS

(a) Force-free response solution

The state equations for the force-free or natural response of a system correspond to setting $\mathbf{u} = \mathbf{0}$ in (1), i.e.

$$\dot{\mathbf{x}} = \mathbf{A}\mathbf{x} \qquad \dots (2)$$

(2) has the solution:

$$\mathbf{x} = e^{\mathbf{A}t}\mathbf{x}_0 \qquad \dots (3)$$

where x_0 is the initial state vector at some specified time t_0 (usually $t_0 = 0$ and corresponds to the instant of switching), $e^{\mathbf{A}t}$ is an exponential function of matrix $\mathbf{A}$ and is a square matrix.

$\quad$ A function $f(\mathbf{A})$ of a matrix $\mathbf{A}$ (in our case of interest $f(\mathbf{A}) = e^{\mathbf{A}t}$) may be determined as follows:

(1) For most linear systems a function of the system matrix $\mathbf{A}$ (where $\mathbf{A}$ is taken as an $n \times n$ matrix) may be expanded as

$$f(\mathbf{A}) = a_0 \mathbf{1} + a_1 \mathbf{A} + a_2 \mathbf{A}^2 + \dots a_k \mathbf{A}^k + \dots a_{n-1} \mathbf{A}^{n-1} \dots (4)$$

where $\mathbf{1}$ is the unit matrix and $a_0, a_1, \dots a_{n-1}$ are constants to be determined, as shown in (2) below.

(2) The eigenvalues* (λ_k, $k = 1, 2, 3 \dots n$) of matrix $\mathbf{A}$ also satisfy equation (4), i.e. when $\mathbf{A}$ is replaced by λ_k

$$f(\lambda_k) = a_0 + a_1 \lambda_k + a_2 \lambda_k^2 + \dots + a_{n-1} \lambda_k^{n-1} \qquad \dots (5)$$

---

* The eigenvalues of a square matrix $\mathbf{A}$ are the roots of the characteristic equation:

$$g(\lambda) = |\mathbf{A} - \lambda \mathbf{1}| = 0$$

which on evaluation of the $|\mathbf{A} - \lambda \mathbf{1}|$ determinant, will reduce to the form:

$$g(\lambda) = \lambda^n + k_{n-1}\lambda^{n-1} + \dots + k_1 \lambda + k_0$$

Thus if we first find the eigenvalues λ_1, $\lambda_2 \ldots \lambda_n$ of $\mathbf{A}$ and substitute each in turn in (5) we obtain n simultaneous equations, which we may solve to determine a_0, a_1, $a_2 \ldots a_{n-1}$.

(b) *Response with driving function*
The solution of

$$\dot{\mathbf{x}} = \mathbf{A}\mathbf{x} + \mathbf{B}\mathbf{u}$$

is

$$\mathbf{x} = e^{\mathbf{A}(t-t_0)}\mathbf{x}_0 + \int_{t_0}^{t} e^{\mathbf{A}(t-\tau)}\mathbf{B}\mathbf{u}(\tau)\,d\tau \qquad \ldots (6)$$

where $\mathbf{x}_0$ is the initial state vector at $t = 0$.

The means of evaluating the solution is illustrated in the following problem. The state equations for the circuit of fig. 6.1C are:

$$d/dt \begin{bmatrix} i_L \\ v_C \end{bmatrix} = \begin{bmatrix} -R/L & -1/L \\ 1/C & -1/[C(R_1+R_2)] \end{bmatrix}$$
$$\begin{bmatrix} i_L \\ v_C \end{bmatrix} + \begin{bmatrix} 1/L & 0 \\ 0 & R_2/[C(R_1+R_2)] \end{bmatrix} \begin{bmatrix} v_1 \\ i_2 \end{bmatrix}$$

$$= \begin{bmatrix} -1 & -1 \\ 2 & -4 \end{bmatrix} \begin{bmatrix} i_L \\ v_C \end{bmatrix} + \begin{bmatrix} 1 & 0 \\ 0 & 3/2 \end{bmatrix} \begin{bmatrix} 4 \\ 8 \end{bmatrix}$$

$$\equiv \mathbf{A}\mathbf{x} + \mathbf{B}\mathbf{u}$$

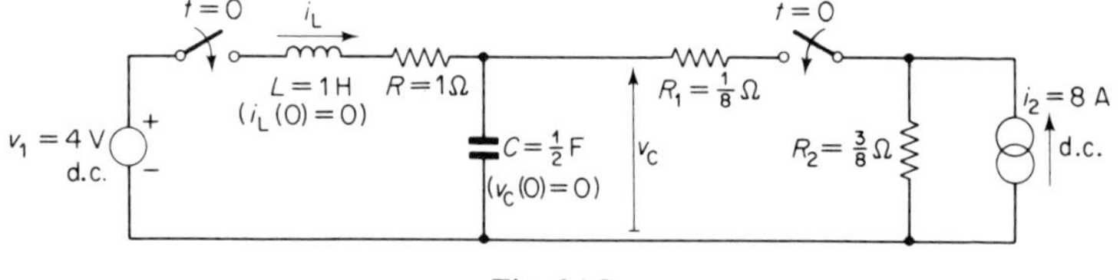

Fig. 6.1C

The solution is

$$\mathbf{x} = e^{\mathbf{A}t}\mathbf{x}_0 + \int_{0}^{t} e^{\mathbf{A}(t-\tau)}\mathbf{B}\mathbf{u}(\tau)\,d\tau$$

$$= \int_{0}^{t} e^{\mathbf{A}(t-\tau)}\mathbf{B}\mathbf{u}(\tau)\,d\tau$$

since the circuit is initially inert, i.e. $i_L(0) = v_C(0) = 0$, and therefore $\mathbf{x}_0 = \mathbf{0}$.

208

To evaluate the integral we first require $e^{\mathbf{A}t}$. The system matrix

$$\mathbf{A} = \begin{bmatrix} -1 & -1 \\ 2 & -4 \end{bmatrix}$$

and its characteristic equation,

$$|\mathbf{A} - \lambda\mathbf{1}| = \begin{bmatrix} -1-\lambda & -1 \\ 2 & -4-\lambda \end{bmatrix}$$

$$= \lambda^2 + 5\lambda + 6 = (\lambda+2)(\lambda+3) = 0$$

so that the eigenvalues of $\mathbf{A}$ are $\lambda_1 = -2$ and $\lambda_2 = -3$. Since $\mathbf{A}$ is 2×2 matrix the expansion for $e^{\mathbf{A}t}$ is

$$e^{\mathbf{A}t} = a_0\mathbf{1} + a_1\mathbf{A}$$

and since λ_1 and λ_2 also satisfy this equation we have

$$e^{\lambda_1 t} = a_0 + a_1\lambda_1 \, , \quad e^{\lambda_2 t} = a_0 + a_1\lambda_2$$

On solving for a_0 and a_1 we obtain:

$$a_0 = 3e^{-2t} - 2e^{-3t}, \quad a_1 = e^{-2t} - e^{-3t}$$

$$\text{so } e^{\mathbf{A}t} = \begin{bmatrix} a_0 & 0 \\ 0 & a_0 \end{bmatrix} + \begin{bmatrix} -a_1 & -a_1 \\ 2a_1 & -4a_1 \end{bmatrix}$$

$$= \begin{bmatrix} 2e^{-2t} - e^{-3t} & -e^{-2t} + e^{-3t} \\ 2(e^{-2t} - e^{-3t}) & -e^{-2t} + 2e^{-3t} \end{bmatrix}$$

and the complete integrand

$$e^{\mathbf{A}t'}\mathbf{Bu} = \begin{bmatrix} 2e^{-2t'} - e^{-3t'} & -e^{-2t'} + e^{-3t'} \\ 2(e^{-2t'} - e^{-3t'}) & -e^{-2t'} + 2e^{-3t'} \end{bmatrix} \begin{bmatrix} 4 \\ 12 \end{bmatrix}$$

$$= \begin{bmatrix} -4e^{-2t'} + 8e^{-3t} \\ -4e^{-2t'} + 16e^{-3t} \end{bmatrix}$$

where $t' = t - \tau$ and as $\mathbf{Bu} = \begin{bmatrix} 4 \\ 12 \end{bmatrix}$

Finally,

$$\mathbf{x} = \begin{bmatrix} i_L \\ v_C \end{bmatrix} = \int_0^t \begin{bmatrix} -4e^{-2(t-\tau)} + 8e^{-3(t-\tau)} \\ -4e^{-2(t-\tau)} + 16e^{-3(t-\tau)} \end{bmatrix} d\tau$$

i.e. $i_L = \int_0^t \{-4e^{-2(t-\tau)} + 8e^{-3(t-\tau)}\} \, d\tau$

$$= \left(\tfrac{2}{3} + 2\,e^{-2t} - \tfrac{8}{3}\,e^{-3t}\right) A$$

$$v_C = \int_0^t \left\{ -4\,e^{-2(t-\tau)} + 16\,e^{-3(t-\tau)} \right\} d\tau$$

$$= \left(\tfrac{10}{3} + 2\,e^{-2t} - \tfrac{16}{3}\,e^{-3t}\right) V.$$

6.2C. Worked Problems

1. Write down the state space equations for the circuit shown in fig. 6.2C.

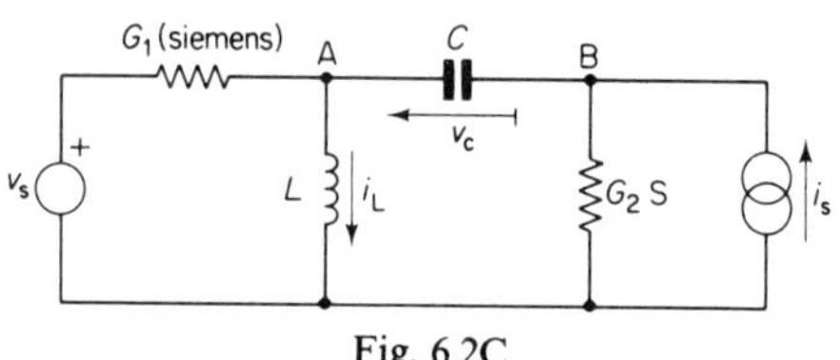

Fig. 6.2C

Solution

The state space equations for the circuit of fig. 6.2C have the form:

$$\frac{d}{dt} \begin{bmatrix} i_L \\ v_C \end{bmatrix} = [A] \begin{bmatrix} i_L \\ v_C \end{bmatrix} + [B] \begin{bmatrix} v_S \\ i_S \end{bmatrix} \quad \dots (1)$$

where $\begin{bmatrix} i_L \\ v_C \end{bmatrix}$, $\begin{bmatrix} v_S \\ i_S \end{bmatrix}$

are, respectively the state and driving function vectors, and $[A]$, $[B]$ are the system matrix for the above state vector and the associated matrix of the driving function vector. $[A]$ and $[B]$ may be determined by writing down the nodal equations and arranging these in the form of (1). Thus on applying the current law at nodes A and B we obtain:

$$(v_L - v_S)G_1 + i_L + i_C = 0 \quad \dots (2)$$

$$-i_C + G_2(v_L - v_C) = i_S \quad \dots (3)$$

and on substituting $v_L = L\dfrac{di_L}{dt}$, $i_C = C\dfrac{dv_C}{dt}$,

(2) becomes: $G_1 L \dfrac{di_L}{dt} + C\dfrac{dv_C}{dt} = -i_L + G_1 v_S$

(3) becomes: $G_2 L \dfrac{di_L}{dt} - C\dfrac{dv_C}{dt} = G_2 v_C + i_S$

210

which may be written in matrix form as

$$\begin{bmatrix} G_1 L & C \\ G_2 L & -C \end{bmatrix} \frac{d}{dt} \begin{bmatrix} i_L \\ v_C \end{bmatrix} = \begin{bmatrix} -1 & 0 \\ 0 & G_2 \end{bmatrix} \begin{bmatrix} i_L \\ v_C \end{bmatrix} + \begin{bmatrix} G_1 & 0 \\ 0 & 1 \end{bmatrix} \begin{bmatrix} v_S \\ i_S \end{bmatrix}$$

and on multiplying throughout by the inverse matrix,

$$\begin{bmatrix} G_1 L & C \\ G_2 L & -C \end{bmatrix}^{-1} = \frac{-1}{LC(G_1 + G_2)} \begin{bmatrix} -C & -C \\ -G_2 L & G_1 L \end{bmatrix}$$

we obtain

$$\frac{d}{dt} \begin{bmatrix} i_L \\ v_C \end{bmatrix} = \frac{-1}{LC(G_1 + G_2)} \left\{ \begin{bmatrix} C & -CG_2 \\ G_2 L & G_1 G_2 L \end{bmatrix} \begin{bmatrix} i_L \\ v_C \end{bmatrix} \right.$$
$$\left. + \begin{bmatrix} -CG_1 & -C \\ -G_1 G_2 L & G_1 L \end{bmatrix} \begin{bmatrix} v_S \\ i_S \end{bmatrix} \right\}$$

2. Write down the state equations for the circuit shown in fig. 6.3C when the switch S is in (a) position 1, (b) position 2. If $v_S = 10$ V d.c., $L = 1$ H, $R = 5\,\Omega$, $C = \frac{1}{6}$ F and the circuit has reached its steady state variation with S in position 1, find the subsequent variation of i_L and v_C when S is switched to position 2 at $t = 0$.

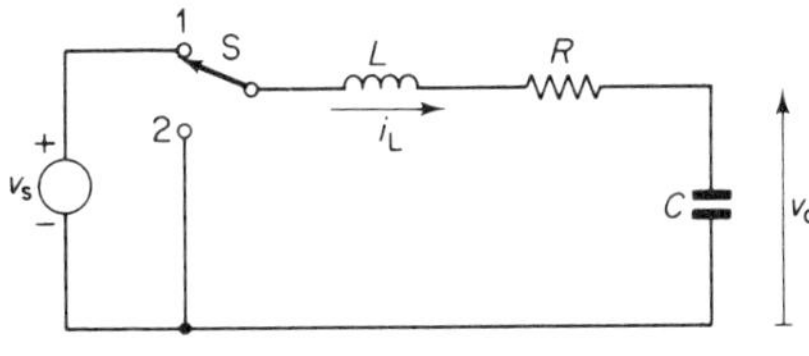

Fig. 6.3C

Solution
(a) In position 1, the differential equations of the circuit are

$$v_S = L\frac{di_L}{dt} + Ri_L + v_C, \text{ i.e. } \frac{di_L}{dt} = -\frac{R}{L}i_L - \frac{1}{L}v_C + \frac{1}{L}v_S$$

$$i_L = C\frac{dv_C}{dt}, \text{ i.e. } \frac{dv_C}{dt} = \frac{1}{C}i_L$$

hence $\dfrac{d}{dt} \begin{bmatrix} i_L \\ v_C \end{bmatrix} = \begin{bmatrix} -R/L & -1/L \\ 1/C & 0 \end{bmatrix} \begin{bmatrix} i_L \\ v_C \end{bmatrix} + \begin{bmatrix} 1/L \\ 0 \end{bmatrix} \begin{bmatrix} v_S \end{bmatrix}$

(b) When S is in position 2, $v_S = 0$ and hence the state equations reduce to

$$\frac{d}{dt}\begin{bmatrix} i_L \\ v_C \end{bmatrix} = \begin{bmatrix} -R/L & -1/L \\ 1/C & 0 \end{bmatrix}\begin{bmatrix} i_L \\ v_C \end{bmatrix} \qquad \ldots (1)$$

i.e. of the form $\dot{\mathbf{x}} = \mathbf{A}\mathbf{x}$.

The solution of equation (1) is

$$\mathbf{x} = e^{\mathbf{A}t}\mathbf{x}(0)$$

$$\text{where } \mathbf{A} = \begin{bmatrix} -R/L & -1/L \\ 1/C & 0 \end{bmatrix} = \begin{bmatrix} -5 & -1 \\ 6 & 0 \end{bmatrix}$$

$$\mathbf{x}(0) = \begin{bmatrix} i_L(0) \\ v_C(0) \end{bmatrix} = \begin{bmatrix} 0 \\ 10 \end{bmatrix} \ldots \text{ the initial value of the state vector at } t = 0.$$

Now the exponential matrix may be expanded in the form,

$$e^{\mathbf{A}t} = a_0[1] + a_1[A] \qquad \ldots (2)$$

where the coefficients a_0 and a_1 are to be determined as follows. The eigenvalues, λ_1 and λ_2, of $\mathbf{A}$ also satisfy equation (2), so

$$e^{\lambda_1 t} = a_0 + a_1 \lambda_1$$

$$e^{\lambda_2 t} = a_0 + a_1 \lambda_2$$

$$\text{hence } a_0 = \frac{\lambda_2 e^{\lambda_1 t} - \lambda_1 e^{\lambda_2 t}}{\lambda_2 - \lambda_1}, \quad a_1 = \frac{e^{\lambda_1 t} - e^{\lambda_2 t}}{\lambda_1 - \lambda_2}$$

The eigenvalues themselves being found by solving the characteristic equation,

$$|\mathbf{A} - \lambda\mathbf{1}| = \begin{vmatrix} -5 - \lambda & -1 \\ 6 & -\lambda \end{vmatrix} = 0$$

i.e. $\lambda^2 + 5\lambda + 6 = 0$, so $\lambda_1 = -2$, $\lambda_2 = -3$.

Thus $a_0 = 3e^{-2t} - 2e^{-3t}$, $a_1 = e^{-2t} - e^{-3t}$,

and $e^{\mathbf{A}t} = (3e^{-2t} - 2e^{-3t})\begin{bmatrix} 1 & 0 \\ 0 & 1 \end{bmatrix} + (e^{-2t} - e^{-3t})\begin{bmatrix} -5 & -1 \\ 6 & 0 \end{bmatrix}$

$$= \begin{bmatrix} -2e^{-2t} + 3e^{-3t} & -e^{-2t} + e^{-3t} \\ 6e^{-2t} - 6e^{-3t} & 3e^{-2t} - 2e^{-3t} \end{bmatrix}$$

and the solution

$$\mathbf{x} = \begin{bmatrix} i_L \\ v_C \end{bmatrix} = e^{\mathbf{A}t}\mathbf{x}(0) = \begin{bmatrix} -2e^{-2t} + 3e^{-3t} & -e^{-2t} + e^{-3t} \\ 6e^{-2t} - 6e^{-3t} & 3e^{-2t} - 2e^{-3t} \end{bmatrix}\begin{bmatrix} 0 \\ 10 \end{bmatrix}$$

i.e. $i_L = 10(-e^{-2t} + e^{-3t})\,\text{A}$, $v_C = 10(3e^{-2t} - 2e^{-3t})\,\text{V}$.

6.3C. Exercise Problems

1. Write down the state equations for the circuits of fig. 6.4C(a) and (b).

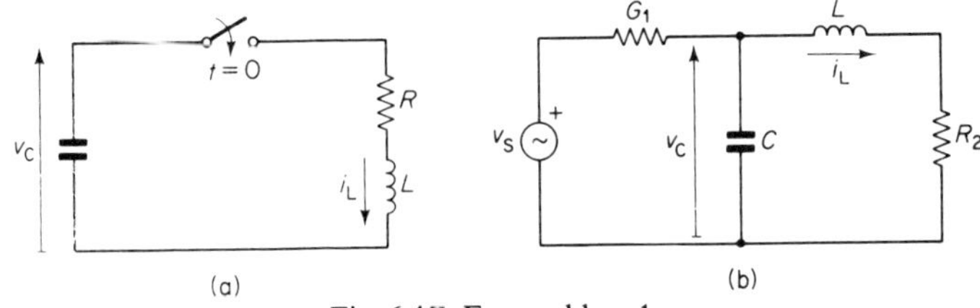

(a)   (b)

Fig. 6.4C. For problem 1

2. Write down the state equations of the circuits shown in fig. 6.5C(a) and (b).

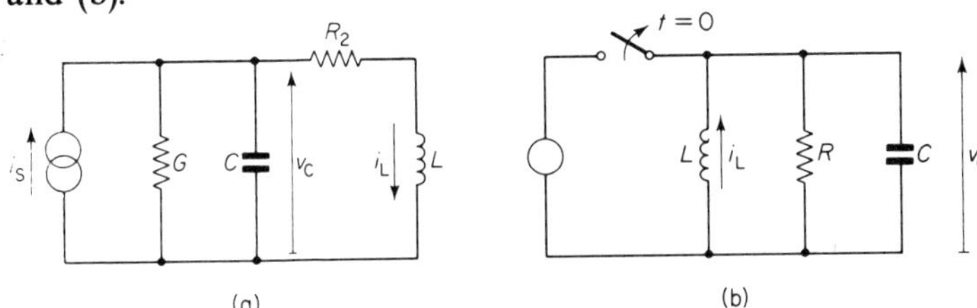

(a)   (b)

Fig. 6.5C. For problem 2

3. Write down the state equations of the circuit of fig. 6.6C.

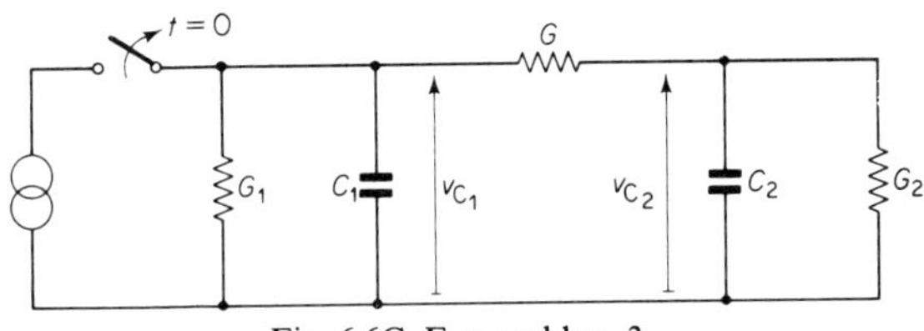

Fig. 6.6C. For problem 3

4. Write down the state equations of the circuit shown in fig. 6.7C at $t \geqslant 0$.

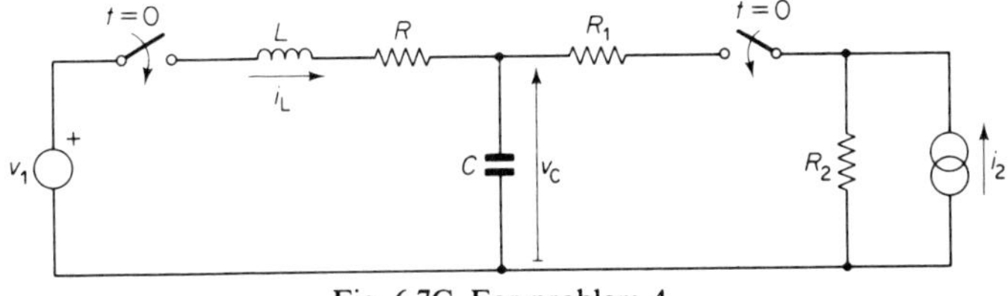

Fig. 6.7C. For problem 4

5. Prior to switching at $t=0$, in the circuit of fig. 6.8C the capacitor voltage $x_2 = 10\,\text{V}$ and the inductor current $x_1 = 0$. Write down the state equations of the circuit for $t>0$ and solve them.

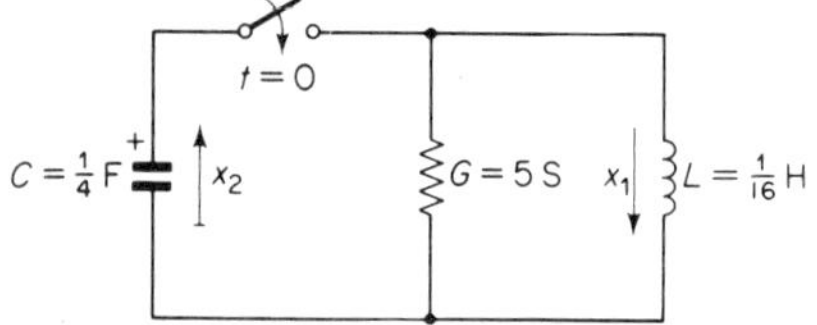

Fig. 6.8C. For problem 5

6. Determine the state equations for the circuit of fig. 6.9C for $t \geqslant 0$.

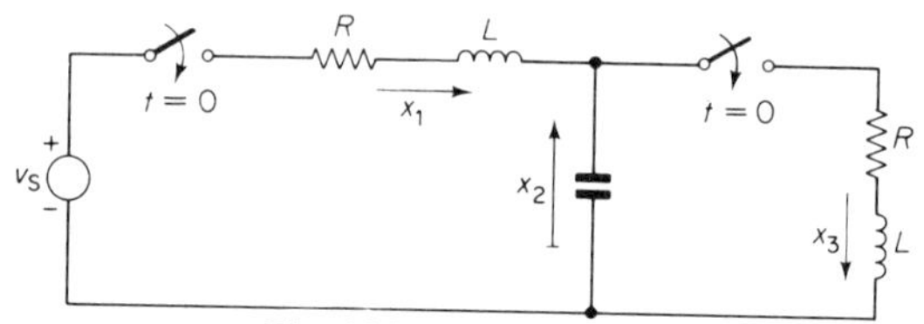

Fig. 6.9C. For problem 6

7. Write down the state equations for the circuit of fig. 6.10C for $t \geqslant 0$. Given that $x_1(0) = x_2(0) = 0$, determine the state vector $\mathbf{x}(t)$.

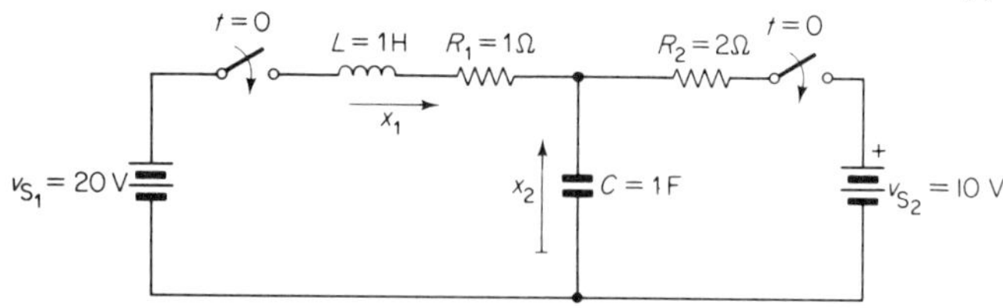

Fig. 6.10C. For problem 7

8. Write down the state equations for the circuit of fig. 6.11C for $t \geqslant 0$. If $R_1 = 20\,\Omega$, $R_2 = 100\,\Omega$, $L_1 = 1\,\text{H}$, $L_2 = 2\,\text{H}$, $M = 1\,\text{H}$ and $x_1(0) = x_2(0)$ determine the state vector $\mathbf{x}(t)$ when the input driving function is $40\,\text{e}^{-5t}$ volts.

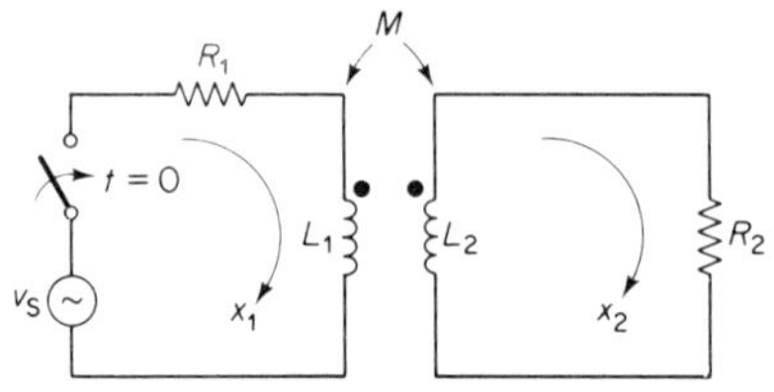

Fig. 6.11C. For problem 8

214

9. Write down the state equation of the controlled source circuit shown
in fig. 6.12C.

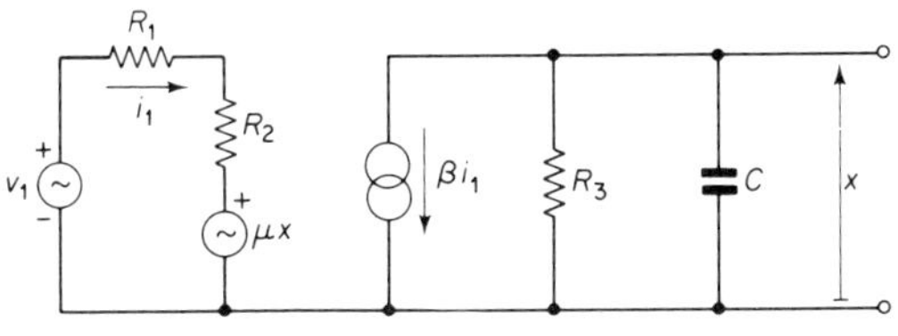

Fig. 6.12C. For problem 9

Answers

1. $v_C = \dfrac{R_2 V}{R_1 + R_2}[1 - e^{-(R_1 + R_2)t/(CR_1 R_2)}]u(t)$ volts

2. $v_2 = [Ke^{-t/(CR)}u(t) - Ke^{-(t-a)/(CR)}u(t-a)]$ volts

3. (a) $i = \dfrac{1}{L}e^{-(R/L)t}u(t)$

   (b) $i = \dfrac{1}{R}(1 - e^{-(R/L)t})u(t)$

   (c) $i = \dfrac{1}{R}\left(t - \dfrac{L}{R} + \dfrac{L}{R}e^{-(R/L)t}\right)u(t)$

4. The transformed circuit is shown in fig. A1

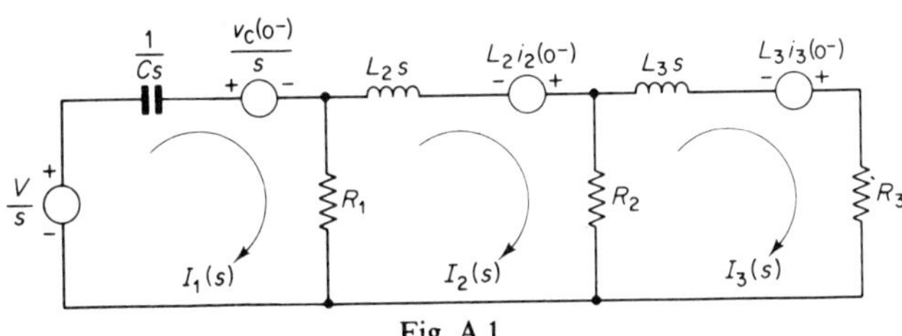

Fig. A.1

The mesh equations of fig. A1 are

$$
\begin{aligned}
(R_1 + 1/(Cs))I_1 \qquad\qquad -R_1 I_2 \qquad\qquad\qquad\qquad &= V/s - v_C(0^-)/s \\
-R_1 I_1 + (R_1 + R_2 + L_2 s)I_2 \qquad\qquad -R_2 I_3 &= L_2 i_2(0^-) \\
-R_2 I_2 + (R_2 + R_3 + L_3 s)I_3 &= L_3 i_3(0^-)
\end{aligned}
$$

5. The mesh equations for the transformer circuit are
$$
\begin{aligned}
(R_1 + L_1 s)I_1 \qquad\qquad + MsI_2 &= Vs/(s^2 + \omega^2) \\
MsI_1 \qquad + (R_2 + L_2 s)I_2 &= 0
\end{aligned}
$$
$$
I_2(s) = -MVs^2/[(s^2 + \omega^2)\{(R_1 + L_1 s)(R_2 + L_2 s) - M^2 s^2\}]
$$

6. The transformed circuit is shown in fig. A2.
   The nodal equations of fig. A2 are:
$$
\left[\frac{1}{R_1} + \frac{1}{R_3} + \frac{1}{R_2 + 1/(Cs)}\right]V_1 - \left[\frac{1}{R_3} + \frac{1}{R_2 + 1/(Cs)}\right]V_2 = I(s) + \frac{v_C(0^-)}{s(R_2 + 1/(Cs))}
$$

$$-\left[\frac{1}{R_3}+\frac{1}{R_2+1/(Cs)}\right]V_1+\left[\frac{1}{R_3}+\frac{1}{R_4}+\frac{1}{R_2+1/(Cs)}\right]V_2=\frac{V(s)}{R_4}-\frac{v_c(0^-)}{s(R_2+1/(Cs))}$$

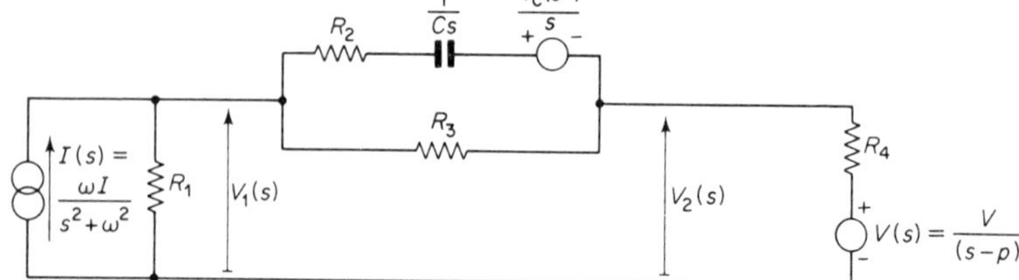

Fig. A.2

7. $$v_C=\frac{V}{L_1C\omega}\left[\frac{\omega}{\alpha^2+\omega^2}\,e^{-\alpha t}-\frac{\omega}{\beta^2+\omega^2}\,e^{-\beta t}\right.$$

$$\left.+\frac{\beta-\alpha}{\sqrt{\{(\alpha^2+\omega^2)(\beta^2+\omega^2)\}}}\,\sin(\omega t-\phi)\right]\text{volts}$$

where $\omega=\sqrt{\left(\dfrac{L_1+L_2}{L_1L_2C}\right)}$, $\phi=\tan^{-1}\left\{\dfrac{\omega(\alpha+\beta)}{\alpha\beta-\omega^2}\right\}$

8. $v_C=[50-e^{-t}(30\cos t+10\sin t]$ volts

9. $i_1=0\!\cdot\!8[1\!\cdot\!65e^{-5t}+1\!\cdot\!25e^{-123\cdot8t}-2\!\cdot\!90e^{-16\cdot2t}]$ amps
   $i_2=10^{-3}[-1\!\cdot\!83e^{-5t}+8\!\cdot\!93e^{-123\cdot8t}-7\!\cdot\!10e^{-16\cdot2t}]$ amps

10. The transformed circuit is shown in fig. A3

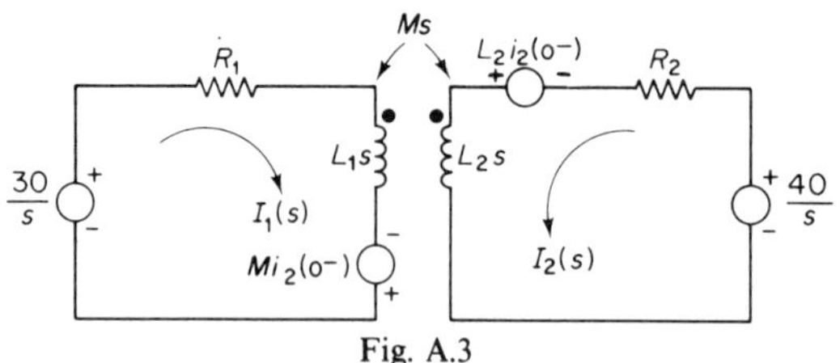

Fig. A.3

$i_1=3-2e^{-5(/2)t}-e^{-10t}$ A
$i_2=2-e^{-5(/2)t}+e^{-10t}$ A

11. $i_1=\frac{2}{3}-\frac{1}{3}e^{-(15/6)t}$ A

12. $i_1=-\frac{3}{5}e^{-25t}$ A, $i_2=\frac{3}{5}e^{-25t}$ A

13. $v_R=\frac{1}{5}CR(1-e^{-t/(CR)})$ for $0<t<5\,\mu\text{s}$
    $=(1-v_1)e^{-(t-5)/(CR)}$ for $5<t<15\,\mu\text{s}$
    $=(1-v_2)e^{-(t-15)/(CR)}-\frac{1}{5}CR(1-e^{-(t-15)/(CR)})$ for $15<t<20\,\mu\text{s}$

where $v_1 = 1 - \frac{1}{5}CR(1 - e^{-5/(CR)})$, $v_2 = 1 - (1 - v_1)e^{-10/(CR)}$, t in μs. Sketch of variation of v_R with time for $C = 1000\,\text{pF}$, $R = 10\,\text{k}\Omega$ is shown in fig. A4.

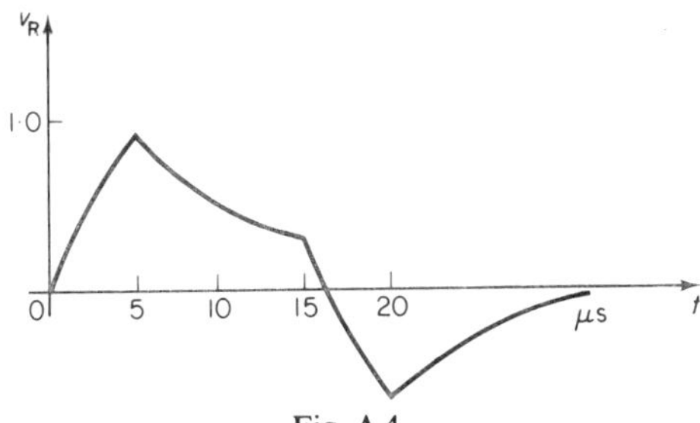

Fig. A.4

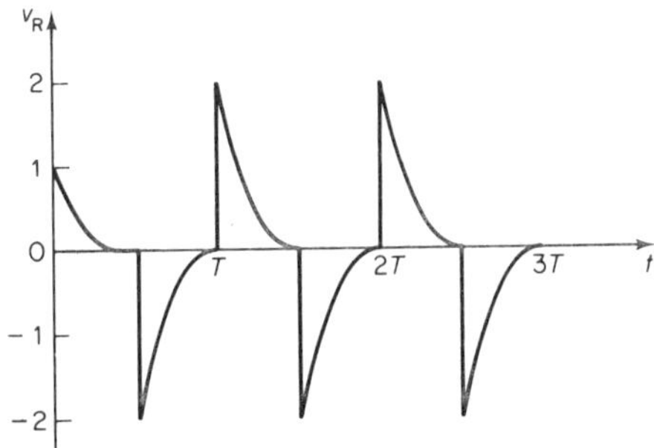

Fig. A.5

14. $v_R = e^{-\alpha t} - 2e^{-\alpha(t - 1/2T)}u(t - \frac{1}{2}T) + 2e^{-\alpha(t - T)}u(t - T) - \ldots$

where $\alpha = 1/(CR)$. The variation of v_R with time when $T = 0.1\,CR$ is shown in fig. A5

15. $i_1 = 1.25 - 0.625(e^{-25t} + e^{-100t})$

$i_2 = 0.62(e^{-25t} - e^{-100t})$

and is a maximum at $t = \dfrac{\ln 4}{75} = 0.0185s$

16. $v_2 = \dfrac{L_2 V}{2L + L_2}\left[1 - \cos\sqrt{\left(\dfrac{2L + L_2}{CL(L + L_2)}\right)t}\right]$ volts

17. $Z(s) = \dfrac{1}{Cs + G + 1/(Ls)} = \dfrac{1}{C}\dfrac{s}{s^2 + (G/C)s + 1/(LC)}$

(a) $V(s) = Z(s)\dfrac{s}{s^2 + \omega^2}$

(b) $V(s) = Z(s)I \dfrac{1}{s}(1 - e^{-Ts})$

(c) $V(s) = Z(s)I \dfrac{1}{s^2}(1 - e^{-Ts})$

18. $Y_{21}(s) = \dfrac{R_1 + L_1 s}{(R + Ls)(R_1 + R_2 + L_1 s) + R_2(R_1 + L_1 s)}$

$$I_2(s) = Y_{21}(s)\left[\dfrac{V_1}{s} + Li(0^-)\right] - \dfrac{L_1 i_1(0^-)(R + Ls)}{R_1(R + R_2 + Ls) + R_2(R + Ls)}$$

19. $I_2(s) = \dfrac{Ms}{(R + Ls)(R + Ls + 1/(Cs)) - M^2 s^2}$

$i_2(t) = \sqrt{2}[2e^{-2t} - e^{-t}(\cos t + \sin t)]$

20. $Y(s) = \dfrac{Ls^2 + 2Rs + 1/C}{2RLs^2 + (3R^2 + L/C)s + 2R/C}$

The poles of $Y(s)$:

$$s_{1,2} = \dfrac{1}{4RL}\left[-(3R^2 + L/C) \pm \sqrt{\left\{(3R^2 + L/C)^2 - 16\dfrac{L}{C}R^2\right\}}\right]$$

The condition for the response to be oscillatory is

$$16\dfrac{L}{C}R^2 > (3R^2 + L/C)^2$$

The radian frequency ω_n and the damping factor α of the natural response are

$$\omega_n = \dfrac{1}{4RL}\sqrt{\left\{16\dfrac{L}{C}R^2 - \left(3R^2 + \dfrac{L}{C}\right)\right\}}, \quad \alpha = -\dfrac{1}{4RL}(3R^2 + L/C)$$

When $R = L = C = 1$, $s_1 = s_2 = -1$ so the response is not oscillatory

21. $Z(s) = 3R \dfrac{s^2 + \dfrac{1}{3RL}(11R^2 + L/C)s + \dfrac{4}{3LC}}{s^2 + \dfrac{5R}{L}s + \dfrac{1}{LC}}$

$$\equiv 3R \dfrac{(s - z_1)(s - z_2)}{(s - p_1)(s - p_2)}$$

The poles of $Z(s)$ are:

$$p_{1,2} = \frac{1}{2}\left[-\frac{5R}{L} \pm \sqrt{\left\{\left(\frac{5R}{L}\right)^2 - \frac{4}{LC}\right\}}\right]$$

and the zeros

$$z_{1,2} = \frac{1}{2}\left[-\frac{1}{3RL}(11R^2 + L/C) \pm \sqrt{\left\{\frac{1}{9R^2L^2}(11R^2 + L/C)^2 - \frac{16}{3LC}\right\}}\right]$$

(a) $i_2(t) = Ae^{p_1 t} + Be^{p_2 t}$

(b) $i(t) = Ce^{z_1 t} + De^{z_2 t}$

22. $\dfrac{V_0(s)}{V_i(s)} = \dfrac{-\mu s}{R_2 C_2(s + 1/\tau_1)(s + 1/\tau_2)}$

where $\tau_1 = R_1 C_1$, $\tau_2 = R_2 R_3 C_2/(R_2 + R_3)$

$$v_0(t) = \frac{-\mu}{R_2 C_2}\left(\frac{\tau_1 \tau_2}{\tau_1 - \tau_2}\right)[e^{-t/\tau_1} - e^{-t/\tau_2}] \text{ volts}$$

23. $\dfrac{V_0(s)}{V_i(s)} = \dfrac{ARCs}{(1 - A)(1 + RCs)^2 + RCs}$

$$v_0(t) = -\frac{1}{RC}e^{-t/(RC)}\left[1 - \frac{1}{RC}t\right] \text{ volts}$$

24. $H(s) = \dfrac{(s + 3)^2}{s^2 + 5s + 7}$

$$y(t) = \frac{7}{9}u(t) + e^{-5/2t}\left(\frac{2}{9}\cos\frac{\sqrt{3}}{2}t + \frac{28}{9\sqrt{3}}\sin\frac{\sqrt{3}}{2}t\right)u(t)$$

25. $H(s) = \dfrac{A(s + 1)}{s^2 + As + A}$

26. $Y(s) = \dfrac{s^2 + 2s + 1}{s(s^2 + 4s + 5)}$

$y(t) = (\frac{1}{5} + \frac{4}{5}e^{-2t}\cos t)u(t)$

27. (a) zeros at 0 and $\pm j\sqrt{[(L + L_1)/(LL_1 C_1)]}$; poles at $\pm j1/\sqrt{(L_1 C_1)}, \infty$

(b) zeros at $\pm j1/\sqrt{[C_2(L_2 + L)]}, \infty$; poles at 0 and $\pm j1/\sqrt{(L_2 C_2)}$

28. Zeros at 0 and $\pm j\sqrt{[(L_1 + L_3)/\{L_1 L_3(C_1 + C_3)\}]}$, ∞; poles at $\pm j1/\sqrt{(L_1 C_1)}$, $\pm j1/\sqrt{(L_3 C_3)}$.

Zeros and poles occur alternately (interlace) along the $j\omega$ axis

29. $H(s) = \dfrac{1}{LC[s^2 + (R/L)s + 1/(LC)]}$

with poles at $-\dfrac{R}{2L} \pm j\sqrt{\left[\dfrac{1}{LC} - \dfrac{R^2}{4L}\right]}$

$|H(j\omega)|$ is at a maximum when $\omega = \sqrt{(\beta^2 - \alpha^2)}$

30. Maximum amplitude $= 20\,\text{dB}$ (at $\omega = 1$); the low frequency asymptote is $0\,\text{dB}$ and the high frequency (for $\omega \gg 1$) is $-40\log_{10}\omega\,\text{dB}$

CHAPTER 2 : FOURIER METHODS IN WAVEFORM AND CIRCUIT ANALYSIS

1. (a) $v(t) = \dfrac{2}{\pi}\sum_{n=1}^{\infty}\dfrac{(-1)^{n+1}}{n}\sin n\omega t$

$= \dfrac{2}{\pi}\left(\sin\omega t - \dfrac{1}{2}\sin 2\omega t + \dfrac{1}{3}\sin 3\omega t - \dots\right)$

r.m.s. value $= 1/\sqrt{3}$

(b) $v(t) = 1 + \dfrac{2}{\pi}\sum_{n=1}^{\infty}\dfrac{(-1)^{n+1}}{n}\sin n\omega t,$

r.m.s. value $= \sqrt{[1^2 + (1/\sqrt{3})^2]} = 2/\sqrt{3}$

(c) $v(t) = \dfrac{1}{2} - \dfrac{4}{\pi^2}\sum_{m=0}^{\infty}\dfrac{1}{(2m+1)^2}\cos(2m+1)\omega t,$

r.m.s. value $= 1/\sqrt{3}$

2. $v(t) = \dfrac{2}{\pi} + \sum_{m=1}^{\infty}\dfrac{-4}{\pi(4m^2 - 1)}\cos 2m\omega t$

$V_R(2\omega) = \dfrac{4}{3\pi(4\omega^2 LC)}, \quad V_R(4\omega) = \dfrac{4}{15\pi(16\omega^2 LC)}$

3. $i(t) = \dfrac{2}{T}\left(\dfrac{1}{2} + \cos\omega t + \cos 2\omega t + \cos 3\omega t + \dots\right)$ where $\omega = 2\pi/T$

The frequency spectrum of the train is shown in fig. A.6

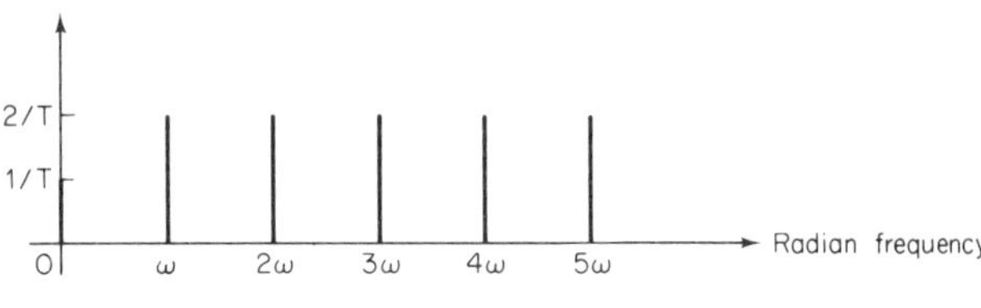

Fig. A.6

4. $v_i(t) = \dfrac{4}{\pi}\left(\cos\omega t - \dfrac{1}{3}\cos 3\omega t + \dfrac{1}{5}\cos 5\omega t - \dots\right)$

$$= \dfrac{4}{\pi}\sum_{m=0}^{\infty}\dfrac{(-1)^m}{2m+1}\cos(2m+1)\omega t$$

$$v_0(t) = \dfrac{4}{\pi}\sum_{m=0}^{\infty}\dfrac{(-1)^m}{(2m+1)}\,\mathrm{Re}\left[\dfrac{1}{\mathrm{j}(2m+1)\omega C}\dfrac{1}{R+1/[\mathrm{j}(2m+1)\omega C]}e^{\mathrm{j}\omega t}\right]$$

$$\approx \dfrac{4}{\pi}\sum_{m=0}^{\infty}\dfrac{(-1)^m}{(2m+1)}\,\mathrm{Re}\left[\dfrac{1}{\mathrm{j}(2m+1)\omega CR}e^{\mathrm{j}\omega t}\right]\text{ when }RC \gg T$$

$$= \dfrac{4}{\pi RC}\left[\dfrac{\sin\omega t}{\omega} - \dfrac{1}{3}\dfrac{\sin 3\omega t}{3\omega} + \dfrac{1}{5}\dfrac{\cos 5\omega t}{5\omega} - \dots\right]$$

5. $i(t) = \dfrac{8}{\pi^2}\left(\sin\omega t - \dfrac{1}{9}\sin 3\omega t + \dfrac{1}{25}\sin 5\omega t - \dots\right),\ \omega = 2\pi/T$

Power dissipated in $R = \frac{1}{3}R$ watts
Power dissipated if filter present

$$= R \times \dfrac{1}{2}\left(\dfrac{8}{\pi}\right)^2\left[1^2 + \dfrac{1}{9^2} + \dfrac{1}{25^2}\right] = 0.3331R$$

6. Amplitude of d.c. component $= \dfrac{\Delta t}{T}$

$$\text{Amplitude of } n^{\text{th}}\text{ harmonic} = 2\dfrac{\Delta t}{T}\dfrac{\sin x}{x},\ x = \tfrac{1}{2}n\omega\Delta t$$

where $\Delta t = 10\mu s$, $T = 100\mu s$, $\omega = 2\pi/T = 2\pi \times 10^4$ rad/s. Fraction of total power contained in harmonics 0 to $100\,\text{kHz} = 0.89$ Voltage outputs at $10\,\text{kHz}$ and $40\,\text{kHz} = 18.9\,\text{V}$, $14.0\,\text{V}$, respectively

7. (a) r.m.s. voltage $= \sqrt{(63.7^2 + 70.7^2 + 30^2)} = 99.8\,\text{V}$
   r.m.s. current $= \sqrt{(10^2 + 5.0^2 + 3.0^2)} = 11.6\,\text{A}$
   (b) $P = 63.7 \times 10 + 70.7 \times 5\cos 40° + 30 \times 3\cos 27° = 985.5\,\text{W}$
   (c) power factor $= \dfrac{P}{99.8 \times 11.6} = 0.8513$

8. Frequency components present in output: d.c., ω, 2ω, 19ω, 20ω, 21ω, 40ω and their respective powers dissipated in R:
   $(bV_m^2)^2/R$, $\frac{1}{2}(aV_m)^2/R$, $\frac{1}{2}(\frac{1}{2}bV_m^2)^2/R$, $\frac{1}{2}(bV_m^2)^2/R$,
   $\frac{1}{2}(aV_m)^2/R$, $\frac{1}{2}(bV_m^2)^2/R$, $\frac{1}{2}(\frac{1}{2}bV_m^2)^2/R$ watts

9. Amplitudes of frequency components in current wave:
   at ω_c: A, at $\omega_c - \omega_m$: $\frac{1}{2}mA$, at $\omega_C + \omega_m$: $\frac{1}{2}mA$

Amplitudes of output voltage:
at ω_c: AR, at $\omega_c - \omega_m$ and $\omega_c + \omega_m$:

$$\tfrac{1}{2}mAR \left/ \sqrt{\left(1 + 4Q^2\, \frac{\omega_m^2}{\omega_c^2}\right)}\right.$$

10. Detector output voltage $= A[(\cos\omega_m t + 2\cdot5)^2 + \sin\omega_m t]^{1/2}$
$$= A\sqrt{(7\cdot25)}[1 + 5/7\cdot25\cos\omega_m t]^{1/2}$$

$$= A\sqrt{7\cdot25}\left[1 + \frac{1}{2}\times 0\cdot69\cos\omega_m t + \frac{1}{2}\times -\frac{1}{2}\times\frac{1}{2!}(0\cdot69\cos\omega_m t)^2 + \ldots\right]$$

$$= A\sqrt{7\cdot25}[1 + 0\cdot345\cos\omega_m t - 0\cdot0298(1 + \cos 2\omega_m t) + \ldots]$$

Amplitude of second harmonic
$$(2\omega_m) = 0\cdot298 \times \sqrt{7\cdot25}\,A = 0\cdot0793\,A$$

11. P.A.M. waveform $v(t) = (1 + 0\cdot5\sin\omega_m t)$

$$\left[\frac{\Delta t}{T} + \frac{2\Delta t}{T}\sum_{n=1}^{\infty}\frac{\sin\left(\tfrac{1}{2}n\omega\Delta t\right)}{\tfrac{1}{2}n\omega\Delta t}\cos n\omega_c t\right]$$

where $\Delta t = 5\mu s$, $T = 50$, $\omega_c = 2\pi \times 20\,\text{krad/s}$.
On expanding this series and using the fact that
$$\sin\omega_m t\cos n\omega_c t = \tfrac{1}{2}[\sin(n\omega_c + \omega_m)t - \sin(n\omega_c - \omega_m)t]$$
we may obtain the frequency components:

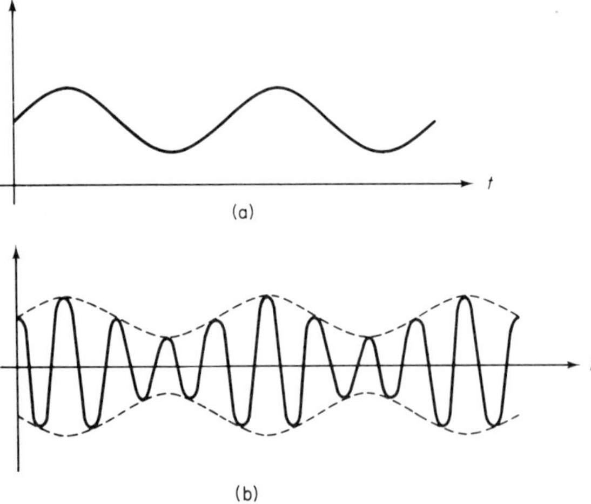

(a)

(b)

Fig. A.7

d.c. 0·1 V, 5 kHz 0·05 V, 15 kHz 0·049 V, 20 kHz 0·197 V, 25 kHz 0·049 V, 35 kHz 0·0468 V, 40 kHz 0·187, 45 kHz 0·0468 V
The voltage outputs when the waveform is passed through a filter are shown in fig. A.7

12. Peak amplitudes of 1st, 3rd, 5th harmonics are 9·542 A, 4·415 A, 2·55 A, respectively

13. $f = 5f_0 = (1/4\pi)10^4$ Hz

$$i = \frac{200}{Z_1}\sin\left(\omega_0 t + \tan^{-1}\frac{1}{5}\right) + \frac{100}{Z_3}\sin\left(3\omega_0 t - \tan^{-1}0·467\right)$$

$$+ \frac{50}{Z_5}\sin\left(5\omega_0 t - \tan^{-1}0·92\right)\text{A}$$

where $Z_1 = |50 - j10|$, $Z_3 = |50 + j23·33|$, $Z_5 = |50 + j46|$

$$P = \frac{1}{2}\left[200 \times \frac{200}{Z_1}\cos\left(\tan^{-1}\frac{1}{5}\right) + 100 \times \frac{100}{Z_3}\cos\left(\tan^{-1}0·467\right)\right.$$

$$\left. + 50 \times \frac{50}{Z_5}\cos\left(\tan^{-1}0·92\right)\right]\text{W}$$

14. (b) 0·6861   (c) 0·8

15. $i = 0·2 + \dfrac{1}{3\sqrt{5}}\cos\left(2\omega t + 153·5°\right)$

$$+ \frac{2}{15\sqrt{17}}\cos\left(4\omega t + 104°\right)\text{A}$$

$$i_{\text{r.m.s.}} = \sqrt{\left[0·2^2 + \frac{1}{2}\left(\frac{1}{9 \times 5} + \frac{4}{225 \times 17}\right)\right]}$$

$$= \sqrt{(0·05138)} = 0·2267\text{ A}$$

Power dissipated in load $= 100 \times i^2_{\text{r.m.s.}} = 5·138$ W

16. (a) Amplitudes of 1st and 3rd harmonics $= 9·10$ A, $3·90$ A respectively

(b) $P = 5 \times 7^2 = 245$ W

17. (a) $F(\omega) = \Delta t\,\dfrac{\sin\left(\frac{1}{2}\omega\Delta t\right)}{\left(\frac{1}{2}\omega\Delta t\right)}$

(b) $F(\omega) = \Delta t\,\dfrac{\sin\left(\frac{1}{2}\omega\Delta t\right)}{\left(\frac{1}{2}\omega\Delta t\right)}\,e^{-j1/2\omega\Delta t}$

Graphs of the magnitude and phase spectra are shown in fig. A.8

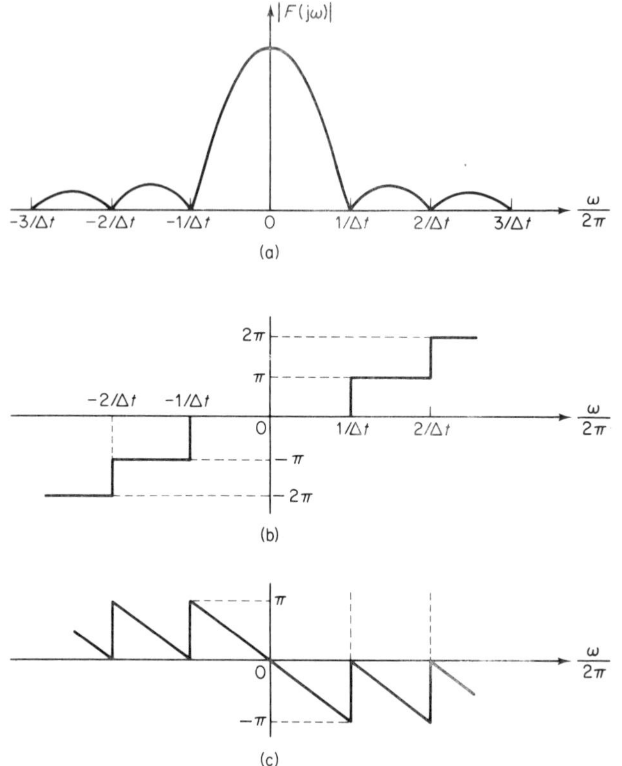

Fig. A.8 (a) $|F(j\omega)|$ vω. (b) Phase vω for symmetrical pulse. (c) For pulse O to Δt

18. (a) $F(\omega) = \dfrac{4V\pi^2 \sin(\omega T/4)}{\omega(4\pi^2 - \omega^2 T^2/4)}$ (b) $F(\omega) = \dfrac{V \sin(\omega T/2)}{\omega(1 - \omega^2 T^2/(4\pi^2))}$

19. (a) $F(\omega) = VT\left(\dfrac{\sin\frac{1}{2}\omega T}{\frac{1}{2}\omega T}\right)^2$ (b) $F(\omega) = j\omega V T^2\left[\dfrac{\sin(\frac{1}{2}\omega T)}{\frac{1}{2}\omega T}\right]^2$

The magnitude spectra are shown in fig. A.9

20. Fraction of total energy $= 2\Delta t \displaystyle\int_0^{1/\Delta t} \left|\dfrac{\sin \pi f \Delta t}{\pi f \Delta t}\right|^2 df = \dfrac{2}{\pi}\int_0^{\pi}\left|\dfrac{\sin x}{x}\right| dx$

21. (a) $i(t) = \dfrac{1}{R}\left[\delta(t) - \dfrac{1}{RC}e^{-t/(RC)}\right]$, (b) $v_0(t) = \dfrac{1}{RC}e^{-2t/(RC)}$

226

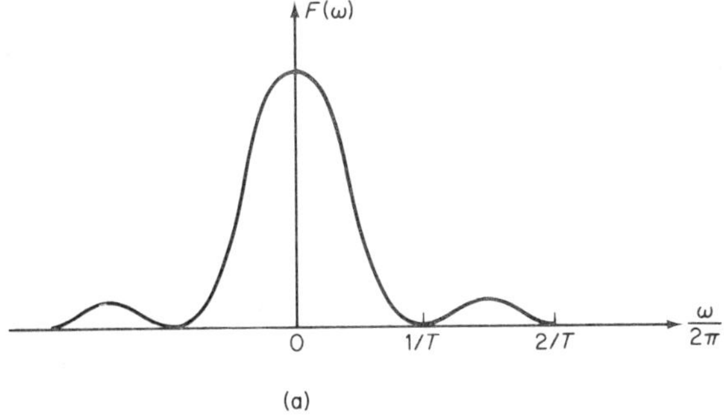

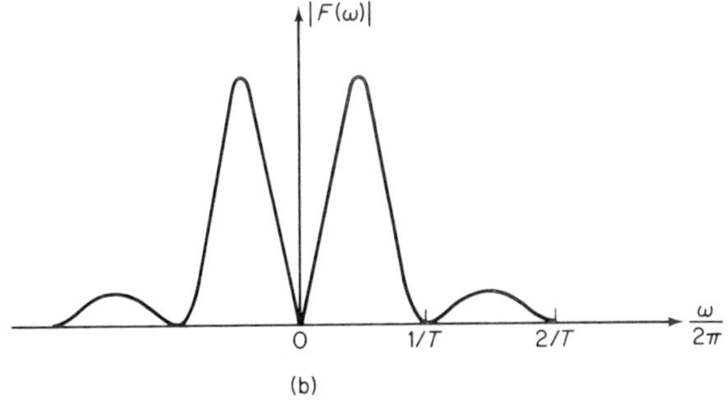

Fig. A.9 (a) $F(\omega)\,\mathrm{v}f$ for triangular pulse. (b) $|F(\omega)|\,\mathrm{v}f$ for double pulse

22. Maximum height $= \dfrac{2V}{\pi}\,\mathrm{Si}(0{\cdot}8) = 0{\cdot}492\,V$; heights at $\pm 0{\cdot}3\Delta t$ are

$$\frac{1}{\pi}\left[\mathrm{Si}(1{\cdot}28)+\mathrm{Si}(0{\cdot}32)\right]\approx 0{\cdot}477\,V$$

CHAPTER 3: TRANSMISSION LINE THEORY

1. $Z_{\mathrm{C}} = 314{\cdot}4\,\angle\,-21{\cdot}06^\circ\,\Omega$
   $\alpha = 0{\cdot}2130\,\mathrm{Np/km}$, $\beta = 0{\cdot}5531\,\mathrm{rad/km}$

(a) $V_{\mathrm{OC}} = \dfrac{10}{|\cosh \gamma l|}$

$$= \frac{20}{\sqrt{[\mathrm{e}^{2\alpha l}+\mathrm{e}^{-2\alpha l}+2\cos 2\beta l]}}\approx 20\mathrm{e}^{-\alpha l}=0{\cdot}662\,\mathrm{V}$$

(b) $I_{SC} = \dfrac{10}{|Z_C \sinh \gamma l|}$

$$= \dfrac{20}{|Z_C|\sqrt{[e^{2\alpha l}+e^{-2\alpha l}-2\cos 2\beta l]}} = 2{\cdot}11\,\text{mA}$$

2. $R = 26{\cdot}3\,\Omega/\text{km}$, $L = 0{\cdot}625\,\text{mH/km}$,
$C = 0{\cdot}0388\,\mu\text{F/km}$, $G = 0{\cdot}938\,\mu\text{S/km}$.

Elements of nominal Π for 1 km of line: 26Ω and $0{\cdot}625\,\text{mH}$ in series for series arm; $0{\cdot}469\,\mu\text{S}$ and $0{\cdot}0194\,\mu\text{F}$ in parallel, comprising each the two shunt arms.

Elements of equivalent π for length l of line:

Series element $= Z_C \sinh \gamma l$
$$= |Z_C|\tfrac{1}{2}\sqrt{[e^{2\alpha L}+e^{-2\alpha L}-2\cos 2\beta l]}\angle -40{\cdot}6^\circ + \phi_S$$
where $\phi_S = \tan^{-1}[\{\sin \beta l(e^{\alpha l}+e^{-\alpha l})\}/\{\cos \beta l(e^{\alpha l}-e^{-\alpha l})\}]$

Shunt elements $= Z_C \coth \tfrac{1}{2}\gamma l$
$$= [|Z_C| X_1/X_2]\angle -40{\cdot}6^\circ + \phi_1 - \phi_2$$

where $X_1 = \sqrt{[(1+e^{-\alpha l}\cos \beta l)^2 + \sin^2 \beta l\, e^{-2\alpha l}]}$,

$X_2 = \sqrt{[(1-e^{-\alpha l}\cos \beta l)^2 + \sin^2 \beta l\, e^{-2\alpha l}]}$

$$\phi_1 = \tan^{-1}\left[\dfrac{-\sin \beta l\, e^{-\alpha l}}{1+e^{-\alpha l}\cos \beta l}\right],$$

$$\phi_2 = \tan^{-1}\left[\dfrac{\sin \beta l\, e^{-\alpha l}}{1-e^{-\alpha l}\cos \beta l}\right]$$

3. $Z_C \approx \sqrt{\dfrac{L}{C}} = 82{\cdot}2\,\Omega$ (since $\omega L \gg R$, $\omega C \gg G$)

$\alpha \approx \tfrac{1}{2}[R/Z_C + GZ_C] = 0{\cdot}478\,\text{Np/km} = 4{\cdot}15\,\text{dB/km}$.

Minimum gain of repeater $= 43\,\text{dB}$.

4. $\alpha \approx \tfrac{1}{2}[R\sqrt{(C/L)}+G\sqrt{(L/C)}]$, $\beta \approx \omega\sqrt{(LC)}$.
Condition for α to be a minimum is $L = RC/G$; $\alpha_{\min} = \sqrt{(RG)}$, $\beta = \omega\sqrt{(LC)}$

5. (a) $\lambda = 154{\cdot}76\,\text{km}$, $v_P = 154\,758\,\text{km/s}$
(b) $|V_T| = 17{\cdot}78\,\text{mV}$, $|I_T| = 0{\cdot}0198\,\text{mA}$, $P = 0{\cdot}3108\,\mu\text{W}$

6. Series elements of nominal T: $27{\cdot}5\,\Omega$, $2{\cdot}37\,\text{mH}$ in series; shunt elements: $0{\cdot}01\,\mu\text{F}$, $6{\cdot}26\,\mu\text{S}$ in parallel.
At $10\,\text{kHz}$ $\alpha = 0{\cdot}0208\,\text{Np/km}$, $\beta = 0{\cdot}2188\,\text{rad/km}$ and so if αl and $\beta l < 0{\cdot}1$, maximum length of line $= 0{\cdot}1/0{\cdot}2188 = 0{\cdot}457\,\text{km}$

228

7. Attenuation of cable before loading

$$= 0.1121 \text{ Np/km} = 0.9735 \text{ dB/km}.$$

After adding coil, attenuation coefficient

$$= 0.0115 \text{ Np/km} = 0.0999 \text{ dB/km}.$$

8. (a) 12 km from input
   (b) $Z_C = 1180 \,\Omega$, $L = 5.90 \,\mu\text{H/m}$, $C = 4.236 \,\text{pF/m}$

9. (a) $C = 0.0101 \,\mu\text{F/km}$

   (b) $Z_C = 330 \,\Omega$, natural load $= \dfrac{(330 \text{ kV})^2}{Z_C} = 330 \text{ MW}$

   (c) 333.3 km

10. (a) $387.2 \angle -8.83°$, 50.02 MW
    (b) 139.98 kV (exact), 140.54 kV (nominal approx)

11. Percentage increase $= 4.34\%$, phase angle $= 1.84°$

12. (a) Solution obtained from $|Z|^2 = (\tfrac{1}{2}Rl)^2 + \{\tfrac{1}{2}Ll\omega - 1/(Cl\omega)\}^2$
    where $|Z| = \{120 \times 10^3/\sqrt{3}\}/300$. This gives $l = 65.35 \text{ km}$
    (b) $I_R = 311.5 \text{ A}$, $P_R = 35.51 \text{ MW}$

13. The nominal πs for both parts of the questions are shown in fig. A.10

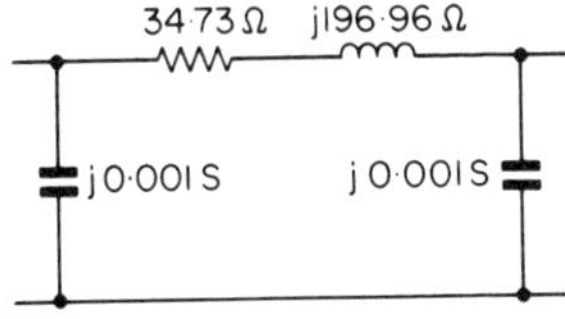

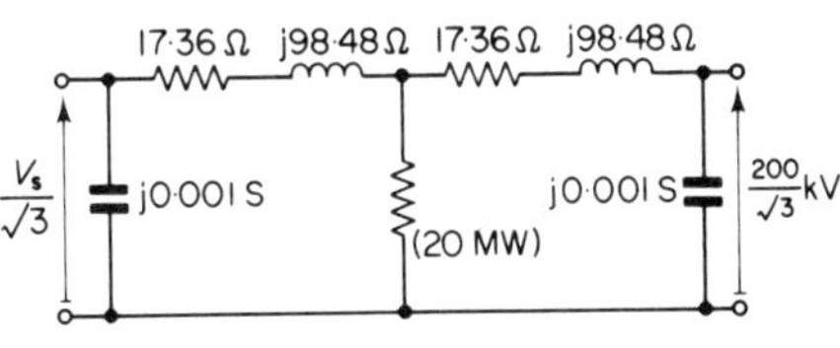

Fig. A.10

The sending end voltage $V_S = 171.6 \text{ kV}$

14. (a) $S = 2$,   (b) $S = 2.618$,   (c) $S = 2$
    (d) $S = 1$,   (e) $S = 1.143$,   (f) $S = 1$

15. (a) $Z_{\text{in}} = \dfrac{jX_1(\tfrac{1}{2}Z_C + j(X_2 + X_3))}{\tfrac{1}{2}Z_C + j(X_1 + X_2 + X_3)}$,

    (b) $Z_{\text{in}} = Z_C(3n_1^2 + \tfrac{1}{2}n_2^2)$

16. Power dissipated in $Y_2 = 0.1 \, I_G^2/Y_C = 125 \text{ W}$
    Power dissipated in $Y_3 = 0.025 \, I_G^2/Y_C = 31.25 \text{ W}$

17. (a) $\rho = 0.2$, $S = 1.5$

    (b) Voltage minima occur at $\frac{1}{4}\lambda$, $\frac{3}{4}\lambda$ ... from load
        Current minima occur at $0, \frac{1}{2}\lambda, \lambda$ ... from load; $\lambda = 0.3$ m. Power
        supplied to $75\,\Omega$ load $= 9.6$ W; $V_{max} = 37.95$ V, $I_{min} = 0.759$ A

18. $r_T = S(1 + \tan^2 \beta d)/(S^2 + \tan^2 \beta d)$,
    $x_T = (1 - S^2) \tan \beta d/(S^2 + \tan^2 \beta d)$.
    $Z_T = (166.6 - j76.3)\,\Omega$

19. $Y_{in} = Y_C \left[-j \cot \beta l_1 + \dfrac{G_2 + jY_C \tan \beta l_2}{Y_C + jG_2 \tan \beta l_2} \right]$

20. $\qquad \alpha = G \sin^2 \beta l/(lY_C)$ nepers per metre,

$$\varepsilon_r = \left[\frac{\lambda_0}{2\pi l} \cot^{-1}(\omega_0 C/Y_C) \right]$$

where $\lambda_0 = $ free space wavelength at radian frequency ω_0

21. Using $Z_{in} = Z_C \tanh \gamma l$,

    show $Z_{in}(\frac{1}{8}\lambda) \approx Z_C(2\alpha l + j) \approx \frac{1}{8}R\lambda + j\sqrt{\left(\dfrac{L}{C}\right)}$

    and $Z_{in}(\frac{3}{8}\lambda) \approx Z_C(2\alpha l - j) \approx \frac{3}{8}R\lambda - j\sqrt{\left(\dfrac{L}{C}\right)}$;

    where $Z_C \approx \sqrt{\dfrac{L}{C}}$, $\alpha \approx R/(2Z_C)$

22. On line l_1: $\rho_1 = 0.3$, $S_1 = 1.86$. On line l_2: $\rho_2 = 0.6$, $S_2 = 4$

23. $y = \dfrac{\lambda}{2\pi} \tan^{-1}\left(\dfrac{1}{\sqrt{S}}\right)$, $B = \dfrac{S-1}{\sqrt{S}} Y_C$

24. (a) Power delivered to load $= I_G^2/(4Y_C)$ watts

    (b) On first section $S = 1$,

        on second (load) section $S = \dfrac{B + \sqrt{(4Y_C^2 + B^2)}}{\sqrt{(4Y_C^2 + B^2)} - B}$

    (c) Power (when jB removed) $= I_G^2 Y_C/(4Y_C^2 + B^2)$ watts

25. $y = \dfrac{\lambda}{2\pi} \tan^{-1} 2 = 0.176\lambda$, $x = 1$

26. $S = \dfrac{S' + 1 + (S' - 1)e^{2\alpha L}}{S' + 1 - (S' - 1)e^{2\alpha L}}$

    (a) $\varepsilon_r = 2.25$ (b) $\alpha = 0.0155$ Np/m $= 0.135$ dB/m

27. $Y = -j5.657$ mS

28. $Z_L = 150\,\Omega$

29. Maximum power $= V_B^2/(SZ_C)$ watts

30. $l = 0{\cdot}111\lambda$, $X = 16{\cdot}92\,\Omega$

31. Input impedance at AB $= Z_{C_1}^2 Z_{C_3}^2/(Z_{C_2}^2 Z_L)$ ohms

32. The connection between generator and load is shown in fig. A.11.
    Two $\frac{1}{4}\lambda$ sections are constructed from the $100\,\Omega$ and $400\,\Omega$ lines and
    the $50\,\Omega$ line used as the main feed line

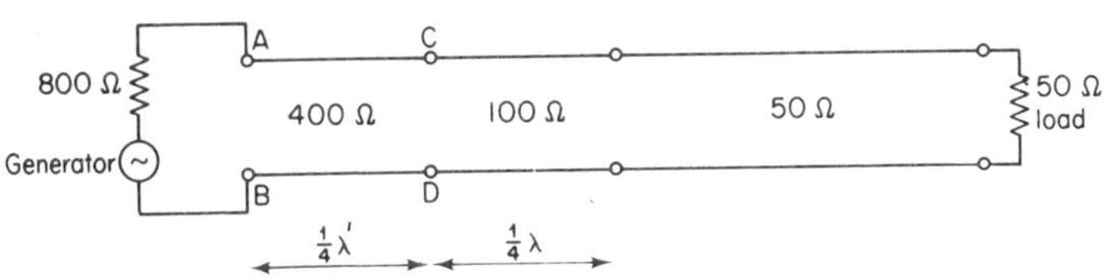

Fig. A.11. Input impedance at CD $= \dfrac{100^2}{50} = 200\,\Omega$,

input impedance at AB $= \dfrac{400^2}{200} = 800\,\Omega$,

which matches to generator

33. $B = -\dfrac{1}{490}\,\text{S} = -2{\cdot}041\,\text{mS}$, $Z_{C_2} = 156{\cdot}5\,\Omega$

34. $y_L = 0{\cdot}8 + j0{\cdot}6$; (a) $d = \frac{1}{8}\lambda$, (b) $Z_C' = \dfrac{1}{\sqrt{2}} Z_C$

35. Load impedance $Z_L = (26{\cdot}5 - j34)\,\Omega$
    Position of short-circuited stub from load $= 0{\cdot}026\lambda$
    Length of this stub $= 0{\cdot}114\lambda$ (to cancel normalized $j1{\cdot}15$
    susceptance)

36. Stub position from load $= 0{\cdot}059\lambda$
    (a) Stub length $= 0{\cdot}109\lambda$ (λ wavelength on $50\,\Omega$ line)
    (b) Stub length $= 0{\cdot}084\lambda'$ (λ' wavelength on $70\,\Omega$ line)

37. $Z_C = 75{\cdot}83\,\Omega$, $l = 0{\cdot}374\lambda$

38. Input admittance at AB $= 100^{-1}(2 - j0{\cdot}6) + 50^{-1}(1 + j0{\cdot}35)$
    $$= (0{\cdot}04 + j0{\cdot}001)\text{S}$$
    V.S.W.R. on input line ≈ 2. Stub position from AB $= 0{\cdot}101\lambda$, length
    of stub $= 0{\cdot}347\lambda$

39. $Z_C = 59{\cdot}16\,\Omega$; $S \approx 1{\cdot}85$ (at $480\,\text{MHz}$), $S \approx 1{\cdot}5$ (at $520\,\text{MHz}$)

40. $\alpha = 0.01055$ Nepers per wavelength. Position of open-circuit stub $=$ 0.166λ from load, stub length $= 0.131\lambda$

41. $l_1 = 0.1\lambda$, $l_2 = 0.138\lambda$; or $l_1 = 0.19\lambda$, $l_2 = 0.362\lambda$

42. (a) $S = 2.77$
    (b) $l_1 = 0.062\lambda$, $l_2 = 0.056\lambda$
    or $l_1 = 0.155\lambda$, $l_2 = 0.349\lambda$

43. (a) Load impedance $= 50(0.7 - j0.93) = (35 - j46.5)\,\Omega$
    Load admittance $= 50^{-1}(0.52 + j0.68)\text{S} = (10.4 + j13.6)\text{mS}.$
    (b) Stub position $= 0.057\lambda$ from load, length $= 0.114\lambda$
    (c) Matching may be obtained when first stub position is at 0 and 0.25λ from load, not when 0.125λ from load.

CHAPTER 4: 2-PORT NETWORKS

1. For the Π: $y_{11} = G + j\omega C$, $y_{12} = y_{21} = -G$,
$$y_{22} = G - j\frac{1}{\omega L}$$

For the T: $y_{11} = \dfrac{C/L + j\omega CG}{G + j(\omega C - 1/(\omega L))}$

$$y_{12} = y_{21} = \frac{-1}{LG/C + j\omega L(1 - 1/(\omega^2 LC)}$$

$$y_{22} = \frac{C/L - jG/(\omega L)}{G + j(\omega C - 1/(\omega L))}$$

2. $z_{11} = 73.64\,\Omega$, $z_{12} = z_{21} = 3.24\,\Omega$, $z_{22} = 29.84\,\Omega$
   $A = 22.73$, $B = 675.1\,\Omega$, $C = 0.3086$ S, $D = 9.210$

3. $A = 1.667$, $B = 73.33\,\Omega$, $C = 0.01667$ S, $D = 1.333$

4. (b) $Z_L = 4.115(1 - j)\,\Omega$

5. (a) $\begin{bmatrix} 1 & Z \\ 0 & 1 \end{bmatrix}$ (b) $\begin{bmatrix} 1 & 0 \\ Y & 1 \end{bmatrix}$ (c) $\begin{bmatrix} 1 + ZY & Z \\ Y & 1 \end{bmatrix}$

   (d) $\begin{bmatrix} 1 + ZY_2 & Z \\ Y_1(1 + ZY_2) + Y_2 & Y_1 Z + 1 \end{bmatrix}$

6. (a) $[z] = 1/\Delta \begin{bmatrix} Y_A + Y_C & Y_A \\ Y_A & Y_A + Y_B \end{bmatrix}$
   where $\Delta = Y_A Y_B + Y_B Y_C + Y_C Y_A$

(b) $[z] = \begin{bmatrix} (Y_A + Y_C)/\Delta + Z & Y_A/\Delta + Z \\ Y_A/\Delta + Z & (Y_A + Y_B)/\Delta + Z \end{bmatrix}$

7. $\dfrac{I_G}{I_2} = \dfrac{(G_G + y_{11})(G_L + y_{12} y_{21})}{y_{21} G_L}$; $I_G = -10\,\text{A}$

8. $h_{11} = r_e + \dfrac{r_b r_c}{r_b + r_c}\,(1 - \alpha)$, $h_{22} = \dfrac{1}{r_b + r_c}$

$h_{12} = \dfrac{r_b}{r_b + r_c}$, $h_{21} = -\dfrac{r_b + \alpha r_c}{r_b + r_c}$

9. $R_{in} = \dfrac{G_L h_{ie} + (h_{oe} h_{ie} - h_{fe} h_{re})}{G_L + h_{oe}}$; $\dfrac{I_2}{I_1} = \dfrac{h_{fe} G_L}{G_L + h_{oe}}$

10. $Z_{in} = (815 \cdot 9 - j500 \cdot 5)\,\Omega$; voltage gain $V_0/V_i = 218 \cdot 6 \angle -148 \cdot 5°$

11. Power output $= 2 \cdot 60\,\text{W}$

12. Voltage gain $V_2/V_1 = \left[1 + \dfrac{R_S(Z_L + R_a)}{Z_L R_a(1 + g_m R_S)} \right]^{-1}$

$Z_{in} = R_S + \dfrac{Z_L R_a(1 + g_m R_S)}{R_a + Z_L}$; $Z_{out} = [1/R_S + 1/R_a + g_m]^{-1}$

13. Voltage gain $V_2/V_1 = \dfrac{-\mu(\mu R_1 + R_a)}{2R_a + R_1(1 + \mu)}$;

$Z_{out} = \dfrac{R_a(R_a + R_1)}{R_1(1 + \mu) + 2R_a}$. $-20 \cdot 4$, $1105\,\Omega$

14. (a) The equivalent circuit is shown in fig. A.12
    (b) $Z_1 = z_{11} - z_{12}$, $Z_2 = z_{12}$,

    $Z_3 = z_{22} - z_{12}$, $\alpha = \dfrac{z_{21} - z_{12}}{z_{22} - z_{12}}$

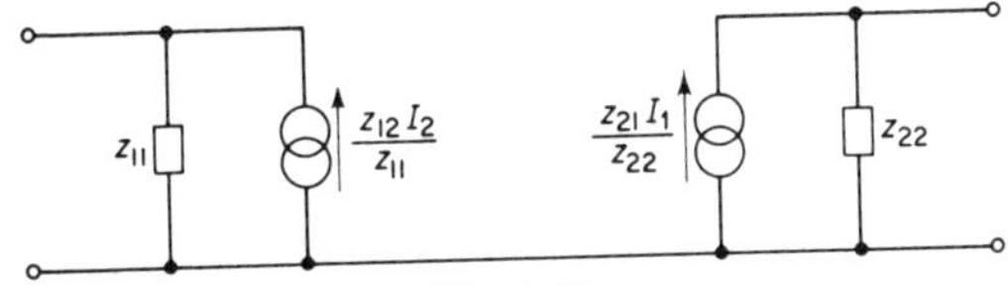

Fig. A.12

15. $Y_A = -y_{12}$, $Y_B = y_{11} + y_{12}$,

$Y_C = y_{22} + y_{12}$, $\mu = \dfrac{y_{21} - y_{12}}{y_{22} + y_{12}}$

16. $\begin{bmatrix} h_{ie} & h_{re} \\ h_{fe} & h_{oe} \end{bmatrix} = \dfrac{1}{1+h_{fb}+\Delta-h_{rb}} \begin{bmatrix} h_{ib} & \Delta-h_{rb} \\ -(h_{fb}+\Delta) & h_{ob} \end{bmatrix}$

$\begin{bmatrix} h_{ic} & h_{rc} \\ h_{fc} & h_{oc} \end{bmatrix} = \dfrac{1}{1+h_{fb}+\Delta-h_{rb}} \begin{bmatrix} h_{ib} & 1+h_{fb} \\ -(1-h_{rb}) & h_{ob} \end{bmatrix}$

where $\Delta = h_{ib}h_{ob} - h_{rb}h_{fb}$

17. $h_{ib} \approx r_\varepsilon + r_{bb'}(1-\alpha)$ $h_{ie} \approx [r_\varepsilon + r_{bb'}(1-\alpha)]/(1-\alpha)$
$h_{rb} \approx \mu_0 + r_{bb'}(g_c + j\omega C_c)$ $h_{re} \approx [\Delta - h_{rb}]/(1-\alpha)]$
$h_{fb} \approx -\alpha \equiv \alpha_0/(1+j\omega/\omega_a)$ $h_{fe} \approx \alpha/(1-\alpha)$
$h_{ob} \approx g_C + j\omega C_C$ $h_{oe} \approx (g_C + j\omega C_C)/(1-\alpha)$
                                 $\Delta = h_{ib}h_{ob} - h_{rb}h_{fb}$

18. (a) $289{\cdot}47\,\text{kV}$ (b) $5504{\cdot}7\angle -90\,\text{A}$

19. The resultant transmission matrix is

$\begin{bmatrix} A & B \\ C & A \end{bmatrix} \begin{bmatrix} 1 & Z_1 \\ 1/Z_m & 1+Z_1/Z_m \end{bmatrix} = \begin{bmatrix} A+B/Z_m & AZ_1+B(1+Z_1/Z_m) \\ C+A/Z_m & CZ_1+A(1+Z_1/Z_m) \end{bmatrix}$

Phase voltage at line input $= 83{\cdot}59\,\text{kV}$

20. $C_1 = 0{\cdot}00135 \angle 91{\cdot}2°\,\text{S};\ 293{\cdot}9\,\text{A}$
$y_{11} = y_{22} = (11{\cdot}1 - j33{\cdot}3)\,\text{mS},$
$y_{12} = y_{21} = -(11{\cdot}1 - j34{\cdot}9)\,\text{mS}$
    Admittance elements of the equivalent Π: $(11{\cdot}1 - j34{\cdot}9)\,\text{mS}$ in series and $j1{\cdot}6\,\text{mS}$ as the two shunt elements

21. (a) $\dfrac{AR_L + B + CR_G R_L + DR_G}{R_L + R_G},$

(b) $\dfrac{|AR_L + B + R_G R_L C + DR_G|^2}{4R_G R_L}$

22. (a) $8{\cdot}686\alpha l\,\text{dB}$, (b) $20\log_{10}(e^{\alpha l} + \tfrac{1}{3}e^{-\alpha l})\,\text{dB}$

23. (a) $10\,\text{dB}$ (b) $20\,\text{dB}$ (c) $5\,\text{dB}$

24. (a) $[z] = \begin{bmatrix} 0 & -b \\ b & 0 \end{bmatrix}$ $[A] = \begin{bmatrix} 0 & b \\ 1/b & 0 \end{bmatrix}$

(b) $Z_{in} = j\omega(b^2 C)$
(c) (i) $Z_1 = b^2/Z$ (ii) $Z_2 = b^2/Z$

234

(d) $[A] = \begin{bmatrix} 0 & b_1 \\ 1/b_1 & 0 \end{bmatrix} \begin{bmatrix} 0 & b_2 \\ 1/b_2 & 0 \end{bmatrix} = \begin{bmatrix} b_1/b_2 & 0 \\ 0 & b_2/b_1 \end{bmatrix}$

(e) $[A] = \begin{bmatrix} 0 & b \\ 1/b & 0 \end{bmatrix} \begin{bmatrix} a & 0 \\ 0 & 1/a \end{bmatrix} = \begin{bmatrix} 0 & b/a \\ a/b & 0 \end{bmatrix}$

25. (a) $Z_C = \sqrt{(\tfrac{1}{4}Z_1^2 + Z_1 Z_2)}$, $\gamma = \log_e\left(1 + \dfrac{Z_1}{2Z_2} + \dfrac{Z_C}{Z_2}\right)$

   (b) $Z_C = Z_1 Z_2/[\sqrt{(\tfrac{1}{4}Z_1^2 + Z_1 Z_2)}]$, $\gamma = \log_e\left(1 + \dfrac{Z_1}{2Z_2} + \dfrac{Z_1}{Z_C}\right)$

   Note $\gamma_T = \gamma_\pi$

26. $Z_C = \sqrt{(Z_A Z_B)}$, $\gamma = \log_e\left[\dfrac{\sqrt{Z_A} + \sqrt{Z_B}}{\sqrt{Z_A} - \sqrt{Z_B}}\right]$. $Z_A = Z_B$

27. $R_1 = R_G\left(\dfrac{n^2 + 1}{n^2 - 1}\right) - 2\sqrt{(R_L R_G)}\left(\dfrac{n}{n^2 - 1}\right)$

   $R_2 = 2\sqrt{(R_L R_G)}\left(\dfrac{n}{n^2 - 1}\right)$

   $R_3 = R_L\left(\dfrac{n^2 + 1}{n^2 - 1}\right) - 2\sqrt{(R_L R_G)}\left(\dfrac{n}{n^2 - 1}\right)$

where $R_G = 50\,\Omega$, $R_L = 600\,\Omega$ where $n = \text{antilog}\,\dfrac{20}{20} = 10$

Hence $R_1 = 50 \times \dfrac{101}{99} - 2\sqrt{(50 \times 600)}\dfrac{10}{99}$

   $= 51{\cdot}01 - 36{\cdot}79 = 14{\cdot}2\,\Omega,$

   $R_2 = 36{\cdot}8\,\Omega$, $R_3 = 600 \times \dfrac{101}{99} - 36{\cdot}8$

   $= 61{\cdot}2 - 36{\cdot}8 = 24{\cdot}4\,\Omega$

28. Attenuation produced by individual Π section $= 5\,\text{dB}$
   Equivalent circuit for ladder problem is shown in fig. A.13

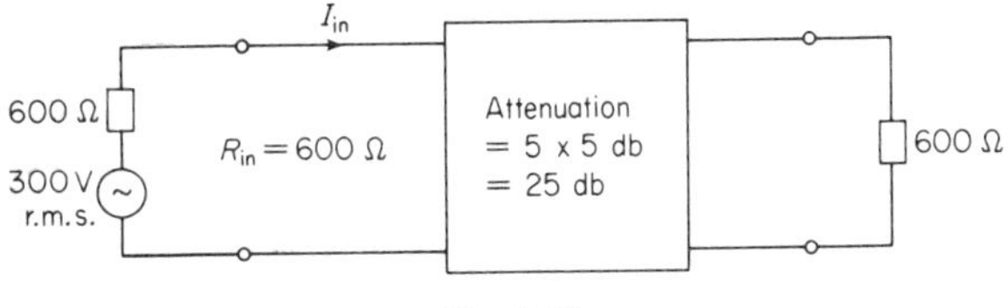

Fig. A.13

Power supplied to ladder at input

$$= R_{in} I_{in}^2 = 600 \times \left(\frac{300}{600 + 600}\right)^2 = \frac{600}{16} = 37{\cdot}5\,\text{W}$$

Let P = power dissipated in load,

then $10 \log_{10} \dfrac{87 \cdot 5}{P} = 25 \text{ dB}$, $P = \dfrac{37 \cdot 5}{316 \cdot 2} = 0 \cdot 118 \text{ W}$

29. $Z_{\text{in}} = \dfrac{AZ_L + B}{CZ_L + D}$; $Z_{01} = \sqrt{\left(\dfrac{AB}{CD}\right)}$, $Z_{02} = \sqrt{\left(\dfrac{BD}{AC}\right)}$

$\theta = \log_e [\sqrt{(AD)} + \sqrt{(BC)}]$

30. $Z_C = \sqrt{\left[\dfrac{Z_1 Z_3 (\frac{1}{4}Z_1 + Z_2)}{Z_1 + Z_3}\right]}$. Design of bridged-T network:

$Z_1 = 100 \, \Omega$, $Z_2 = \dfrac{50}{x} \, \Omega$, $Z_3 = 50x \, \Omega$

31. $X_2 = \pm \sqrt{\left(\dfrac{R_1}{R_2 - R_1}\right)}$, $X_1 = \dfrac{-R_1 R_2}{X_2}$

32. $C_1 = 4131 \text{ pF}$ (series element), $L_2 = 173 \cdot 5 \, \mu\text{H}$ (shunt element); phase lead of output with respect to input $= 70° \, 2'$.

Or $L_1 = 153 \cdot 3 \, \mu\text{H}$ (series), $C_2 = 3650 \text{ pF}$ (shunt); phase lag $= 70° \, 2'$.

33. $Z_{01} = \sqrt{[(Z_1 + Z_2)(Z_1 Z_2 + Z_2 Z_3 + Z_3 Z_1)/(Z_2 + Z_3)]}$,

$Z_{02} = \sqrt{[(Z_2 + Z_3)(Z_1 Z_2 + Z_2 Z_3 + Z_3 Z_1)/(Z_1 + Z_2)]}$

$X_1 = X_3 = -\sqrt{(R_1 R_2)}$

34. (a) $Z_{\text{it}_1} = [-(A - D) \pm \sqrt{\{(A - D)^2 + 4BC\}}]/2C$

$Z_{\text{it}_2} = [-(D - A) \pm \sqrt{\{(D - A)^2 + 4BC\}}]/2C$

(b) $Z_{\text{it}_1} = \sqrt{(\frac{1}{16}Z_1^2 + Z_1 Z_2)} - \frac{1}{4}Z_1$

$Z_{\text{it}_2} = \frac{1}{4}Z_1 + \sqrt{(\frac{1}{16}Z_1^2 + Z_1 Z_2)}$

35. (a) $Z_{01} = [-\frac{1}{2}X_1(\frac{1}{2}X_1 + 2X_2)]^{1/2}$, $Z_{02} = \left[\dfrac{-2X_1 X_2^2}{\frac{1}{2}X_1 + 2X_2}\right]^{1/2}$

(b) $0 \leqslant \omega \leqslant 2/\sqrt{(LC)}$ (c) $1/\{2\sqrt{(LC)}\} \leqslant \omega < \infty$

36. $Z_{01} = \sqrt{(Z_{\text{SC}_1} Z_{\text{OC}_1})} = \sqrt{[(\frac{1}{2}mZ_1)(\frac{1}{2}mZ_1 + 2Z_2/m + (1 - m^2)Z_1/2m]}$

$= \sqrt{(\frac{1}{4}Z_1^2 + Z_1 Z_2)} = \sqrt{(Z_1 Z_2)}\sqrt{(1 + Z_1/(4Z_2))} = R_0 \sqrt{(1 - x^2)}$

$Z_{02} = \sqrt{(Z_{\text{SC1}} Z_{\text{OC2}})}$

$= \left[\left(\dfrac{\frac{1}{2}mZ_1 \{2Z_2/m + (1 - m^2)Z_1/(2m)\}}{\frac{1}{2}mZ_1 + 2Z_2/m + (1 - m^2)Z_1/(2m)}\right)\right.$

$\left. \{2Z_2/m + (1 - m^2)Z_1/(2m)\}\right]$

$= \dfrac{\frac{1}{2}mZ_1 \{2Z_2/m + (1 - m^2)Z_1/(2m)\}}{R_0 \sqrt{(1 - x^2)}} = \dfrac{R_0[1 - (1 - m^2)x^2]}{\sqrt{(1 - x^2)}}$

The graph of Z_{02} v x for various values of m are given in Chapter 5 Z_{02} remains approximately constant and equal to R_0 when $m=0.6$ for $0<x<0.9$ and also since $Z_{01}=Z_{CT}$, the characteristic impedance of a symmetrical T section, the $m=0.6$ half section effects an approximate match.

CHAPTER 5: FILTER NETWORKS

1. (a) $0\leqslant\omega\leqslant2/\sqrt{(LC)}, f_C = 1/[\pi\sqrt{(LC)}]$
   (b) Z_C is real at all frequencies, hence pass-band extends 0 to ∞.
   (c) $1/[2\sqrt{(LC)}]\leqslant\omega<\infty, f_C = 1/[4\pi\sqrt{(LC)}]$
   (d) $0\leqslant\omega\leqslant2/\sqrt{(LC)}, f_C = 1/[\pi\sqrt{(LC)}]$

2. (a) $\omega_{C_1} = \omega_0[\sqrt{(1+a^2)}-a], \omega_{C_2} = \omega_0[\sqrt{(1+a^2)}+a]$
   where $\omega_0 = 1/\sqrt{(LC)}$
   (b) $\omega_{C_1} = \frac{1}{4}\omega_0[\sqrt{(a^2+16)}-a], \omega_{C_2} = \frac{1}{4}\omega_0[\sqrt{(a^2+16)}+a]$
   where $\omega_0 = 1/\sqrt{(LC)}$

3. Use the condition that the characteristic impedance of the Π must be real, i.e.
$$Z_C = \frac{-X_1 X_2}{\sqrt{[-\frac{1}{4}X_1^2 - X_1 X_2]}} = \frac{k^2}{\sqrt{[-\frac{1}{4}X_1^2 + k^2]}} \text{ is real when } k^2 > \frac{1}{4}X_1^2$$
or $-2k<X_1<2k$

4. The elements of the high-pass section (synthesized via the constant k low-pass, shown in fig. A.14(b), are:
$$C_h = \frac{1}{2\omega_C R_0} = 5305\,\text{pF}, \quad L_h = \frac{R_0}{\omega_C} = 3.820\,\text{mH}.$$
The elements of the input and output $m=0.6$ half sections, shown in fig. A.14(c) are:
$$L_1 = \frac{2m}{1-m^2}\,2L_h = 14.325\,\text{mH}, \quad C_1 = \frac{2}{m}C_h = 17683\,\text{pF}$$
$$L_2 = \frac{2}{m}(2L_h) = 25.47\,\text{mH}$$

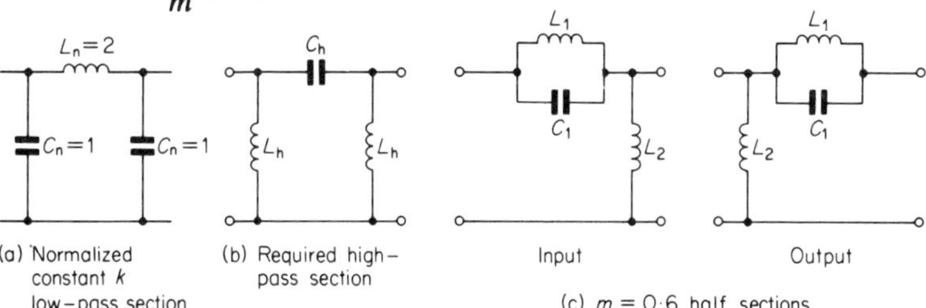

(a) Normalized constant k low-pass section

(b) Required high-pass section

Input

Output

(c) $m=0.6$ half sections

Fig. A.14

5. $m = 0.275$, $L = 76.39\,\text{mH}$, $C = 0.2122\,\mu\text{F}$; so elements of series arms $\tfrac{1}{2}mL = 10.50\,\text{mH}$; elements of shunt arm are $mC = 0.05836\,\mu\text{F}$, $(1-m^2)L/4m = 64.19\,\text{mH}$

6. Pad: T section consisting of two $65.8\,\Omega$ series resistors, with a $4.4\,\Omega$ resistor in shunt.

   Constant k section: $L = 0.1114\,\text{mH}$, $C = 0.02273\,\mu\text{F}$.

7. $m = 0.305$, $L = 12.73\,\text{mH}$, $C = 0.03537\,\mu\text{F}$ and thus series arm elements of m-derived section are $\tfrac{1}{2}mL = 1.941\,\text{mH}$ and shunt arm elements are

$$mC = 0.01079\,\mu\text{F}, \quad \frac{1-m^2}{4m}L = 9.464\,\text{mH}.$$

   Frequency at which section matches

$$300\,\Omega\ \text{terminations} = 12.99\,\text{kHz}.$$

8. $R_0 = 632.5\,\Omega$, $f_C = 16.776\,\text{kHz}$

   Redesigned filter is an m-derived filter with $m = 0.305$,
   series elements $\tfrac{1}{2}mL = 1.83\,\text{mH}$

$$\text{shunt elements } mC = 0.00915\,\mu\text{F}, \quad \frac{1-m^2}{4m}L = 8.921\,\text{mH}$$

   The insertion loss v frequency characteristics are shown in fig. A.15.

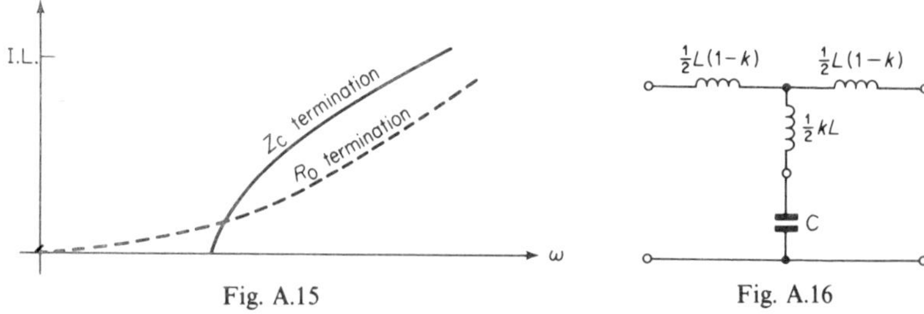

Fig. A.15   Fig. A.16

9. (a) $180°$, $Z_{\text{in}} = 0$, (b) $135°$, $Z_{\text{in}} = \tfrac{1}{5}R_0(1+\text{j}2)\,\Omega$

10. Replace coupled inductors by an equivalent T of uncoupled inductors, as shown in fig. A.16. On comparing this circuit with low-pass m-derived circuit of the standard form,

   we have $\tfrac{1}{2}m'L' = \tfrac{1}{2}L(1-k)$, $\dfrac{1-m'^2}{4m'}L' = \tfrac{1}{2}kL$

   from which $m' = \sqrt{[(1-k)/(1+k)]}$

11. Selecting $Z_\text{B} = \text{j}\omega L_\text{B} + 1/(\text{j}\omega C_\text{B})$ we have

$$Z_\text{C} = \sqrt{\{L_\text{B}/C_\text{A} - 1/(\omega^2 C_\text{A} C_\text{B})\}}.$$

Hence lattice acts as a high-pass filter with $\omega_C = 1/\sqrt{(L_B C_A)}$ and $R_0 = \sqrt{(L_B/C_A)}$

Choosing $C_A = C_B = \dfrac{1}{\omega_C R_0} = 88\cdot4\,\text{pF}$, $L_B = R_0/\omega_C = 31\cdot8\,\mu\text{H}$.

12. $L = 60/\pi = 19\cdot10\,\text{mH}$, $C = 1/(6\pi) = 0\cdot05305\,\mu\text{F}$

Diagrams for m-derived full and half sections are shown in fig. A.17

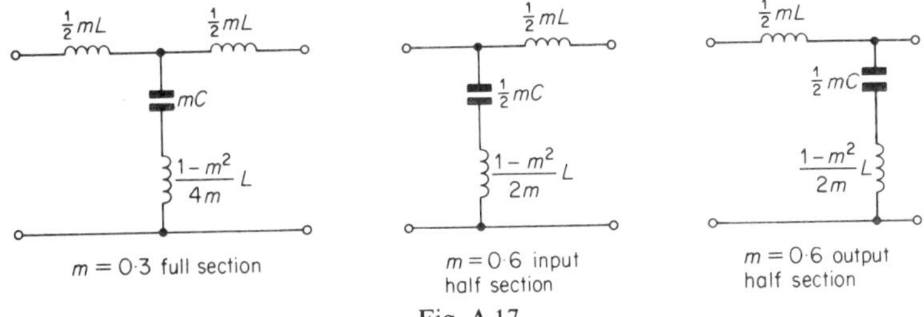

Fig. A.17

13. $\text{I.L.R.} = [1 + Z_1/(2Z_2) + (\tfrac{1}{4}Z_1^2 + Z_1 Z_2 + R_0^2)/(2R_0 Z_2)]$
For $Z_1 = 1/(j\omega C)$, $Z_2 = j\omega L$, $R_0 = \sqrt{(L/C)}$

$$\text{I.L.} = 20\log_{10}\left|1 - \frac{1}{2\omega^2 LC} + j\frac{\left(\dfrac{1}{4}\dfrac{1}{\omega^2 C^2} - 2\dfrac{L}{C}\right)}{2\sqrt{(L/C)}\omega L}\right|$$

$$= 20\log_{10}\left|1 - 2\frac{\omega_C^2}{\omega^2} + j\left(\frac{\omega_C^3}{\omega^3} - 2\frac{\omega_C}{\omega}\right)\right|$$

$$= 10\log_{10}\left(1 + \frac{\omega_C^6}{\omega^6}\right)\,\text{dB}$$

14. Two alternative filters designs are shown in fig. A.18. Attenuation loss $= 0\cdot067\,\text{dB}$ at $25\,\text{kHz}$, $10\cdot93\,\text{dB}$ at $75\,\text{kHz}$.

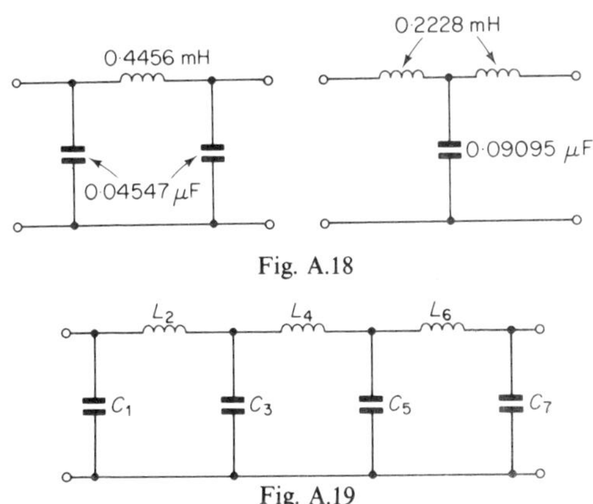

Fig. A.18

Fig. A.19

15. One design is shown in fig. A.19: $C_1 = C_7 = 0.01416\,\mu F$, $L_2 = L_6 = 99.23\,\mu H$ $C_3 = C_5 = 0.05736\,\mu F$, $L_4 = 159.15\,\mu H$. Attenuation loss at $230\,kHz = 50.62\,dB$.

16. One design is shown in fig. A.20: $C_1 = C_3 = 0.02684\,\mu F$, $L_2 = 4.746\,mH$ Attenuation loss at $40\,kHz = 22.46\,dB$.

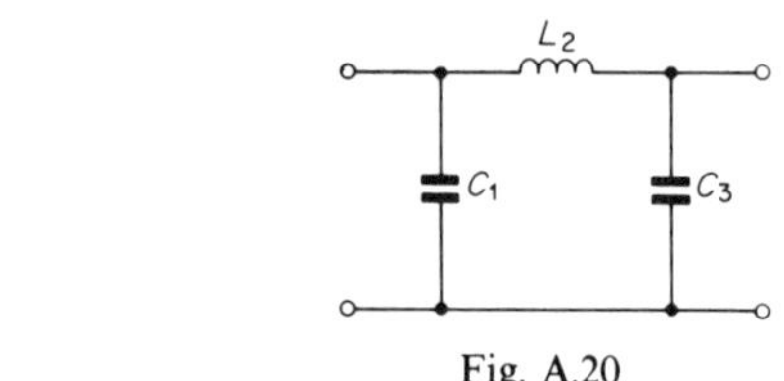

Fig. A.20

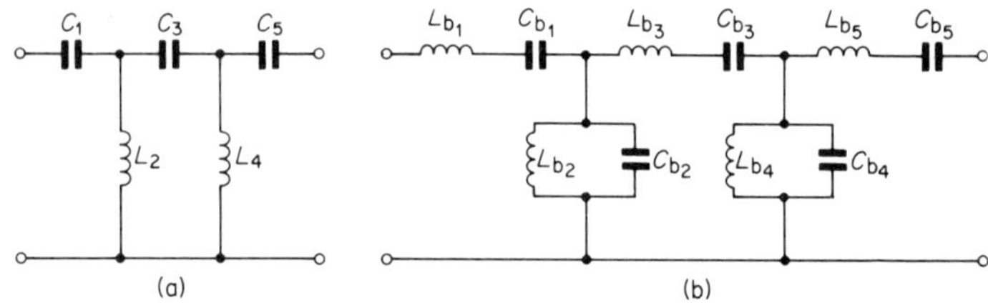

Fig. A.21

17. (a) One design is shown in fig. A.21(a): $C_1 = C_5 = 20316\,pF$, $L_2 = L_4 = 0.5222\,mH$, $C_3 = 15587\,pF$.
   (b) One design is shown in fig. A.21(b): $L_{b_1} = L_{b_5} = 0.3398\,mH$, $C_{b_1} = C_{b_5} = 1491\,pF$, $L_{b_3} = 0.4776\,mH$, $C_{b_3} = 1061\,pF$, $L_{b_2} = L_{b_4} = 7.293\,\mu H$, $C_{b_2} = C_{b_4} = 69462\,pF$.

18. Minimum number of elements is 9; maximum loss in range 0 to 90% of pass-band is $0.607\,dB$.

19. One design is shown in fig. A.22: $L_{S_1} = L_{S_3} = 13.32\,mH$, $C_{S_1} = C_{S_3} = 19803\,pF$, $L_{S_2} = 33.54\,mH$, $C_{S_2} = 7866\,pF$. Frequencies of infinite attenuation $= [2\pi\sqrt{(L_{S_1} C_1)}]^{-1} = 9799.5\,Hz$,

$$[2\pi\sqrt{(L_{S_2} C_{S_2})}]^{-1} = 9798.5\,Hz.$$

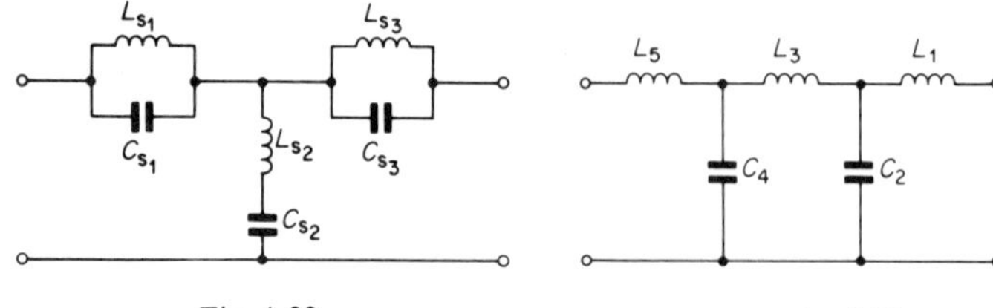

Fig. A.22

Fig. A.23

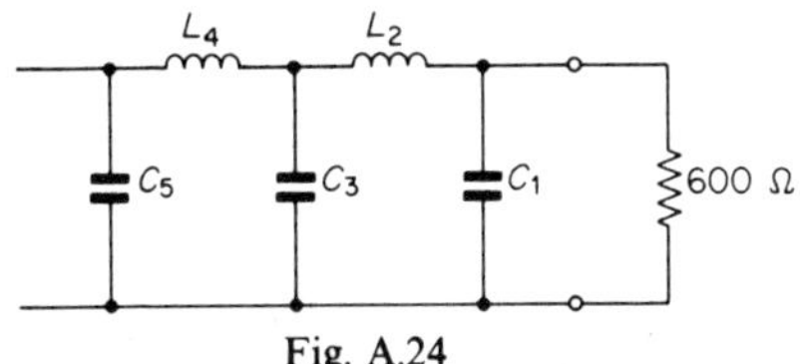

Fig. A.24

20. Insertion loss $= 3.01\,\text{dB}$ (at $\omega = 1$), $23.0\,\text{dB}$ (at $\omega = 2$). Insertion loss at $\omega = 2$ is $4.83\,\text{dB}$ greater.

21. The filter is shown in fig. A.23: $L_1 = 2.459\,\mu\text{H}$, $C_2 = 2847\,\text{pF}$, $L_3 = 11.0\,\mu\text{H}$, $C_4 = 5393\,\text{pF}$, $L_5 = 12.296\,\mu\text{H}$

22. The filter is shown in fig. A.24: $C_1 = 566\,\text{pF}$, $L_2 = 0.2758\,\text{mH}$, $C_3 = 10578\,\text{pF}$, $L_4 = 0.3038\,\text{mH}$, $C_5 = 883\,\text{pF}$

23. Insertion phase shift

$$= [90° + \tan^{-1}(0.4\omega'^2 - 1)/\{\omega'(1 - 0.0668\omega'^2)\}]$$

Good linearity.

CHAPTER 6A. INTRODUCTION TO PASSIVE NETWORK SYNTHESIS

1. $a > b$ and both real numbers.
   The network and sketch of $Z(j\omega)$ v ω are shown in fig. A.25.
   $L = 1$, $L_1 = (a^2 - b^2)/b^2$ henries, $C_1 = (a^2 - b^2)^{-1}$ farads.

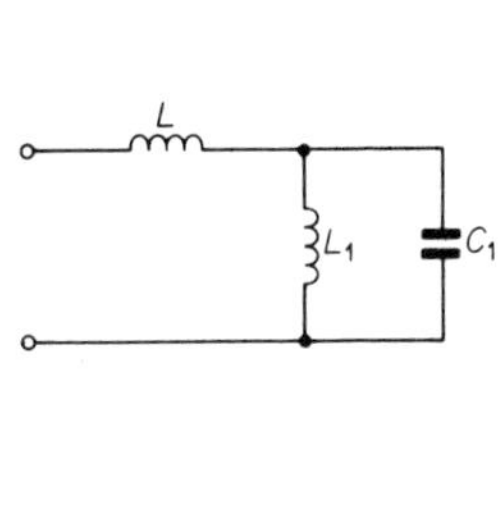
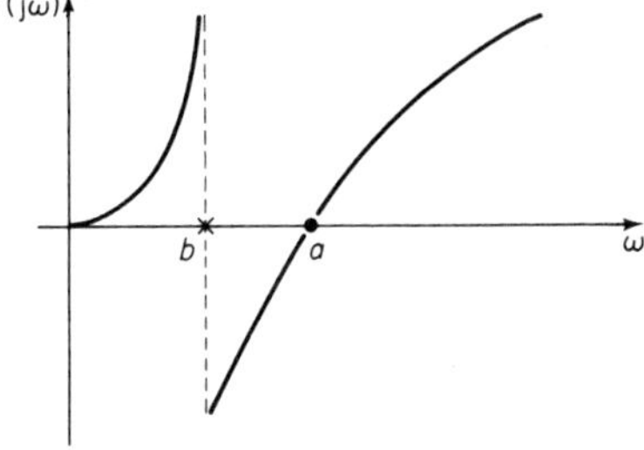

Fig. A.25

2. (a) and (c) are realizable, (b) is not since it has consecutive zeros at $\omega = 0$ and $\omega = 1$ with no interlacing by a pole.

3. (a) $Z(s) = s + \dfrac{1}{\dfrac{1}{4}s + \dfrac{1}{2s}}$,

the network is shown in fig. A.26(a).

241

(b) $Z(s) = 10 + \dfrac{1}{\dfrac{1}{2}s + \dfrac{1}{10}}$,

the network is shown in fig. A.26(b).

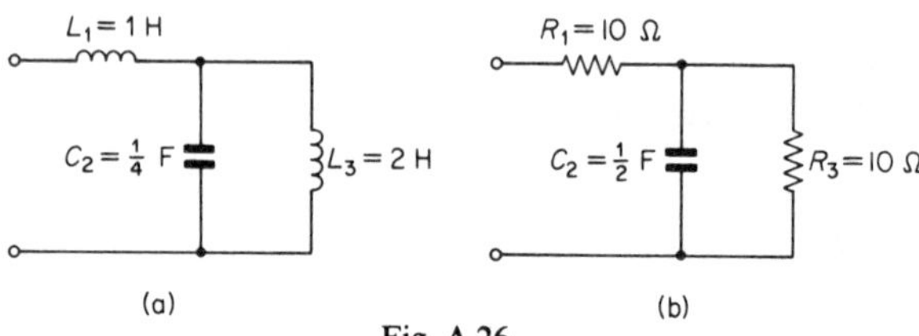

Fig. A.26

4. (a) $Z(s) = \dfrac{\frac{1}{2}}{s} + \dfrac{1}{3s+2}$,

the network is shown in fig. A.27(a).

(b) $Y(s) = \dfrac{s}{s^2+4} + \dfrac{s}{2s^2+3} = \dfrac{1}{s+4/s} + \dfrac{1}{2s+3/s}$,

a network is shown in fig. A.27(b).

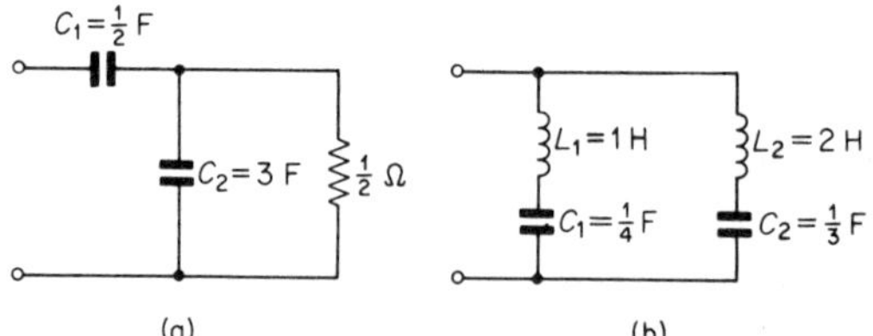

Fig. A.27

5.

$$Z(s) = \dfrac{1}{4s} + \dfrac{1}{\dfrac{1}{3s} + \dfrac{1}{\dfrac{1}{2s} + \dfrac{1}{\dfrac{1}{s} + \dfrac{1}{1}}}} \cdot \dfrac{1}{s}$$

and the network is shown in fig. A.28.

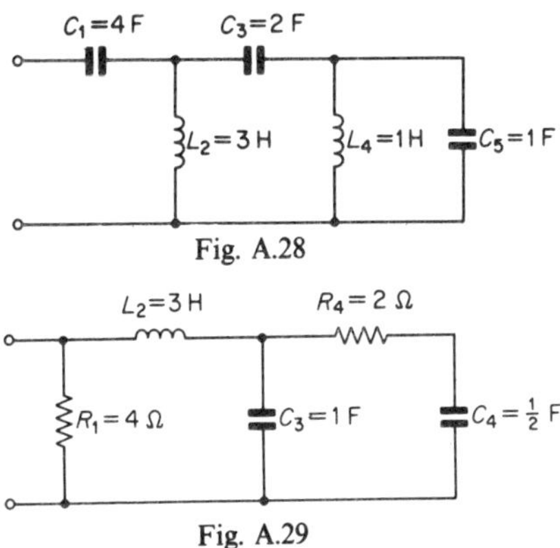

Fig. A.28

Fig. A.29

6. The ladder network is shown in fig. A.29.

7. $L_1 = \frac{1}{2}$H, $L_2 = \frac{1}{4}$H, $C_1 = \frac{1}{3}$F, $C_2 = \frac{1}{2}$F.

8. Two networks giving the required reactance characteristic are shown in fig. A.30.

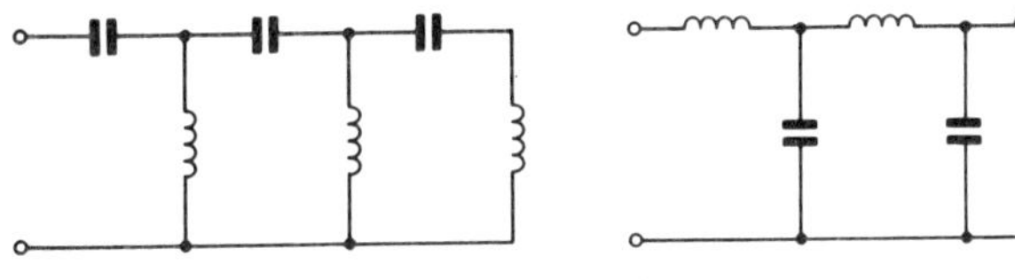

Fig. A.30

9. 2 parallel L–C networks in series with element values $L_1 = 2\cdot103\,$mH, $C_1 = 119\,$pF; $L_2 = 0\cdot449\,$mH, $C_2 = 89\,$pF.

10. The reactance v ω sketch is shown in fig. A.31. The impedance could be synthesized in the form of any one of the networks shown in fig. A.32.

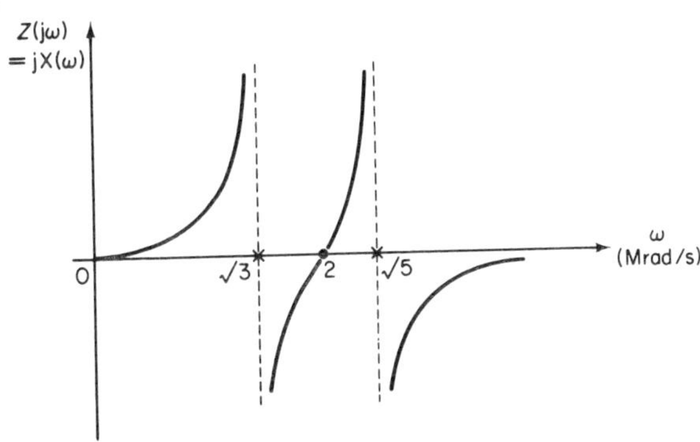

Fig. A.31

243

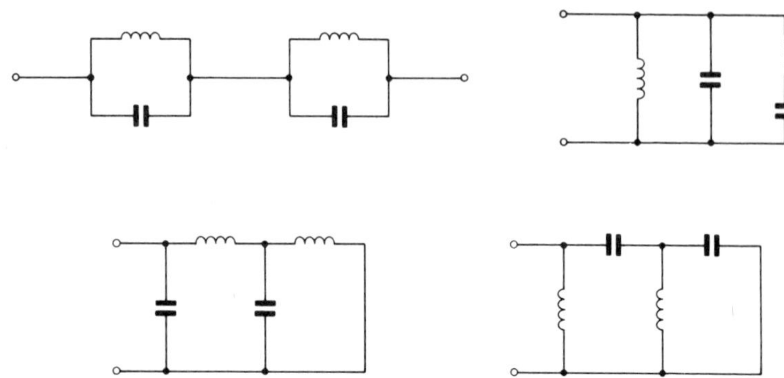

Fig. A.32

11. $Z(s) = 3 + \dfrac{1}{5s} + \dfrac{1}{\frac{1}{4} + 2s}$, the network is shown in fig. A.33.

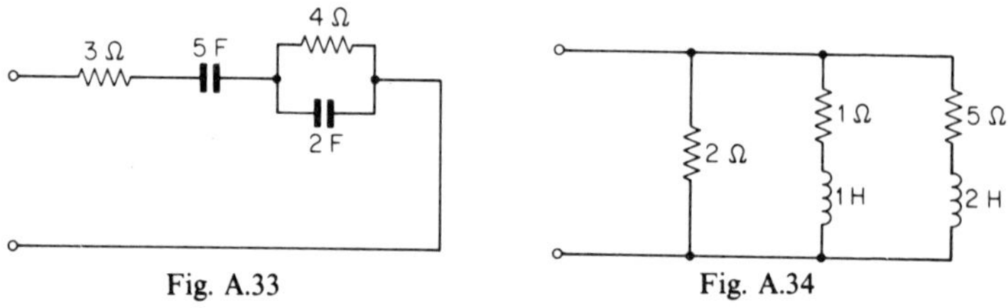

Fig. A.33

Fig. A.34

12. $Y(s) = \frac{1}{2} + \dfrac{1}{(s+1)} + \dfrac{1}{5s+2}$, the network is shown in fig. A.34.

13. (a) $Z(s) = 3 + s + \dfrac{6s}{s+6}$, the network is shown in fig. A.35(a).

(b) $Z(s) = s + \dfrac{1}{\dfrac{1}{9} + \dfrac{1}{\dfrac{9}{2} + \dfrac{1}{\dfrac{9}{2}}}}$, the network is shown in fig A.35(b).

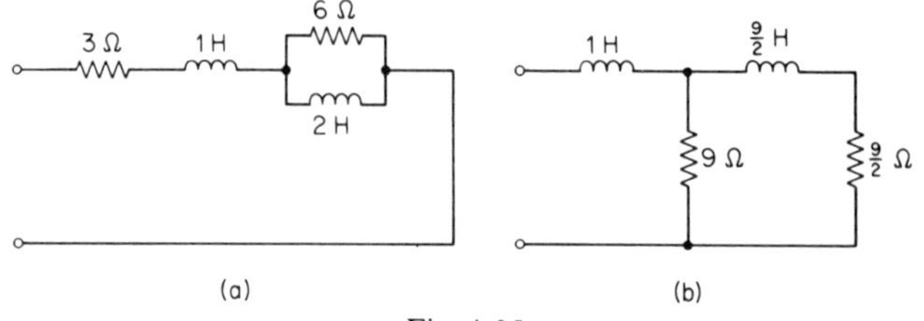

(a)

(b)

Fig. A.35

244

14. (a) R–L network, e.g. see fig. A.36(a) for a possible realization
   (b) R–C network, e.g. see fig. A.36(b)
   (c) R–L–C network, e.g. see fig. A.36(c)
   (d) R–C network, e.g. see fig. A.36(d)

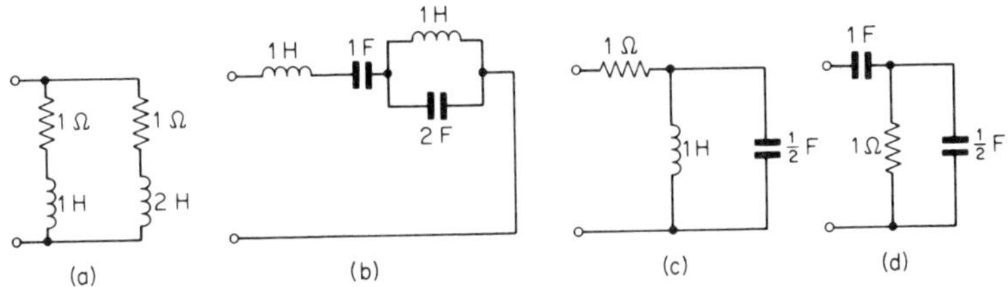

Fig. A.36

15. $L_A = 1\,\text{H}$, $R_B = 1\,\Omega$, $L_C = 3\,\text{H}$, $R_D = 2\,\Omega$, $L_E = 4\,\text{H}$.

16. $y_{22} = \dfrac{20s^2 + 1}{60s^3 + 7s}$, $y_{21} = \dfrac{-1}{60s^3 + 7s}$.

The required network is shown in fig. A.37.

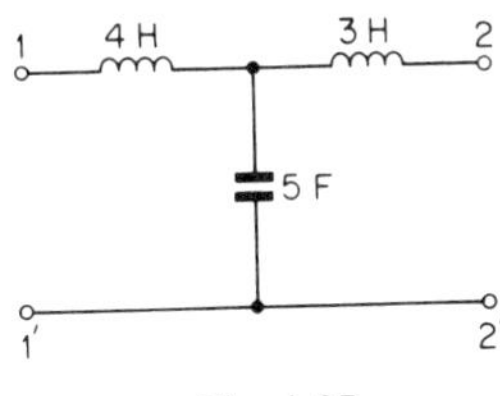

Fig. A.37

17. $S_{11}(s) = \dfrac{s^3}{s^3 + 2s^2 + 2s + 1}$, $Z_{\text{in}}(s) = \dfrac{2s^3 + 2s^2 + 2s + 1}{2s^2 + 2s + 1}$

The realization of the characteristic by a T and Π is shown in fig. A.38

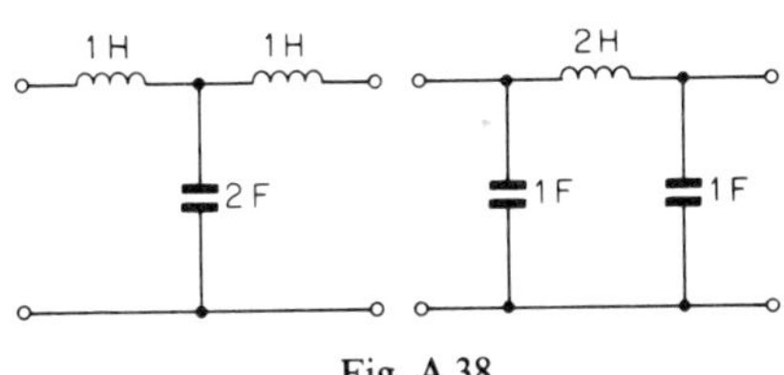

Fig. A.38

1. (a) $a = \dfrac{V_G}{2\sqrt{R_G}}, \quad S_{11} = \dfrac{R_L - R_G}{R_L + R_G}$

(b) $a = \dfrac{V_G}{2\sqrt{Z_C}} = \dfrac{V_G}{2\sqrt{50}}, \quad S_{11} = \dfrac{100 - 50}{100 + 50} = \dfrac{1}{3}$

(c) $a = \dfrac{V_G}{2\sqrt{50}}; \quad S_{11} = \dfrac{Z_{in} - 50}{Z_{in} + 50}, \quad Z_{in} = 2^2 \times 10$ so $S_{11} = -\dfrac{1}{9}$

2. $a = \dfrac{V_G}{2\sqrt{R_0}}; \quad S_{11} = \dfrac{Z_{in} - R_0}{Z_{in} + R_0}$ where $Z_{in} = R_1 + \dfrac{R_2(R_1 + R_0)}{R_1 + R_2 + R_0}$

(a) $S_{11} = 0$ when $Z_{in} = R_0$ which occurs when $R_2 = 26.7\,\Omega$
(b) $P = \frac{1}{2}|a|^2(1 - |S_{11}|^2) = 3V_G^2/(32R_0)$ watts

3. $S_{11}(j\omega) = \dfrac{-\omega^2 LC + j\omega(L - C)}{(2 - \omega^2 LC) + j\omega(L + C)} \cdot \quad P/P_{avail.} = (1 - |S_{11}|^2) = \frac{1}{2}$

4. (a) At $\omega = 0$, $|S_{11}| = 0$, $P/P_{avail.} = 1$
(b) At $\omega = 1$, $|S_{11}| = 1/\sqrt{2}$, $P/P_{avail.} = \frac{1}{2}$
(c) At $\omega = 2$, $|S_{11}| = 0.9923$, $P/P_{avail.} = \frac{1}{65} = 0.1534$

5. (a) $[S] = \begin{bmatrix} 0 & e^{-\gamma l} \\ e^{-\gamma l} & 0 \end{bmatrix}$ in both cases

(b) $[S] = \dfrac{1}{n^2 + 1} \begin{bmatrix} n^2 - 1 & 2n \\ 2n & -(n^2 - 1) \end{bmatrix}$ in both cases

6. (a) $\begin{bmatrix} V_1^- \\ V_2^- \end{bmatrix} = \dfrac{1}{Z_{C_1} + Z_{C_2}} \begin{bmatrix} (Z_{C_2} - Z_{C_1}) & 2Z_{C_1} \\ 2Z_{C_2} & (Z_{C_1} - Z_{C_2}) \end{bmatrix} \begin{bmatrix} V_1^+ \\ V_2^+ \end{bmatrix} \equiv [S] \begin{bmatrix} V_1^+ \\ V_2^+ \end{bmatrix}$

(b) $\begin{bmatrix} b_1 \\ b_2 \end{bmatrix} = \dfrac{1}{Z_{C_1} + Z_{C_2}} \begin{bmatrix} (Z_{C_2} - Z_{C_1}) & 2\sqrt{(Z_{C_1} Z_{C_2})} \\ 2\sqrt{(Z_{C_1} Z_{C_2})} & (Z_{C_1} - Z_{C_2}) \end{bmatrix} \begin{bmatrix} a_1 \\ a_2 \end{bmatrix} \equiv [S] \begin{bmatrix} a_1 \\ a_2 \end{bmatrix}$

7. $|S_{11}| = 0.4$, $|S_{12}| = |S_{21}| = 0.5$, $|S_{22}| = 0.0991$

8. (a) $\begin{bmatrix} V_1^- \\ V_2^- \end{bmatrix} = \dfrac{1}{1 + Z/Z_1 + Z_2/Z_1} \begin{bmatrix} Z/Z_1 + Z_2/Z_1 - 1 & 2 \\ 2Z_2/Z_1 & Z/Z_1 - Z_2/Z_1 + 1 \end{bmatrix} \begin{bmatrix} V_1^+ \\ V_2^+ \end{bmatrix}$

$\begin{bmatrix} b_1 \\ b_2 \end{bmatrix} = \dfrac{1}{1 + Z/Z_1 + Z_2/Z_1} \begin{bmatrix} Z/Z_1 + Z_2/Z_1 - 1 & 2\sqrt{(Z_2/Z_1)} \\ 2\sqrt{(Z_2/Z_1)} & Z/Z_1 - Z_2/Z_1 + 1 \end{bmatrix} \begin{bmatrix} a_1 \\ a_2 \end{bmatrix}$

(b) $\begin{bmatrix} V_1^- \\ V_2^- \end{bmatrix} = \dfrac{1}{1+Y/Y_1+Y_2/Y_1} \begin{bmatrix} 1-Y_2/Y_1-Y/Y_2 & 2Y_2/Y_1 \\ 2 & Y_2/Y_1-Y/Y_1-1 \end{bmatrix} \begin{bmatrix} V_1^+ \\ V_2^+ \end{bmatrix}$

$\begin{bmatrix} b_1 \\ b_2 \end{bmatrix} = \dfrac{1}{1+Y/Y_1+Y_2/Y_1} \begin{bmatrix} 1-Y_2/Y_1-Y/Y_1 & 2\sqrt{(Y_2/Y_1)} \\ 2\sqrt{(Y_2/Y_1)} & Y_2/Y_1-Y/Y_1-1 \end{bmatrix} \begin{bmatrix} a_1 \\ a_2 \end{bmatrix}$

9. $A = \dfrac{1}{2S_{21}} [S_{12}S_{21} - (1+S_{11})(S_{22}-1)]$

$B = \dfrac{1}{2S_{21}} [(1+S_{11})(1+S_{22}) - S_{12}S_{21}]$

$C = \dfrac{-1}{2S_{21}} [S_{12}S_{21} + (1-S_{11})(S_{22}-1)]$

$D = \dfrac{1}{2S_{21}} [S_{12}S_{21} + (1-S_{11})(1+S_{22})]$

I.L.R. $= \dfrac{Az_L + B + Cz_G z_L + Dz_G}{(z_L + z_G)}$

$\qquad = \tfrac{1}{2}(A+B+C+D) = \dfrac{1}{S_{21}}$ if $z_G = z_L = 1$

10. $[S] = \begin{bmatrix} 0 & \sqrt{(1-C^2)} & jC & 0 \\ \sqrt{(1-C^2)} & 0 & 0 & jC \\ jC & 0 & 0 & \sqrt{(1-C^2)} \\ 0 & jC & \sqrt{(1-C^2)} & 0 \end{bmatrix}$

11. $|S_{12}| = |S_{34}| = \sqrt{(1-|S_{14}|^2)} = \sqrt{(1-\alpha^2)}$

12. $|R|^2_{\text{s.c.}} = (0{\cdot}65)^2 = 0{\cdot}4225; \ |R_{\text{o.c.}}|^2 = (-0{\cdot}1)^2 = 0{\cdot}01$
Load for zero reflection $= 4\tfrac{1}{3} \times$ line characteristic impedance.

CHAPTER 6C. INTRODUCTION TO STATE VARIABLE ANALYSIS

1. (a) $\dfrac{d}{dt} \begin{bmatrix} i_L \\ v_C \end{bmatrix} = \begin{bmatrix} -R/L & 1/L \\ -1/C & 0 \end{bmatrix} \begin{bmatrix} i_L \\ v_C \end{bmatrix}$

(b) $\dfrac{d}{dt} \begin{bmatrix} i_L \\ v_C \end{bmatrix} = \begin{bmatrix} -R_2/L & 1/L \\ -1/C & -G_1/C \end{bmatrix} \begin{bmatrix} i_L \\ v_C \end{bmatrix}$

2. (a) $\dfrac{d}{dt} \begin{bmatrix} i_L \\ v_C \end{bmatrix} = \begin{bmatrix} -R_2/L & 1/L \\ -1/C & -G/C \end{bmatrix} \begin{bmatrix} i_L \\ v_C \end{bmatrix} + \begin{bmatrix} 0 \\ 1/C \end{bmatrix} [i_S]$

(b) $\dfrac{d}{dt}\begin{bmatrix} i_L \\ v_C \end{bmatrix} = \begin{bmatrix} 0 & -1/L \\ 1/C & -1/(CR) \end{bmatrix}\begin{bmatrix} i_L \\ v_C \end{bmatrix}$

3. $\dfrac{d}{dt}\begin{bmatrix} v_{C_1} \\ v_{C_2} \end{bmatrix} = \begin{bmatrix} -(G_1+G)/C_1 & G/C_1 \\ G/C_2 & -(G_2+G)/C_2 \end{bmatrix}\begin{bmatrix} v_{C_1} \\ v_{C_2} \end{bmatrix}$

4. $\dfrac{d}{dt}\begin{bmatrix} i_L \\ v_C \end{bmatrix} = \begin{bmatrix} -R/L & -1/L \\ 1/C & -1/[C(R_1+R_2)] \end{bmatrix}\begin{bmatrix} i_L \\ v_C \end{bmatrix}$
$\qquad + \begin{bmatrix} 1/L & 0 \\ 0 & R_2/[C(R_1+R_2)] \end{bmatrix}\begin{bmatrix} v_1 \\ i_2 \end{bmatrix}$

5. $\dfrac{d}{dt}\begin{bmatrix} x_1 \\ x_2 \end{bmatrix} = \begin{bmatrix} 0 & 1/L \\ -1/C & -G/C \end{bmatrix}\begin{bmatrix} x_1 \\ x_2 \end{bmatrix} = \begin{bmatrix} 0 & 16 \\ -4 & -20 \end{bmatrix}\begin{bmatrix} x_1 \\ x_2 \end{bmatrix}$

$x_1 = 13{\cdot}33(e^{-4t}-e^{-16t})\,\text{A}, \; x_2 = (-3{\cdot}33\,e^{-4t}+13{\cdot}33\,e^{-16t})\,\text{V}$

6. $\dfrac{d}{dt}\begin{bmatrix} x_1 \\ x_2 \\ x_3 \end{bmatrix} = \begin{bmatrix} -R/L & -1/L & 0 \\ 1/C & 0 & -1/C \\ 0 & 1/L & -1/C \end{bmatrix}\begin{bmatrix} x_1 \\ x_2 \\ x_3 \end{bmatrix} + \begin{bmatrix} 1/L \\ 0 \\ 0 \end{bmatrix}[v_S]$

7. $\dfrac{d}{dt}\begin{bmatrix} x_1 \\ x_2 \end{bmatrix}=\begin{bmatrix} -R_1/L & -1/L \\ 1/C & -1/(R_2C) \end{bmatrix}\begin{bmatrix} x_1 \\ x_2 \end{bmatrix}$
$\qquad + \begin{bmatrix} 1/L & 0 \\ 0 & 1/(R_2C) \end{bmatrix}\begin{bmatrix} v_{S_1} \\ v_{S_2} \end{bmatrix}$

$$\begin{bmatrix} x_1 \\ x_2 \end{bmatrix} = \begin{array}{l} \dfrac{10}{3}-\dfrac{10}{3}e^{-3/4t}\left(\cos\dfrac{\sqrt{15}}{4}t-\dfrac{21}{\sqrt{15}}\sin\dfrac{\sqrt{15}}{4}t\right) \\[2ex] \dfrac{50}{3}-\dfrac{50}{3}e^{-3/4t}\left(\cos\dfrac{\sqrt{15}}{4}t+\dfrac{9}{5\sqrt{15}}\sin\dfrac{\sqrt{15}}{4}t\right) \end{array}$$

8. $\dfrac{\mathrm{d}}{\mathrm{d}t}\begin{bmatrix} x_1 \\ x_2 \end{bmatrix} = \dfrac{1}{(1 - M^2/L_1 L_2)} \begin{bmatrix} -R_1/L_1 & -MR_2/(L_1 L_2) \\ -MR_1/(L_1 L_2) & -R_2/L_2 \end{bmatrix} \begin{bmatrix} i_1 \\ i_2 \end{bmatrix}$

$\qquad\quad + \dfrac{1}{(1 - M^2/(L^1 L^2))}\ [1/L_1]\ [v_S]$

$\quad \begin{bmatrix} x_1 \\ x_2 \end{bmatrix} = \begin{bmatrix} 0{\cdot}80(1{\cdot}65\,e^{-5t} + 1{\cdot}25\,e^{-123{\cdot}8t} - 2{\cdot}90\,e^{-16{\cdot}2t}) \\ 10^{-3}(-1{\cdot}83\,e^{-5t} + 8{\cdot}93\,e^{-123{\cdot}8t} - 7{\cdot}10\,e^{-16{\cdot}2t}) \end{bmatrix}$

9. $\dfrac{\mathrm{d}x}{\mathrm{d}t} = \left\{ \dfrac{\mu\beta}{(R_1 + R_2)C} - \dfrac{1}{R_3 C} \right\} x - \left\{ \dfrac{\beta}{(R_1 + R_2)C} \right\} v_1$